International Conference of Advanced Research on Renewable Energy for Universal Sustainability (ARUS 2021)

Journal of Physics: Conference Series Volume 2053

Online
22 – 24 June 2021

Editors:

Shamsul Bin Sarip
Mohd Nabil Bin Muhtazaruddin
Hazilah Binti Mad Kaidi
Nurul Aini Binti Bani
Suriani Mohd. Sam

ISBN: 978-1-7138-5024-3
ISSN: 1742-6588

Printed from e-media with permission by:

Curran Associates, Inc.
57 Morehouse Lane
Red Hook, NY 12571

Some format issues inherent in the e-media version may also appear in this print version.

Institute of Physics
Dirac House, Temple Back
Bristol BS1 6BE UK

Phone: 44 1 17 929 7481
Fax: 44 1 17 920 0979

techtracking@iop.org

Additional copies of this publication are available from:

Curran Associates, Inc.
57 Morehouse Lane
Red Hook, NY 12571 USA
Phone: 845-758-0400
Fax: 845-758-2633
Email: curran@proceedings.com
Web: www.proceedings.com

TABLE OF CONTENTS

1st International Conference of Advanced Research on Renewable Energy for Universal Sustainability 2021 (ARUS2021)

PREFACE

The 1st International Conference of Advanced Research on Renewable Energy for Universal Sustainability 2021 (ARUS2021) was organised by Arus Infiniti Sdn Bhd and supported by Razak Faculty of Technology and Informatics, Universiti Teknologi Malaysia (UTM). In light of ongoing developments resulting from the COVID-19 globally and after deliberate discussion and thoughtful consideration with our co-organisers and collaborators, we have decided to move our ARUS 2021 Conference to a **fully online, virtual experience** held on 22-24th June 2021 from Kuala Lumpur, Malaysia. As much as we wish for a meaningful gathering full of respectable researchers, we cannot overlook the danger the COVID-19 might cause. We firmly believe that this was the safest approach for the well-being of our global participants and audiences.

The main objective of the conference is to explore the frontiers of Science, Technology, Engineering and Management in various areas addressing innovative research in the broad field of renewable energy and green technology and discussing and entertaining topics that stretch into the future. It also provides a venue and many opportunities to meet and network with colleagues from this broad field, share personal experiences, and learn from others. We are thrilled that our virtual conference platform makes it possible for the learning opportunities and network engagement to happen during this pandemic in the comfort and safety of their home or office.

Since its inception, ARUS2021 has been attracting many delegates from all over the world at ARUS2021, and all participants are given the opportunity to meet and exchange knowledge from our Keynote speakers, Professor Dr. Yasuyuki Ikegami (Japan), Professor Dato' Ir Dr Abu Bakar Jaafar (Malaysia), Ir. Noraziah Muda (Malaysia) and Ts. Ahmad Hadi Mohamed Rashidi (Malaysia), renowned experts in renewable energy and green technology. With the theme "Clean Energy : Potentials and Trends in Future Development", all experts shared their enriching insights on the current trends and issues in global energy sustainability.

A 45-minutes slot was allocated to each keynote speaker, including 5 minutes for Q&A, while every presenting participant received a 15-minutes slot, including Q&A. Throughout the conference, audiences have demonstrated great interest in the discussed topics with active participations during Q&A sessions. Each session was attended by at least 50 participants with the largest audience of about 150 participants during our Special Session. The whole conference was held online on WEBEX platform, with several dry run sessions conducted by the organizer, prior to the conference to ensure the success delivery of the conference. The organiser also accepted a pre-recorded video presentation as an alternative to an online presence for presenters experiencing internet problems.

On behalf of the organising committee, we extend our heartiest gratitude to all the esteemed delegates that participate in ARUS2021. We hope your experience at ARUS2021 is a fruitful and beneficial to you in many ways. We sincerely hope that this association will be maintained and that you will continue to support us in our future endeavours. It is a great privilege for us to present the proceedings of ARUS2021 to the authors and delegates of the conference. We would like to thank IOP Publishing for this publication opportunity. We hope that you will find it useful and inspiring.

We would like to express our deepest gratitude to our sponsors; Naglus Industries Sdn. Bhd, Enforture Sdn Bhd, Sifutrademark and HALEA. We truly appreciate your support and your very generous donation that we believe has dramatically help us to develop and succeed. Without your involvement and generosity, we would not be able to aim high and reach our potential. Thank you again for your generous and continuous support.

We hope you can join us again and we look forward to seeing all participants in another series of ARUS conferences.

Assoc. Prof. Ir. Dr Shamsul Bin Sarip
Dr. Mohd Nabil Bin Muhtazaruddin
Dr. Hazilah Binti Mad Kaidi
Ts. Dr. Nurul Aini Binti Bani
Ts. Dr. Suriani Mohd. Sam

Editors of Conference proceedings, ARUS 2021

Organising Committee

General Chair
Assoc. Prof. Ir. Dr. Shamsul Sarip
(Universiti Teknologi Malaysia)

Conference Chair
Ts. Dr. Mohamad Zaki Hassan
(Universiti Teknologi Malaysia)

Vice Chair/Publicity Manager
Ts. Dr. Nurul Aini Bani
(Universiti Teknologi Malaysia)

Secretary
Dr. Hazilah Mad Kaidi
(Universiti Teknologi Malaysia)

Treasurer
Dr. Siti Hasliah Salleh
(Universiti Teknologi Malaysia)

Technical Program Chair
Ts. Dr. Suriani Mohd. Sam
(Universiti Teknologi Malaysia)

Publication
Dr. Mohd Nabil Muhtazaruddin
(Universiti Teknologi Malaysia)

Local Arrangement
Dr. Mohamed Azlan Suhot
(Universiti Teknologi Malaysia)

Sponsorship
Dr. Rozzeta Dolah
(Universiti Teknologi Malaysia)
Dr. Rohayah Che Amat
(Universiti Teknologi Malaysia)

ADVISORY COMMITTEE

1. Prof. Dr. Gordon Huang (University of Regina, Canada)
2. Priv.-Doz. Dipl.-Ing. Dr. Amela Ajanovic (Vienna University of Technology, Austria)
3. Assoc. Prof. Dr. Abdullahi Abubakar Mas'ud (Jubail Industrial College, SA)
4. Assoc. Prof. Ts. Dr. Wan Zuha Wan Hasan (Universiti Putra Malaysia, Malaysia)
5. Assoc. Prof. Ir Dr Ungku Anisa Bte. Ungku (Universiti Tenaga Nasional, Malaysia)
6. Asst. Prof. Pirat Khunkitti (Khon Kaen University, Thailand)
7. Dr. Firdaus Muhammad Sukki (Robert Gordon University, UK)
8. Dr. Siti Hawa Abu Bakar (Universiti Kuala Lumpur, Malaysia)
9. Dr. Nur Fadilah Bte. Ab. Aziz (Universiti Tenaga Nasional, Malaysia)
10. Dr. Nazmi Sellami (Robert Gordon University, UK)
11. Dr. Mohammad Ershadul Karim (Universiti Malaya, Malaysia)
12. Dr. Mohammad Nurul Islam (Universiti Teknologi Brunei, Brunei Darussalam)
13. Dr. Ahmad Fazlizan Abdullah (Universiti Kebangsaan Malaysia, Malaysia)
14. Dr. Roslynna Rosli (Universiti Teknologi Brunei, Brunei Darussalam)
15. Dr. Umer Rashid (Universiti Putra Malaysia, Malaysia)
16. Dr. Azrinawati Mohd Zin (Universiti Teknologi Mara, Pulau Pinang)
17. Dr Ridoan Karim (Universiti Malaya, Malaysia)
18. Ts. Ir. Azwin Zailti Abd Razad (TNB Research Sdn. Bhd.)
19. Ts. Ir. Noraziah Omar (TNB Research Sdn. Bhd.)

TECHNICAL COMMITTEE

Abdulhafid Elfaghi	Universiti Tun Hussein Onn, Malaysia
Abdullahi Abubakar Mas'ud	Jubail Industrial College, Saudi Arabia

Afifah Shuhada Rosmi	Universiti Malaysia Perlis, Malaysia
Ahmad Anas Nagoor Gunny	Universiti Malaysia Perlis, Malaysia
Ahmad Hamdan Bin Ariffin	Universiti Tun Hussein Onn Malaysia
Akmal Hadi Bin Ma Radzi	Universiti Malaysia Perlis, Malaysia
Ang Qian Yee	Universiti Malaysia Perlis, Malaysia
Azizul Helmi Sofian	Universiti Malaysia Pahang, Malaysia
Ching Yee Yong	University College Of Technology Sarawak
Chokchai Mueanmas	Thaksin University, Thailand
Cornelius Bavoh	Universiti Teknologi Petronas, Malaysia
Firdaus Muhammad Sukki	Edinburgh Napier University, United Kingdom
Goh Yingxin	University Of Malaya, Malaysia
Hashem Salarzadeh Jenatabadi	University Of Malaya, Malaysia
Jitti Kasemchainan	Chulalongkorn University, Thailand
Kah Hon Leong	Universiti Tun Abdul Razak, Malaysia
Khairatun Najwa Mohd. Amin	Universiti Malaysia Pahang, Malaysia
Khaled Obaideen	Segi University, Malaysia
Kumuthawathe Ananda-Rao	Universiti Malaysia Perlis, Malaysia
Midhat Nabil Ahmad Salimi	Universiti Malaysia Perlis, Malaysia
Mohammod Aminuzzaman	Universiti Tun Abdul Razak, Malaysia
Mohd Fauzi Mamat	Universit Teknikal Malaysia Melaka, Malaysia
Muhamad Razuhanafi Mat Yazid	Universiti Kebangsaan Malaysia
Muhamad Safwan Abd Rahman	Universiti Teknologi Nasional, Malaysia
Mustaqqim Abdul Rahim	Universiti Malaysia Perlis, Malaysia
Noor Hasyimah Rosman	Universiti Kebangsaan Malaysia
Norhusna Mohamad Nor	Universiti Teknologi Mara, Malaysia
Norliza Dzakaria	Universiti Teknologi Mara, Malaysia
Noshin Fatima	Universiti Kebangsaan Malaysia, Malaysia
Nur Azwa Muhamad Bashar	Universiti Teknologi Mara, Malaysia

Nurazuwa Md Noor	Universiti Tun Hussien Onn, Malaysia
Nurul Hanim Razak	Universiti Teknikal Malaysia Melaka
Qian Yee Ang	Universiti Malaysia Perlis, Malaysia
Rafikullah Deraman	Universiti Tun Hussein Onn Malaysia, Malaysia
Ridoan Karim	Universiti Malaya, Malaysia
Roslinda Seswoya	Universiti Tun Hussein Onn, Malaysia
Roslynna Rosli	Universiti Teknologi Brunei, Brunei Darussalam
Siti Nur Aishah Mat Yusuf	Universiti Malaysia Perlis, Malaysia
Slamet Wahyudi	Brawijaya University, Indonesia
Suriani Ibrahim	University Malaya, Malaysia
Syamimi Saadon	University Putra Malaysia, Malaysia
Teoh Yi Peng	Universiti Tun Abdul Razak, Malaysia
Ungku Anisa Ungku Amirulddin	Universiti Teknologi Nasional, Malaysia
Wai Zhe Leow	Universiti Malaysia Perlis, Malaysia
Wan Azani Mustafa	Universiti Malaysia Perlis, Malaysia
Wan Mohd Sabki Wan Omar	Universiti Malaysia Perlis, Malaysia
Wan Zuraida Wan Kamis	Universiti Teknologi Mara Cawangan Pulau Pinang
Wardah Fatimah Mohammad Yusoff	Universiti Kebangsaan Malaysia, Malaysia

Contact person for queries:

Name : Assoc. Prof. Ir. Dr. Shamsul Sarip
Affiliation: Universiti Teknologi Malaysia
Email :shamsuls.kl@utm.my

Journal of Physics: Conference Series **2053** (2021) 011002 doi:10.1088/1742-6596/2053/1/011002

Peer review declaration

All papers published in this volume of Journal of Physics: Conference Series have been peer reviewed through processes administered by the Editors. Reviews were conducted by expert referees to the professional and scientific standards expected of a proceedings journal published by IOP Publishing.

- **Type of peer review: Single-blind / Double-blind / Triple-blind / Open / Other (please describe)**
  - All the review process was done single-blind through Microsoft CMT platform. Average three reviewers involved for each paper. In the CMT system, we provide each reviewer with a Review Questions form.

- **Conference submission management system:**
  - **Microsoft Conference Management Toolkit (CMT)**

- **Number of submissions received:** 36

- **Number of submissions sent for review:** 36

- **Number of submissions accepted:** 26

- **Acceptance Rate (Number of Submissions Accepted / Number of Submissions Received X 100):** 72.2 %

- **Average number of reviews per paper:** 3 reviewers

- **Total number of reviewers involved:** 53

- **Any additional info on review process:** No

- **Contact person for queries:**
  Name : Assoc. Prof. Ir. Dr. Shamsul Sarip
  Affiliation: Universiti Teknologi Malaysia
  Email :shamsuls.kl@utm.my

The readiness of IoT enabled Smart Buildings in Malaysia

Mee San Kiu[1*], Ying Jie Chin[1], Fah Choy Chia[1] and Phui Fung Wong[1]

[1] Department of Surveying, Lee Kong Chian Faculty of Engineering and Science, Universiti Tunku Abdul Rahman, Sungai Long Campus, Bandar Sungai Long, 43000 Kajang, Selangor, Malaysia

*Corresponding author: kiums@utar.edu.my

Abstract. The rapid evolution of the Internet of Things (IoT) has significantly impacted the way of conduct business and influenced current daily lifestyles. Observing this phenomenon, researchers and industry practitioners are increasingly paying attention to the research and development of smart buildings. This research aimed to focus on the readiness of smart buildings in Malaysia and research objectives were set to determine the readiness, potential benefits of implementing smart buildings and the challenges faced in the current state towards adopting smart buildings in Malaysia. Based on the 140 questionnaire survey responses from various construction professionals, descriptive statistics and Kruskal-Wallis test were conducted. This study has contributed to the body of knowledge on understanding on current readiness of Malaysia in the implementation of Smart Building and brought up the main challenges faced towards the adoption of Smart Building. In conclusion, Malaysia has not fully ready to take advantage of the benefits the smart revolution will bring, including greener, safer and more efficient energy use.

1. Introduction

The smart building is very important to the construction industry because it offers a dynamic approach to sustainability of the functioning asset at lower costs and with fewer effort than ever. Besides, the adoption of smart buildings also enhances sustainable development in the country by overcoming environmental issues. However, the construction industry is a conservative industry, so greater efforts are needed to evoke changes in the implementation of smart buildings.

In recent decades, growing numbers of things are being connected to the Internet, thereby enriching the digital world. The Internet of Things (IoT) promotes and improves people's life and work efficiency in many different areas. Thus, the use of IoT devices to make buildings smarter and more efficient is increasing in interest [1]. Cities and urban areas around the world are rapidly evolving as a result of globalisation, emerging smart systems, and the penetration of IoT. Nonetheless, the implementation of smart buildings is still in its infancy for many cities.

According to [2], governments and societies all over the world are increasingly paying attention to the development of smart buildings. In Saudi Arabia, [3] found that the construction industry is gradually integrating smart technology in its buildings. However, Saudi Arabia is still far behind in the use of smart building technologies, only 15% of buildings with appropriate used while 85% are without adequate utilised. In Denmark, [4] observed that the integrated design solutions that different products from different suppliers into one building have become a trend. In China, [5] pointed out that the adoption of smart buildings has widened in the following years.

In Malaysia, considerable attempts are being observed in recent years to integrate smart and sustainable design concepts into future urban planning. [6] has conducted a case study in Malaysia to predict energy consumption in Building Energy Management System (BEMS) for smart buildings by using machine learning. Another case study was also conducted by [7] on awareness of smart energy management in residential buildings in Kajang and Putrajaya, Malaysia. Besides, [8] also conduct a review on the energy efficiency programs and policies in Malaysia. From the previous studies highlighted here, most studies were focused on the building's smart energy management in Malaysia. It is obviously still lacking of studies on IoT enabled smart buildings in Malaysia, including the awareness and readiness of smart buildings in Malaysia.

Hence, this research will focus on the readiness of smart buildings in Malaysia as well as take an integrated approach to determine the potential benefits of implementing smart buildings and the challenges faced in current state.

2. Internet of Things and Smart Buildings

2.1. Smart buildings

Smart building is a structure utilising the computerised procedure to control the building's operations automatically involving ventilation, lighting, security, heating, air conditioning, and other systems. Building owners and designers have to consider the developments of environmental systems, climate control, security, access, and fire detection in the design of smart building [9]. The anticipation of smart building is using interdependency among building and business system to achieve the capability of ameliorated operational and energy efficiency as well as overcome facility management challenges, resulting to automated or optimised facility management processes and provide enrichment of data for analysis [10]. Smart building uses microchips, actuators, and sensors so as to administer diverse building operations dependent on particular business functions and services rules as well as gather data which bring about cost-effective and productive environment.

2.2. IoT Technology

The IoT is an information technology that connects physical devices with internet to empower data transmission, acquisition, and processing as well as enable smart identification and management in cloud server [11]. There are various areas in which IoT can improve human life and work in a significant manner, including conveyance, healthcare, automation, facility management, etc. IoT empowers an object to see, tune in, and communicate simultaneously. IoT thus transforms the devices from being intelligent by incorporating its widespread and pervasive communication technologies, computing, embedded devices, and numerous different applications to influence and revolutionise human life [12].

In the era of IoT, a large amount of data and resources had been offered by internet-connected things which provide boundless possibilities for services and applications. The application and services provided by the IoT can be both residential and commercial ranging from automation, e-health, e-marketing, intelligent transportation system, logistic services, etc. Outlook on the future, the internet will comprise a mass of intelligent objects alongside embedded devices and not only being the network of computers. Therefore, the scope and size of the adoption IoT will increase exponentially which provide new opportunities and challenges. Various nations have grown long haul national strategies for the usage of IoT. For example, smart buildings give the capacity to the building occupants to access intelligent equipment remotely from anyplace over the world. The development initiatives framed the establishment of IoT. The main part of these IoT devices is mounted inside buildings, thus give us impetus to the concept of smart buildings. In smart buildings, diverse electronic gadgets and contraptions are interconnected and are communicating with one another. A large amount of data will be generated as a huge quantity of devices are communicating with one another in such a system. Big data analytics and management are expected to play an important role to enhance IoT-enabled smart buildings.

2.3. IoT enabled Smart Buildings

The fundamental of a smart building is the ability of standalone technology devices operated with the existing network connection to ensure a good working system of a building [13]. Smart buildings enable to enhance the service for its occupants by incorporating a broad assortment of devices, control systems, sensors, and actuators that are interconnected and jointly function. Besides, smart buildings can be monitored or controlled remotely by integrating a correspondence network within the buildings' components. For instance, all the windows must be closed automatically as an action that is a touch off by the change of temperature before the air conditioner is turned on. Hence, a sensor has to trigger the air conditioner, and afterwards, data ought to be interchanged between the windows and air conditioner. The systems controlling the air conditioner and windows are more likely to be made by various manufacturers based on the technical interoperability viewpoint. This management and automating control of the building requires an integration process [14].

Smart buildings incorporate and account for data from various sources or embedded devices for materials, construction, control, and intelligence as the whole building system. This is to upgrade versatility to meet the worth drivers of the smart building which include comfort and satisfaction, life span, as well as energy and efficiency. Smart buildings enable to provide its occupants with a secure, comfortable and productive environment without compromising on energy and operational performance. For example, some of them incorporate occupancy monitoring systems, smart metering (water, gas, and electricity) and even high-efficiency HVAC systems. Worldwide energy consumption in both residential and commercial buildings has risen consistently and far surpassed the other significant areas such as transportation and industrial. Besides, the demand for comfort levels and building services is increasing together with the building occupancy time due to the population growth, which marks an expanding pattern in the demand for energy. Thus, the energy policy is extremely important as it decides the course that enterprises and industries will take with regard to smart building design, construction, operation, and maintenance. In addition, the HVAC systems consume more energy during building operation when contrasted with other building energy systems. IoT enabled smart buildings to enable the use of data to minimise the energy consumption and optimise operational efficiency.

3. Research methodology

In this research, the quantitative approach was utilised to measure the readiness, benefits and challenges of implementing smart buildings in Malaysia. The primary data was collected by distributing questionnaire surveys online. For sampling, a non-probability sampling method which are often appropriate for exploratory research was adopted. Convenience sampling that composed of people who are easy to reach was selected in this study as it was more convenient to collect data and an inexpensive way to gather initial data. After collecting the survey data, statistical analysis is used to analyse the data gathered. SPSS software was utilised to carry out the descriptive statistical analysis and non-parametric test. The results get from the statistical analysis will be interpreted in more generalisable in next section.

The survey area for the research was Klang Valley, Malaysia due to Klang Valley is known as the heartland of Malaysia's commerce and industry and has achieved outstanding construction performance. Moreover, the respondents will be limited to Malaysian construction practitioners such as developers, architects, engineers, quantity surveyor and contractor. They were the right group of respondents who understand the real situation in the construction industry and may have experience in smart building projects. This study was not limited to any specific type of project.

4. Results and Discussion

A total of 400 sets of questionnaire survey's requests were sent out but only 140 responses were received. The respondents' demographic background was summarised in table 1 below.

Table 1. Respondents demographic (N=140)

General Information	Categories	Frequency	Percentage (%)
Working Experience	Less than 5 years	62	44.3
	6 to 10 years	35	25.0
	11 to 15 years	17	12.1
	16 to 20 years	8	5.7
	More than 20 years	18	12.9
Profession	Developer	12	8.6
	Architect	22	15.7
	Engineer	25	17.9
	Quantity Surveyor	33	23.6
	Contractor	38	27.1
	Others	10	7.1

Table 2 indicated results on the readiness in implementing smart buildings in Malaysia. The result on the respondents' experience in smart building project revealed that more than half of Malaysian construction professionals (n=85, 60.7%) are lack of practical experience as they have not participated in any smart building projects before. The results also inferred that the three most common types of smart building projects in Malaysia were *Office buildings* (n=26), *Residential* (n=21) and *Commercial Retail* (n=19). The other smart building project (n=7) experienced by the respondents were such as research and development (R&D) building, Light Rail Transit (LRT) station, Mass Rapid Transit (MRT) project and their own house.

Results also indicated the smart systems that were commonly used by respondents in the building project were *Lighting System* (n=46), *Energy Management System* (n=45), *Access Control System* (n=44), *HVAC Control System* (n=43), and *Safety and Security System* (n=37). Minor respondents had used smart audiovisual systems (n=26) in building project, while the remaining 2.1% of respondents were involved in other systems such as surveillance systems and transporting systems which particularly applied in infrastructure projects. The outcome indicated that there are only partial of construction professionals in Malaysia are familiar with smart buildings system.

Results in Table 2 also inferred most respondents (n=118, 84.3%) were interested in using smart technologies in the buildings. However, less than half of the respondents (n=63, 45%) agreed that Malaysian construction industry is ready for the smart building revolution while 40% (n=56) were unsure whether Malaysian built environment is currently being smart-ready. This revealed that Malaysia is not fully ready to take advantage of the benefits the smart revolution will bring, including greener, safer and more efficient energy use. The knowledge gaps among built environment practitioners in Malaysia need to be bridged as this is one of the major obstacles to the implementation of smart buildings. Stakeholders at all levels may need immediate and effective large-scale capacity building and awareness programs.

Table 2. Results of readiness of Smart Building in Malaysia (N=140)

Item	Categories	Frequency	Percentage (%)
Experience in Smart Building Projects	Yes	55	39.3
	No	85	60.7
Type of project involved for Smart Building	None	85	60.7
	Office	26	18.6
	Residential	21	15.0
	Commercial retail	19	13.6
	Hotel	12	8.6
	Education	11	7.9
	Industrial Warehouse	9	6.4
	Healthcare	8	5.7
	Others	7	5.0
Use of Smart Building Systems in Project.	None	85	60.7
	Lighting System	46	32.9
	Energy Management	45	32.1
	Access Control	44	31.4
	HVAC control system	43	30.7
	Safety and Security System	37	26.4
	Audio visual systems	26	18.6
	Others	3	2.1
Intention to promote smart technologies in Building Project	Extremely interested	52	37.1
	Somewhat interested	66	47.2
	Not sure	16	11.4
	Not very interested	6	4.3
	Not at all interested	0	0.0
Readiness of implementing smart buildings in Malaysia	Yes	63	45.0
	No	21	15.0
	Maybe	56	40.0

The Cronbach alpha value of reliability tests on the 10 statements of benefits of smart building and 12 statements on challenges encountered in smart building implementation are 0.911 and 0.878 respectively as shown in table 3 indicated that the constructs are internally consistent and reliable.

Table 3. Cronbach's Alpha Test Results

Statements	Number of Item	Cronbach's Alpha Coefficient
Benefits of Smart Buildings	10	0.911
Challenges encountered in smart building implementation	12	0.878

Table 4 indicated results of the mean agreement and Kruskal-Wallis test of 10 statements on benefits that drive smart building on the construction projects. The findings revealed that 7 out of 10 statements on the benefits of smart buildings have mean agreement of more than 4.0. The seven most perceived benefits of the smart building were *Improving convenience* (mean = 4.24) ranked first, *Enhance occupant's safety and security* (mean = 4.22) ranked second, *Enhance occupant's health and comfort* (mean = 4.19) ranked third and *Save energy* (mean = 4.16) ranked forth. These results indicated that the top three important benefits of implementing smart building in construction projects lies in social and behavioural dimensions and the forth benefit is more towards environmental sustainability.

Kruskal-Wallis test was carried out to investigate whether there is a significant difference between one's profession and his or her opinion on the benefits of smart buildings. The hypotheses for the test were null hypothesis and alternative hypothesis. The null hypothesis (Ho) assumed that there is no significant difference between one's profession and his or her opinion on the benefits of smart buildings, while the alternative hypothesis (H1) assumed that there is a significant difference between one's profession and his or her opinion on the benefits of smart buildings.

The results shown a statistically significant difference in the opinion of benefit *"improve convenience"* ($\chi 2 = 16.670$, $p = 0.005$) and *"enhance reputation"* ($\chi 2 = 13.188$, $p = 0.022$) in smart building between different profession. In cities with busy momentum, property buyers are pursuing high quality and improved convenience in living when it comes to buying a property. Therefore, developers and designers will try to embed the design of smart building systems within the property to improve the end-users' convenience and quality living experience, and concurrently with the aim to enhance the project reputation. Meanwhile, for quantity surveyors and contractors, smart buildings integrate part of design and construction, thereby achieving cost-saving benefits for the project by providing valuable advice about value engineering and cost management shall be their priority. Moreover, contractor's focus is on the improvement on the efficiency of project management and project schedule rather than concerning on improved convenience of smart building system brought to occupants.

Table 4. Results of Benefits of Smart Buildings.

Benefits	Mean	Chi-Square $\chi 2$	Asymp. Sig.
Improve convenience	4.24	16.670	**0.005***
Enhance occupant's safety and security	4.22	8.080	0.152
Enhance occupant's health and comfort	4.19	4.384	0.496
Save energy	4.16	3.032	0.695
Increase building value	4.08	9.570	0.088
Increase productivity	4.05	5.602	0.347
Enhance reputation	4.01	13.188	**0.022***
Save water	3.99	2.482	0.779
Optimise space utilisation	3.84	3.909	0.563
Reduce operating and maintenance costs	3.65	8.718	0.121

Table 5 summarised the results of mean agreement of 12 statements on challenges of implementing smart buildings in Malaysia. The findings revealed that 4 out of 12 statements are having mean agreement more than 4.0. The four major challenges of implementing smart building were *High initial costs* (mean=4.52) ranked first, *Lack of funding* (mean=4.26) ranked second, *Lack of a regulatory environment to the adoption of smart buildings* (mean=4.11) ranked third and *Lack of skilled and specialist workers in smart buildings concepts, devices and solutions* (mean=4.02) ranked forth.

Table 5. Results of the Challenges of Implementing Smart Buildings in Malaysia.

Challenges	Mean
High initial costs	4.52
Lack of funding	4.26
Lack of a regulatory environment favourable to adopting the concept smart building	4.11
Lack of skilled and specialist workers in smart buildings concepts, devices and solutions	4.02
Privacy and data security	3.95
Limited consumer's demand	3.94
Lack of awareness in smart building concepts and technologies	3.91
Learning how to operate new systems or training staff	3.88
Reliability of smart building components	3.84
Lack of effective data management	3.81
Complexity of devices' connectivity	3.77
Complexity of prediction task	3.72

The most significant challenge of implementing a smart building in Malaysia mentioned above was high initial costs than a conventional building. Although many people perceive that smart buildings were easy to use and can minimise all operating costs of buildings like energy, air conditioning, and control environment, it is difficult for most individuals to pay the initial cost for this type of construction. Therefore, financial capability and smart buildings will be the main concern for developer [15]. The findings of [16] also showed similar results that high cost of purchasing devices ranked first in the challenges in smart building.

Lack of funding opportunities or failure to access finance on acceptable terms is typically one of the most commonly cited obstacles to investment in smart buildings. This was primarily due to lack of confidence of the potential investors, lack of incentive policies in place, insufficient and instable available funding and lack of attractive financing for owners with low to medium incomes. As public spending decreases, funding opportunities will become limited and the uncertainty over proposed schemes also increases. Similar results in [17] indicated that lack of funding ranked second.

Another major challenge to adoption of smart buildings in Malaysia was the lack of adequate of a regulatory environment favourable to adopting the concept smart building. Smart buildings are a trend that gradually takes shape by the day, but it must be ensured that not only will smart buildings happen, but they will happen in the right way. As perceived by [18], a range of standards and regulations were needed to help the smart building meet its potential as these regulations will help address issues at various levels which creating an environment that promotes public confidence and ensures stability, transparency, competition, investment, innovation, and growth in the city.

Among the challenges of implementing smart buildings in Malaysia, the complexity of prediction task was ranked at the last with the mean of 3.72. In general, occupant's behaviors are not likely to remain constant, but to change over time. Besides, they may have unique patterns that overlap in time or space and may also have patterns that only occur when combinations of occupants occupy the space simultaneously.

There are two other challenges of implementing smart buildings in Malaysia which were highlighted by six of the respondents during the questionnaire survey: 1) Lack of government incentives to implement energy-saving and 2) smart building initiative and resistance to change from traditional buildings. The adoption of government incentives can promote the implementation of smart buildings, but this depends on government policy. At the moment, there are lack of incentive policies in place and therefore the cost for such adoption and investment falls on service users [19], including Malaysia. Apart from that, the findings revealed that the existing culture and environment in Malaysia as well as the mindset of building owners and users mostly resist to change from traditional buildings. The willingness of local citizens to accept new technology is low due to concerns on the significant price increase and unproven technologies in Malaysia.

In order to overcome these major challenges, the government should come out with incentives to support building developers to incorporate more IoT enabled smart applications in the building design, which end up being sustainable buildings in our cities. Moreover, there is a strong need for government to come up with enforcement policies to incorporating IoT enabled smart applications into urban planning and the enforcement action should start right away. The building developers should also start to create awareness and marketing on the benefits of IoT enabled the smart home system to the housing buyers. There is also a need for Malaysian higher education to educate more talents and create more researchers who are to come out with Malaysian own smart home application, which end up to be more cost-effective in long run. These shall involve providing more research grants on the researching on IoT enabled smart building application. Finally, there is a need to educate Malaysian citizens on the benefits perceived by the IoT enabled smart building through education institutions. These overcoming strategies which involved government, developers and researchers should be concurrently implemented and no doubt Malaysia can be a step closer towards achieving sustainable cities.

5. Conclusion and Recommendation

It can be concluded that Malaysia has not fully ready to take advantage of the benefits the smart building offer. The Malaysian construction professionals might know the concept and how powerful smart buildings are, but they are lacking opportunity to realise it. This study has contributed to the body of knowledge on understanding on current readiness of Malaysia in the implementation of Smart Building and brought up the main challenges faced towards the adoption of Smart Building in Malaysia. Furthermore, this research has also provided great advantages to identify the area where more efforts are needed and, therefore, the direction of future research. The recommendations for future research are to explore further smart buildings readiness from different aspects such as technology, people, and management and determine the strategies to overcome the challenges of implementing smart buildings in Malaysia.

6. References

[1] Rathore M M, Ahmad A, Paul A and Rho S 2016 *Computer Networks* **101** 63-80
[2] Zhao Q, Xia L and Jiang Z 2018 *Energy Informatics* **1**
[3] Reffat R M 2010 *Proc. of the 10th Int. Conf. for Enhanced Building Operations (Kuwait)*
[4] Ma Z, Badi A and Jørgensen B N 2016 *IEEE Green Energy and Systems Conf. (California)*
[5] Ghaffarianhoseini A, Berardi U, Alwaer H, Chang S, Halawa E, Ghaffarianhoseini A and Clements D C 2016 *Archit. Sci. Rev.* **59** 338-357
[6] Shapi M K M, Ramli N A and Awalin L J 2021 *Developments in the Built Environment* **5**
[7] Ahmed M S, Mohamed A, Homod, R Z, Shareef H and Khalid K 2017 *J. of Eng. Sci. Technol.* **12** 1280-94
[8] Shaikh P H, Nor N B M, Sahito A A, Nallagownden P, Elamvazuthi I and Shaikh M S 2017 *Renew. Sustain. Energy Rev.* **75** 1392-1403
[9] Salleh H, Ali A S, Kamaruzaman S N and Chuing L S 2009 *Malays. Constr. Res. J.* **4** 42-53
[10] Li X, Nie L, Chen S, Zhan D and Xu X 2015 *IEEE Int. Conf. on Mobile Services (New York)* pp 438-445
[11] Giang N K, Lea R, Blackstock M and Leung V C M 2016 *Proc. of 2nd. Int. Workshop on Smart*

Cities (New York) vol 7
[12] Bashir M R and Gill A Q 2017 *IEEE Intelligent Systems Conf.* (London) pp 151-159
[13] Basarudin N A, Yeon A L, Yusoff Z M, Dahlan N H M and Author N M 2017 *Malays. Constr. Res. J.* **2** 209-222
[14] Jia M, Komeily A, Wang Y and Srinivasan R S 2019 *Autom. Constr.* **101** 111-126
[15] Hojjati S N and Khodakarami M 2016 *Int. J. of Adv. Netw. Appl.* **7** 2936-43
[16] Ogunde A O, Emmanuel I, Nduka D, Ayedun C A and Ogunde A 2018 *Int. J. Civ. Eng. Technol.* **9** 577-585
[17] Hsiao J S, Chen J C P and Tsaih L S J 2019 *IOP Conf. Ser: Mater. Sci. Eng. (Bangkok)* vol 652
[18] Wadhwa M 2015 *SSRN Electronic J.*
[19] Gobbo J A, Souza M G Z N and Gobbo S C O 2017 *Int. J. of Sustain. Real Estate and Constr. Economics* **1** 31-50

Effect of mixing temperature in the development of green polypropylene – coal bottom ash composite slab

Md Fauzan Kamal Mohd Yapandi[1*], Mohamad Fetri Zainal[1], Hanafi Yusop[1], Famiza Latif[2], Noor Najmi Bonnia[2], Mohammad Noor Jalil[2] and Nur Zaqira Izzati Sukhairul Zaman[1]

[1] TNB Research Sdn. Bhd., Kawasan Institusi Penyelidikan, Jalan Ayer Itam, 43000 Kajang, Selangor, Malaysia
[2] Faculty of Applied Sciences, Universiti Teknologi MARA, 40450 Shah Alam, Selangor, Malaysia

*Corresponding author: fauzankamal@tnb.com.my

Abstract. Coal bottom ash (CBA) is a residue produced from the burning of pulverized coal for electricity generation. The high porosity and low abrasion resistance of CBA hindered its commercialization; thus tonnes of CBA were discharged in landfills and surface impoundments. Considering the adverse social and environmental issues associated with disposal management, it is imperative to optimize the utilization of CBA. About 30% of CBA were homogeneously blended with Polypropylene (PP) to produce a "green" PP-CBA composite slab. The influences of mixing temperatures on the physical, chemical and mechanical properties of the fabricated slabs were examined in the study. FESEM micrographs illustrated a smoother surface with a low density of voids in composite systems mixed at low temperatures (190°C). Moisture content was retained in the PP-CBA matrix and imparted supplementary strength to the composite. The relatively amorphous state of the composite allowed for easier dispersion of molecules under applied stress. From the tests, PP-CBA composite slabs mixed at 190°C exhibited the largest drop impact energy and ideal physical and structural properties.

1. Introduction

Coal is deemed as a significant source of energy to cater to the surging electricity demand and growing global population. In 2019, 7,921 Mt of coal was produced, with Asia emerging as the largest market for global coal production and consumption [1]. Despite the substantial coal-gas switching observed in other regions, the energy mix strategy in Asia is still reliant on coal power (36% of electricity source), notably due to the challenges associated with the economic growth and transitions to renewable energy [2] [3]. However, the proliferation of coal combustion contributes to the massive production of coal ashes (~780 Mt/year) and other coal combustion by-products such as boiler slag and flue gas desulfurization residue [4]. Approximately 80% of the coal ashes are captured as coal fly ash (CFA) while coal bottom ash (CBA) makes up the remaining proportion [5].

CFA is widely used as supplementary cementitious material in concrete production, road base construction and synthesis of zeolite due to its pozzolanic properties [6]. On the contrary, the porous and granular structure of CBA limits its applications, hence CBA is often discharged into impoundment ponds and results in an abundance of coal ash lagoons [7][8].

The metal oxides of CBA as listed in table 1 are capable of forming inter-molecule crosslinks with polymer to form a robust structure. However, limited studies have been conducted on the feasibility of CBA in polymer composite systems despite its promising benefits [9]. Therefore, green polymer-based composites with CBA as the reinforcement material were developed in this study. Polypropylene (PP) is selected due to its availability, good chemical resistance, high impact and flexural strength and potential as a 'filler' to the porous structure of CBA [10]. Thus, the incorporation of both PP and CBA is expected to produce a sustainable composite capable of tackling issues related to the dumping of CBA.

The effects of different mixing temperatures were explored to determine the ideal process condition in developing green PP-CBA composite slab with desirable mechanical strength. PP particles and CBA samples were heated to a high enough temperature to achieve sufficient mixed with no degradation on the quality of composite pellets and slabs. The physical, chemical and mechanical properties of each slab were tested to establish the optimal mixing temperature.

Table 1. Chemical compositions of CBA in Malaysia power plants [11][12]

Chemical contents	**Composition of coal bottom ash**		
	Tanjung Bin Power Plant [13]	Jimah Power Plant	Kapar Power Plant [14]
SiO2	42.7	49.4	9.78
Al2O3	23.0	22.3	20.75
Fe2O3	17.0	13.7	37.1
CaO	9.80	9.00	11.1
K2O	0.96	1.00	-
TiO2	1.64	2.20	-
MgO	1.54	0.87	3.2
P2O5	1.04	0.65	-
Na2O	0.29	0.13	-
SO3	1.22	0.68	-
BaO	0.19	-	-
MnO	-	0.08	-
ZnO	-	-	1.8

2. Methodology

2.1. Fabrication of PP-CBA Composite

The CBA samples were crushed and ground to ~250 μm before mixing with PP in Muriyama Dispersion Mixer at different temperatures of 190°C, 200°C and 220°C. The amount of CBA incorporated in the mixture was fixed at 30% to satisfy the minimum requirement of a 'green' product requirement as stated in the SIRIM Eco-Label certification. Homogeneous putty-like blends from the mixer were left to harden at room temperature (T) before undergoing further crushing. The PP-CBA pellets were placed in a pre-greased mould (size: 15 cm x 15 cm x 4mm) and pressed for 10 minutes under constant pressure (1000 psi) at 190°C using hot compression moulding to produce homogeneously blended PP-CBA composite slabs.

2.2. Characterization of PP-CBA Composites

The composite specimens were pre-coated with gold prior to Field Emission Scanning Electron Microscope (FESEM) analysis as they were non-conductive. The FESEM analysis was performed to examine the morphologies of the composites. Fourier- Transform Infrared Spectroscopy (FTIR) analysis were carried out using Perkin Elmer Spectrum One Series with 4 cm^{-1} resolution and 16 times scan rate to visualize the changes in chemical structures of the composites. Significant functional groups in each composite sample were observed within the range of 4000-450 cm^{-1}.

The XRD patterns for all composites were recorded at room temperature from 2θ= 10° to 90° at a scanning speed of 2.5°θ/2θ per minute by X-ray diffractometer, XRD PAN analytical X'Pert PRO using CuKα radiation (λ=1.5418 Å). Drop impact test conforming to ASTMD5628 was conducted using Dynatup 8250 drop- weight impact machine. Specimens of size 5mm x 5mm were supported on two 12.5 mm diameter of stainless steel cylindrical movable support. A 3.0 kg hemispherical tip impactor (diameter of 16 mm) was released from a height of 75 mm.

2.3. Modification of Mixing Temperature
The compatibility of PP and CBA in the fabricated composite slabs were evaluated by varying the mixing temperature at 190°C, 200°C and 220°C. Similar characterization procedures were carried out to determine the mixing temperature that exhibited the ideal mechanical properties.

3. Results and Discussions
3.1. FESEM Analysis of PP-CBA Composites
CBA has successfully integrated into the polymer matrix, as suggested by the lack of phase separation and relatively smooth surface of composite slabs mixed at 190°C and 200°C. On the contrary, a porous and irregular texture was detected from the FESEM analysis of composite slab mixed at 220°C, as depicted figure 1(c). This discrepancy can be explained by the creation of a large gel macrostructure from the heightened gelation process which in turn, obstructed the melting of the composite slab.

Gelation behaviour was observed in PP-CBA composites mixed at high temperature due to strong hydrogen bonding between the hydrogen atoms of PP chains and CBA oxides. The viscosity of the polymer was significantly reduced at high mixing temperature and the energy generated was sufficient to overcome the intermolecular forces of PP. Thus, the surge of molecular mobility of PP and high content of alumina, crystalline quartz, hematite and other CBA metal oxides enhanced the cross-linkage in PP-CBA composite slab.

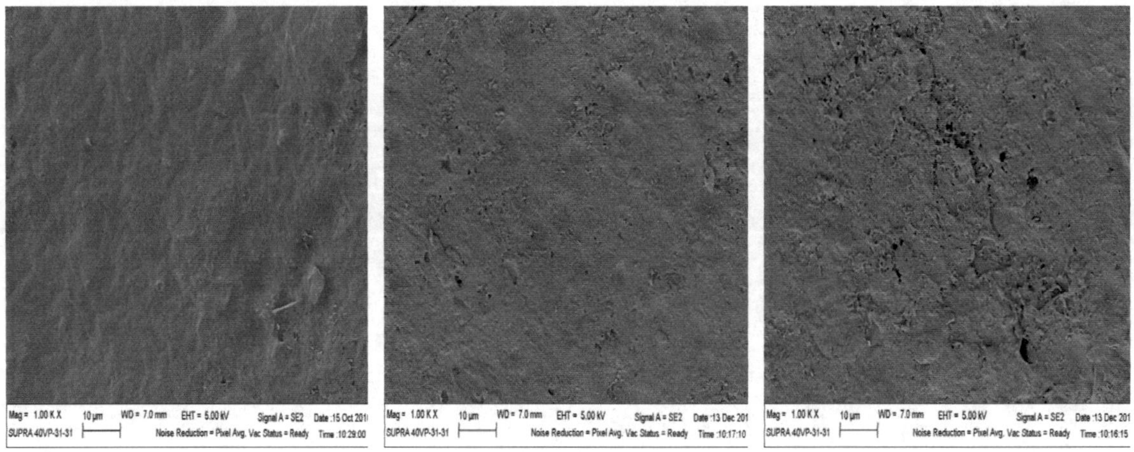

(a) (b) (c)

Figure 1. FESEM micrographs for 70:30 PP-CBA composite slabs mixed at (a) 190°C, (b) 200 °C and (c) 220°C.

3.2. FTIR Analysis of PP-CBA Composites

Moisture escaped from silanol groups of CBAs and created additional free silica groups at high mixing temperatures (220°C). These reactions were verified by the absence of silanol peak at 3300 cm^{-1} and an increase in silica's peak intensity at ~800 cm^{-1} in FTIR spectra shown in figure 2. The ejection of volatile components from the composite at high mixing temperature has notably created a higher density of unoccupied space on the composites.

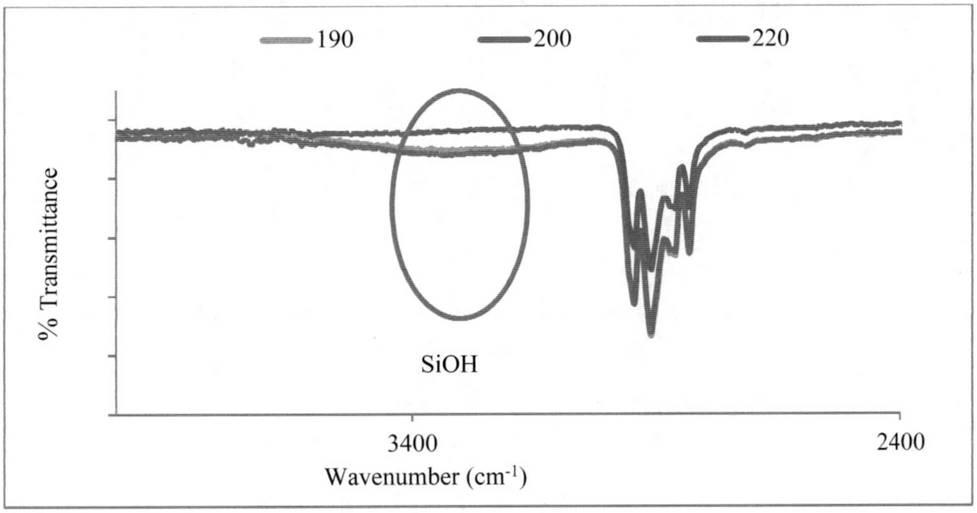

(a)

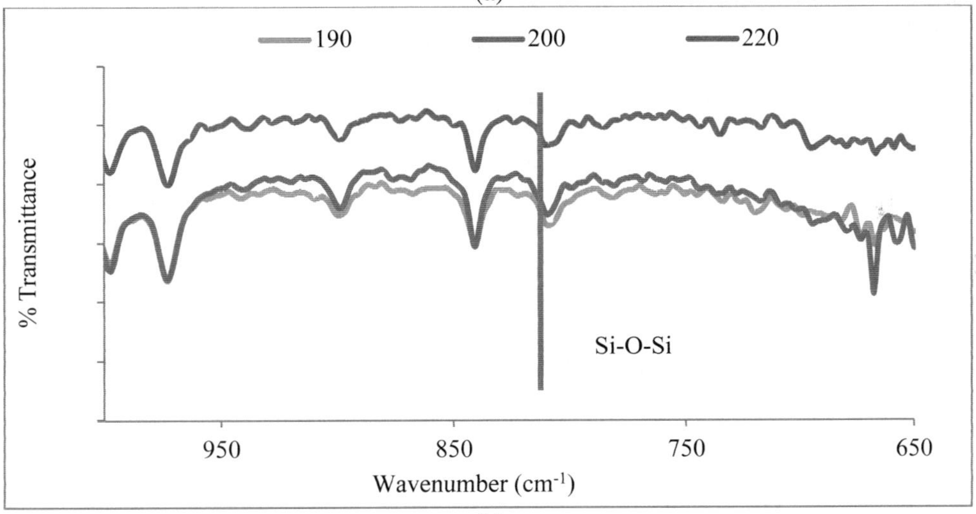

(b)

Figure 2. (a) Si-OH and (b) Si-O-Si peaks observed in FTIR spectrum for 70:30 PP-CBA composite slabs at different mixing temperatures.

Additionally, these vacant spaces assisted the rotational movement of PP chains to form a more ordered isotactic arrangement. The increase in crystallinity of the composite was verified by the greater isotactic peak of PP observed at ~1160 cm^{-1} in figure 3.

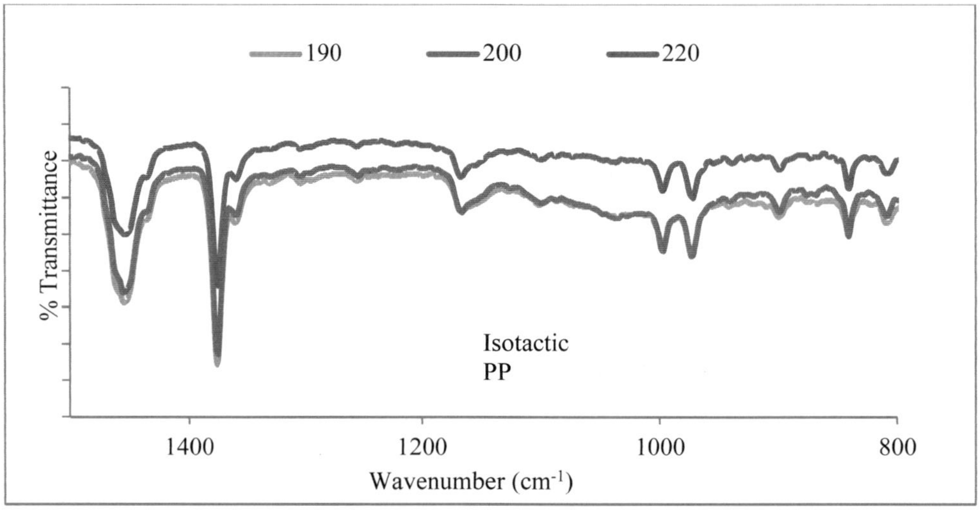

Figure 3. Isotactic peak of PP in FTIR spectra for PP-CBA composite slabs mixed at different temperatures.

3.3. XRD Analysis of PP-CBA Composites

The XRD diffractogram in figure 4 displayed an increase in intensity peak at $2\theta = 14° - 21°$ which proved that the degree of crystallinity of composite slab intensified at higher mixing temperature. The increase in mobility motion prompted the realignment of polymer into a less viscous state (crystalline phase) and well-ordered arrangement.

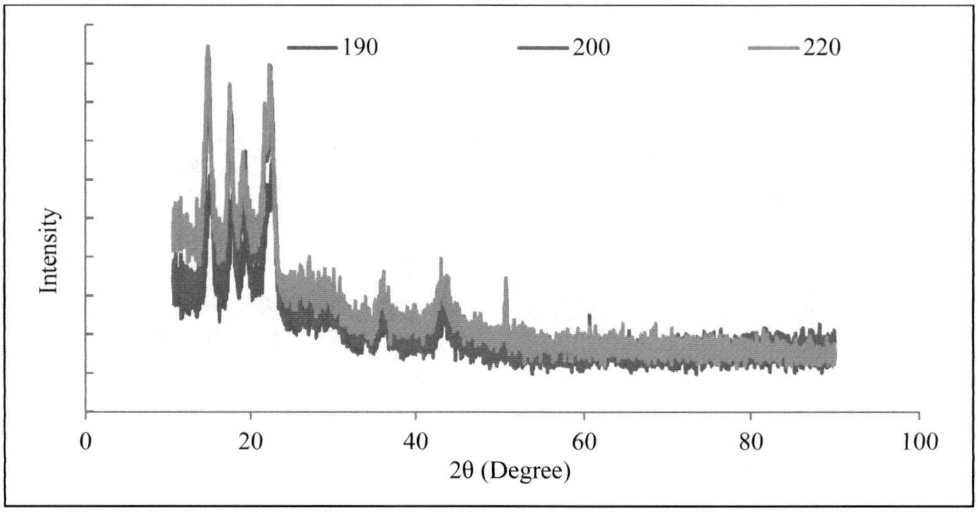

Figure 4. XRD diffractogram of 70:30 PP-CBA composite slabs mixed at 190°C, 200°C and 220°C.

3.4. Drop Impact Studies

Small and large gaps were observed in the composite slabs mixed at high temperatures, as shown in figure 5, due to the high density of large pores caused by the elimination of moisture from the specimens. These gaps were unable to withstand high impact from the applied load and thus, initiated micro-cracks and further crack propagations. The mechanical strength of the specimens were compromised from these stress-induced cracks and subsequently resulted in low drop impact energies. The average drop impact energies for all the PP-CBA composites are presented in table 2.

Table 2. Average drop impact energies recorded at various mixing temperatures.

Mixing Temperature (°C)	Average Drop Impact Energy (J)
190	2.6
200	2.5
220	2.2

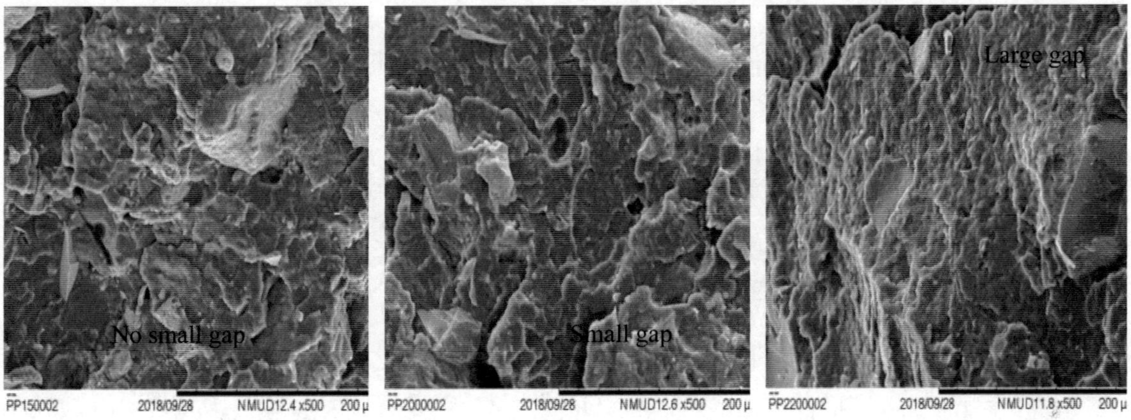

(a) (b) (c)

Figure 5. FESEM micrographs of 70:30 PP-CBA composite slabs mixed at (a) 190°C, (b) 200 °C and (c) 220°C after drop impact test at 75 mm high.

Conversely, the composite slab mixed at 190°C demonstrated the greatest impact energy value (2.6 J) and quality upon drop impact test at 75mm high (figure 5(a)). The performance could be attributed to the presence of Si-OH group which provided supplementary mechanical strength to the fabricated slab. Furthermore, the slab recorded a lower intensity peak and exhibited a relatively amorphous phase. In this state, the intermolecular forces were slightly weaker than those in PP-CBA composites mixed at high temperatures. PP particles were able to move freely and filled the porous structure of CBA when the composite was subjected to external stress. Consequently, the even distribution of molecules aided in reducing stress impact and formed a homogeneous and smooth surface of the composite.

4. Conclusion

A "green" PP-CBA composite slab was successfully fabricated by incorporating 30% coal bottom ash waste. PP-CBA composite slab mixed at 190°C demonstrated the ideal mechanical, structural and chemical properties. At this temperature, integration of CBA into polymer matrix occurred due to intermolecular bonding between the CBA oxides and hydrogen atoms in PP chains. As a result, the crystalline phase of PP shifted to a more amorphous state in the PP-CBA composite. The phase transition enabled freely moving PP particles to occupy the porous structure of CBA under the application of force and form a smoother PP-CBA composite. On the contrary, the ejection of moisture from silanol groups of CBAs at 220°C led to the formation of free silica groups and more vacant spaces in the composite. Hence, the porous PP-CBA composite slab was unable to withstand high drop impact, as opposed to the slab mixed at 190°C.

5. References

[1] International Energy Agency (IEA). Coal information overview and statistical report (2020 Edition) International Energy Agency Report. 2020.
[2] Harris D, Heidrich C, Feuerborn J. Global aspects on Coal Combustion Products. VGB Powertech. Germany; 2019.
[3] International Energy Agency (IEA). Coal Fired Power. Paris; 2020.
[4] Izquierdo M, Querol X. Leaching behaviour of elements from coal combustion fly ash: An overview. *Int J Coal Geol*. 2012;**94**:54–66.
[5] Singh M, Siddique R. Effect of coal bottom ash as partial replacement of sand on workability and strength properties of concrete. *J Clean Prod*. 2016;**112**:620–30.
[6] Sakir S, Raman SN, Safiuddin M, Amrul Kaish ABM, Mutalib AA. Utilization of by-products and wastes as supplementary cementitious materials in structural mortar for sustainable construction. *Sustainability*. 2020;**12**(9):1–35.
[7] Ramzi NIR, Shahidan S, Maarof MZ, Ali N. Physical and Chemical Properties of Coal Bottom Ash (CBA) from Tanjung Bin Power Plant. *IOP Conf Ser Mater Sci Eng*. 2016;**160**(1).
[8] Sathonsaowaphak A, Chindaprasirt P, Pimraksa K. Workability and strength of lignite bottom ash geopolymer mortar. *J Hazard Mater*. 2009;**168**(1):44–50.
[9] Asokan P, Saxena M, Asolekar SR. Coal combustion residues - Environmental implications and recycling potentials. *Resour Conserv Recycl*. 2005;**43**(3):239–62.
[10] Hisham A. Maddah. Polypropylene as a Promising Plastic: A Review. *Am J Polym Sci* 2016;**6**(1):1–11.
[11] Itam Z, Beddu S, Mohammad D, Kamal NLM, Zainoodin MM, Syamsir A, et al. Extraction of metal oxides from coal bottom ash by carbon reduction and chemical leaching. *Mater Today Proc*. 2019;**17**:727–35.
[12] Abubakar AU, Baharudin KS. Potential Use of Malaysian Thermal Power Plants. *Int J Sustain Constr Eng Technol*. 2012;**3**(2):25–37.
[13] Muhardi A, Marto A, Kassim KA, Makhtar AM, Wei LF, Lim YS. Engineering characteristics of Tanjung Bin coal ash. *Electron J Geotech Eng*. 2010;**15**:1117–29.
[14] Naganathan S, Subramaniam N, Nasharuddin Bin Mustapha K. Development of brick using thermal power plant bottomash and fly ash. *Asian J Civ Eng*. 2012;**13**(2):275–87.

Acknowledgements

The authors would like to acknowledge the contribution of the Faculty of Applied Sciences, UiTM and TNB Research Sdn. Bhd. for their support in providing the research facilities to carry out this project.

Journal of Physics: Conference Series **2053** (2021) 012003 doi:10.1088/1742-6596/2053/1/012003

Static concentrating photovoltaic modelling using MATLAB

Firdaus Muhammad-Sukki [1,2,*], Haroon Farooq [3], Siti Hawa Abu-Bakar [4], Jorge Alfredo Ardila-Rey [5], Nazmi Sellami [6], Ciaran Kilpatrick [6], Mohd Nabil Muhtazaruddin [2], Nurul Aini Bani [2] and Muhammad Zulkipli [7]

[1] Edinburgh Napier University, United Kingdom
[2] Universiti Teknologi Malaysia, Malaysia
[3] University of Engineering and Technology, Pakistan
[4] Universiti Kuala Lumpur, Malaysia
[5] Universidad Técnica Federico Santa María, Chile
[6] Robert Gordon University, United Kingdom
[7] Universiti Tun Hussein Onn Malaysia, Malaysia

* Corresponding author's email : f.muhammadsukki@napier.ac.uk

Abstract. The world has recorded an increasing interest and staggering investment in renewable technology in the last two decades, specifically in solar photovoltaic (PV). Concentrating PV (CPV) is one of PV's technology advancements and is gaining popularity for integration in a building. Various CPV designs are currently being investigated by researchers. The aim of this paper is to design and develop a MATLAB programme that can predict the electrical properties of a static concentrator that is designed with a ±40° acceptance angle. The programme was utlizied to determine the angular characteristics of the static concentrator between acceptance angle of -50° and 50°. It is proposed that the optoelectronic gain, $C_{opto\text{-}e}$ values be incorporated into the model to simulate a CPV design. The incident angle values (within ±50°) were chosen to demonstrate that the static concentrator could collect solar energy within its designed acceptance angle of ±40°. The current-voltage and power-voltage characteristics are generated for each simulation, and critical parameters such as the maximum power, open-circuit voltage, short-circuit current, and optoelectronic gain were identified and measured. The programme was found to be able to determine the electrical properties for the static concentrator.

1. Introduction

Solar photovoltaic (PV) technology has drawn major investment and research interest since the early 2000s, with the focus on improving the effectiveness of solar photovoltaic cells and addressing the economy feasibility of producing energy with the solar technology. The interaction of light photons with electrons in a solar cell that gives rise to the photovoltaic effect, primarily constructed of a doped semiconductor, mainly silicon to produce a p-n junction, is the working principle of the solar PV system. The electrons within the junction are freed as a result of this interaction, resulting in a hole to be occupied by any free electron moving in the p-n junction. With a constant supply of light source, these electrons contacts aggregate, resulting in an electron flow and, thus producing current. A commercially available solar cell made from silicon typically generates 0.5–0.6 V with resulting current of 28–35 mA/cm^2 [1]. For practical application, solar cells are grouped to form solar panels, allowing the generation of solar energy, and solar arrays are built by combining these panels, offering a more extensive surface area and generating much more electricity.

One way to improve the efficiency of a solar panel is by increasing the irradiance of sun energy that arrive at the surface area of the solar cell. This can be achieved through the use of a solar concentrator. If properly configured, a solar panel can be built with a relatively less number of solar cells than a regular solar panel of equal surface area would require [2]. The concentrating PV (CPV) technique is the name for this method. Furthermore, depending on the geometry of these lenses, the panel may absorb a large fraction of sun energy, even when it is not precisely oriented towards the light source, enhancing its efficiency even further in comparison to a regular solar panel.

With the advancement of PV systems, the world is seeing a growing demand to estimate solar PV performance and anticipate the total amount of energy produced under a particular set of conditions. Consequently, equivalent circuits for PV cells and other mathematical models have been developed to forecast a PV cell's performance characteristics and possible output power. It is worth noting that these models are not identical; each one has its own set of parameters. Experiments are frequently used to confirm the efficiency of these models.

Li et al. [3] detailed the development and testing of a MATLAB simulation for obtaining the model's parameters utilising open circuit currents, closed circuit currents, and maximum power, via an iterative calculation until convergence. The models' performance were compared to the projected features from the datasheet of the PV cells and experimental findings of different PV modules, yielding an average inaccuracy of 5.53%. Mahmoud et al. [4] thoroughly presented the design and testing of a new method for calculating PV cell parameters using the data provided within the manufacturer's datasheet and the single-diode equivalent circuit. When tested against experimental results, it had a normalised root mean square deviation of less than 4.91%, and for most tests, the error was around 1%, therefore, making this an excellent mathematical simulation model.

Mehta et al. [5] designed and implemented a single-diode model using MATLAB to tackle the noise in the measurements and results in a highly accurate model where the errors recorded were between 2.5938×10^{-5} and 1.6299×10^{-3} A when compared with experimental results. Vinod et al. [6] created a single-diode model using MATLAB/Simulink. Their stepwise model requires three parameters to be identified analytically, namely the photocurrent, the saturation current and the reverse saturation current. The simulation results achieved a maximum relative error of 1.65% when compared with the experimental result.

This work seeks to forecast the current-voltage (I-V) and power-voltage (P-V) characteristics of a static concentrator. The following is how the paper is organised: Section 1 provides an overview of the static concentrator and solar photovoltaic modelling, Section 2 details the processes to model the concentrator's features, Section 4 shows the findings from the bespoke model, and Section 5 concludes.

2. Modelling of the static concentrator

This study uses analytical methods in the MATLAB programming environment to offer a modelling of a single-diode equivalent circuit for a CPV configuration. When Vinod et al. [6] used Simulink to model a solar PV module, they provided a stepwise technique, which is used in this model. During modelling, the datasheet for a solar PV cell from National Renewable Energy Centre was used, as well as results from earlier studies characterising the static concentrator. Simulation-generated characteristics of I-V and P-V are examined to experiment-derived values.

Ramirez-Iniguez et al. [7] designed the static concentrator to enhance electrical yield, reducing the amount of expensive PV material used, lowering the PV system's cost [8]. Ramirez-Iniguez et al. [7] have provided a detailed description of the static concentrator creation process. It is proposed that the value of optoelectronic gain should be incorporated in the modelling of any CPV. All CPV concentrates sun energy from a vast region into a smaller concentrated area where a solar PV cell is installed [9]. Equation 1 shows the relationship between the solar irradiance, G, with the irradiance at reference of 1,000 W/m^2, G_{ref}, the optoelectronic gain, C_{opto-e} as well as the angle of incidence with the integration of CPV, G_{con},. C_{opto-e} is equivalent to 1 for a non-CPV.

$$G_{con} = C_{opto-e}\left(\frac{G}{G_{ref}}\right)\cos\theta \tag{1}$$

Muhammad-Sukki et al. presented their indoor tests to characterise the static concentrator-PV configuration in [10]. The static concentrator is 3 cm in height and has a 1 cm^2 square exit aperture area. It also has a geometrical concentration gain of 4.91. Along its x-axis, the concentrator is designed to attain a half-acceptance angle of ±40 degrees. At varied incidence angles between -50 degrees and 50 degrees, the angular response of the static concentrator -PV configuration and the plain PV cell was examined. The angular response allows the static concentrator's optoelectronic gain to be measured at various incidence angles. The static concentrator's optoelectronic gain at various incidence angles is presented in the following Table 1.

Table 1. The static concentrator's optoelectronic gain at various angles of incidence.

The Angle of Incidence (°)	Opto-electronic Gain, C_{opto-e}
0	4.17143
±5	4.00000
±10	3.89205
±15	3.60465
±20	3.37313
±25	3.14642
±30	2.78105
±35	1.84722
±40	1.21561
±45	0.95142
±50	0.69643

A solar PV cell's current and voltage output can be predicted by converting it to an equivalent electrical circuit. A diode coupled in parallel with a photocurrent, I_{ph}, is ideal for representing a PV cell's equivalent circuit as a light generated current source. However, in practice, factors such as the material's resistivity, the influence of shunt resistance and the ohmic losses must all be considered. Series resistance, R_s and shunt resistance, R_{sh} are two factors that can be simulated. The PV cell's equivalent circuit can be represented using a single-diode model. Equation (2) [11] gives the following formula for the output current of a single-diode model:

$$I = I_{ph} - I_d - I_{sh} \quad (2)$$

where I_d is the diode currents. Equation (1) can be expressed as [12]:

$$I = I_{ph} - I_s\left[\exp\left(\frac{q(V + IR_s)}{n_d kT} - 1\right)\right] - \frac{V + IR_s}{R_{sh}} \quad (3)$$

where I_s is the diode's saturation current, n_d is the diode ideality factor, q is the electron charge (1.602 x 10^{-19} C), k is the Boltzman constant (1.38 x 10^{-23} J/K), T is the p-n junction temperature (in Kelvin), and V is the the PV cell's voltage.

It is necessary to calculate I_s, I_{ph}, and I_{rs} in this model. The following equations will be used to calculate the values. I_{ph} is unaffected by V (or R_s), is linearly dependent on solar irradiation, and is temperature-dependent. Equation 4 [6] illustrates this link.

$$I_{ph} = (I_{sc} + \alpha(T - T_{ref})) * G_{con} \quad (4)$$

where T_{ref} is 25 °C and α is the current temperature coefficient.

Saturation current I_s and reverse saturation current, I_{rs} are calculated using both Equations (5) and (6), with the value of the energy bandgap, E_g, as reported by Vinod et al. [6]. The essential parameters values of the solar PV cell are listed in Table 2. The solar PV cell's manufacturer, Narec, gave these parameters (National Renewable Energy Centre).

$$I_{rs} = I_{sc} / \left[\exp\left(\frac{qV_{oc}}{n_{d1}kT} \right) - 1 \right] \tag{5}$$

$$I_s = I_{rs} \left[\frac{T}{T_{ref}} \right]^3 \exp\left[\left(\frac{qE_g}{n_d k} \right) \left(\frac{1}{T} - \frac{1}{T_{ref}} \right) \right] \tag{6}$$

Table 2. Essential parameters of the solar cell

Component	Value
Solar radiation intensity, G_{ref}	1,000 W/m^2
Reference Temperature, T_{ref}	25 °C
Square solar cell area, A	1 cm^2
Short circuit current, I_{sc} (at 0°)	0.0350 A
Open-circuit voltage, V_{oc} (at 0°)	0.586 V
Ideality factor, n	1.109
Energy bandgap, E_g	1.12
Charge of an electron, q	1.6 x 10^{-19} C
Boltzman constant, k	1.38 x 10^{-23} m^2 kg s^{-2} K^{-1}
Series resistance, R_s	0.047994 Ω
Shunt resistance, R_{sh}	2148.53 Ω
Short circuit current temperature coefficient, α	0.00024 A/°C

3. Simulation process

Equation (1)- was used to create a MATLAB algorithm to simulate the electrical power output generated by the static concentrator-PV construction (6). The code was created to emulate the static concentrator-PV configuration's angular properties from -50 degrees to 50 degrees with a 5 degrees increment. Figure 1 shows the stages involved in creating the I-V and P-V curves for each incident angle. The solar beams are perpendicular to the static concentrator at a 0 incidence angle. The incident angle range (±50 degrees) was chosen to show that the static concentrator can capture solar rays within its acceptance angle of ±40 degrees. The code considers a number of factors, including solar irradiation, optoelectronic gain (as shown in Equation (1)), and the solar PV cell's key characteristics. The values of I_{ph}, I_{rs}, and I_s are then calculated analytically by the programme. The equivalent I value for voltages that range from 0 to V_{oc} is then derived using the Newton- Raphson iteration method by substituting Equations (4)–(6) into Equation (3). P is derived by multiplying the matching values of I and V. Finally, the I-V and P-V curves are plotted. The I-V and P-V characteristics were produced in each simulation, and maximum power (P_{max}), the short circuit current (I_{sc}), open-circuit voltage (V_{oc}), and optoelectronic gain (C_{opto-e}) were calculated and recorded.

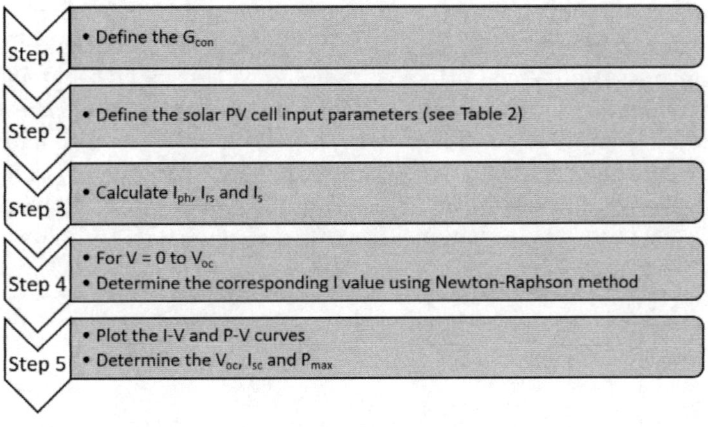

Figure 1. The process to obtain the angular response of the static concentrator at all incidence angles.

4. Results and discussions

The simulated characteristics of I-V and P-V for the static concentrator -PV configuration and a plain PV cell at normal incidence were run at a temperature of 25 °C with sun irradiation of 1,000 W/m^2 as shown in Figure 3. According to calculations, the plain PV cell produced the maximum power output of 0.0165 W, open-circuit voltage of 0.5860 V, and short circuit current of 0.0350 A. The addition of static concentrator boosts the maximum output power to 0.0740 W, the open-circuit voltage to 0.0627 V, and the short circuit voltage to 0.1460 A. A value of 4.1714 is calculated as the simulated optoelectronic gain.

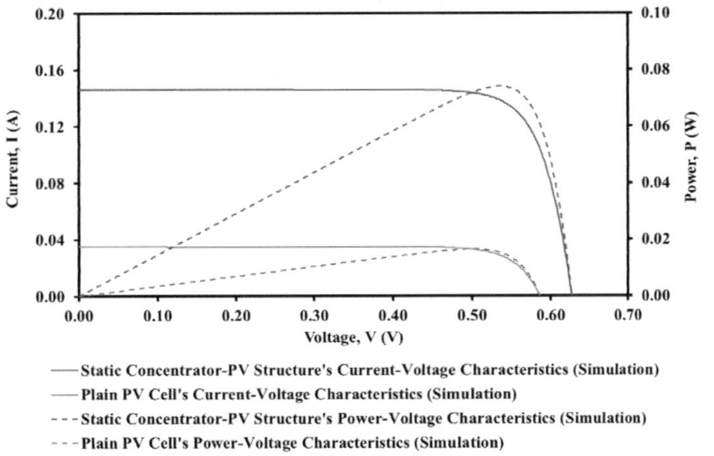

Figure 2. Simulated characteristics of I-V and P-V of the static concentrator-PV configuration and the plain PV cell at STCs at normal incidence.

In order to analyse the angular response of the static concentrator-PV configuration and the plain PV cell, it was tested at various incidence angles ranging from -50 degree to 50 degree at STCs. Each simulation determined and recorded the maximum power, short circuit current, open-circuit voltage, and optoelectronic gain.

Figure 3 shows the simulated short circuit currents of the static concentrator-PV configuration and a plain PV cell at different incidence angles. The short circuit current for the static concentrator-PV configuration dropped from 0.1460 A at average to 0.1219 A at ±15° of incidence. When the incident angle is greater than ±30°, the short circuit current dropped to less than half of its highest value. Outside of the intended half-acceptance angle of 40, the static concentrator-PV configuration produces a lower short circuit current than a plain PV cell. Moreover, these rays escaped through the side profile of the concentrator and did not reach the exit aperture at angles outside of the half-acceptance angle. Instead, they emitted through the concentrator's side profile [13]. Within the 'design' half-acceptance angles, the static concentrator-PV configuration provides a considerably larger short circuit current than a plain PV cell. As the incident angle increases from 0° to ±50°, the plain PV cell shows a similar trend, with the simulated short circuit current at normal incidence drops from 0.0350 A to lower values.

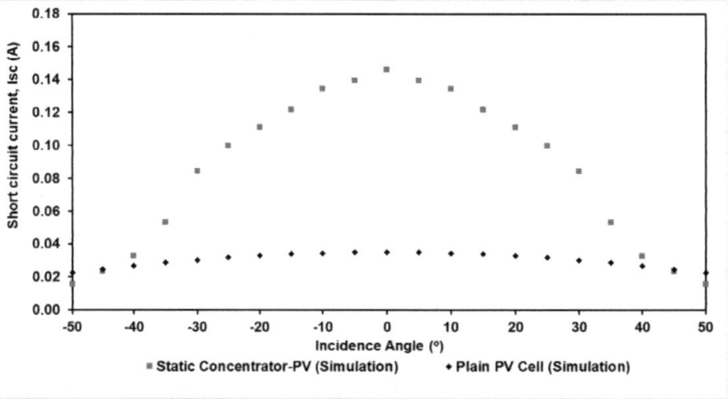

Figure 3. The simulated short circuit currents for the static concentrator-PV configuration and a plain PV cell at STCs.

Figures 4 demonstrates the simulated open-circuit voltages generated for the static concentrator-PV configuration vs a plain PV cell for various angles of incidence at STCs. For the static concentrator-PV configuration at normal incidence, the value of open circuit voltage gradually decreased from 0.6270 V to 0.5440 V at a ±50 degrees angle of incidence. Meanwhile, the corresponding value for the plain PV cell at normal incidence gradually decreased from 0.5860 V to 0.5740 V at a ±50 degrees angle of incidence.

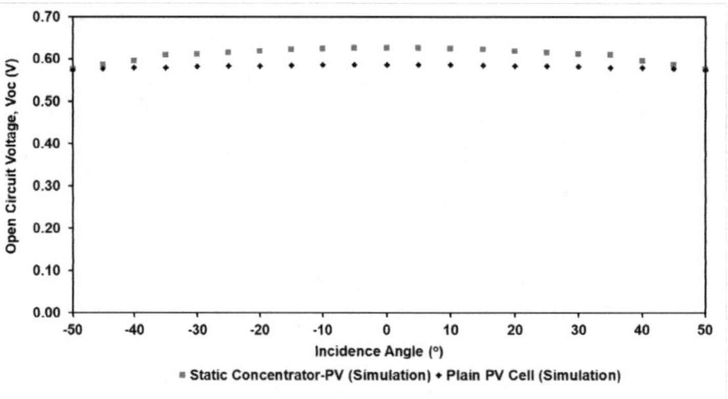

Figure 4. Comparison of the simulated open circuit voltages for the static concentrator-PV configuration and a plain PV cell at STCs.

Figure 5 compares the numbers derived from simulations for the static concentrator-PV configuration with a plain PV cell at STCs for different incident angles at maximum power generation. According to simulations, the max value of 0.0740 W was attained at normal incidence for the static concentrator-PV configuration. At a ±15 degrees angle of incidence, the value dropped to 0.0613 W. However, as the incident angle becomes greater than ±30 degrees, the maximum power value drops to lower than 50%

of the maximum value. Outside of the design half-acceptance angle of 40 degrees, the maximum power value generated by the static concentrator-PV configuration is lower than that produced by a plain PV cell. The plain PV cell exhibits a similar trend, where the corresponding value drops from 0.0165 W at normal incidence to lesser values when the incident angle is increased from 0 to ±50 degrees.

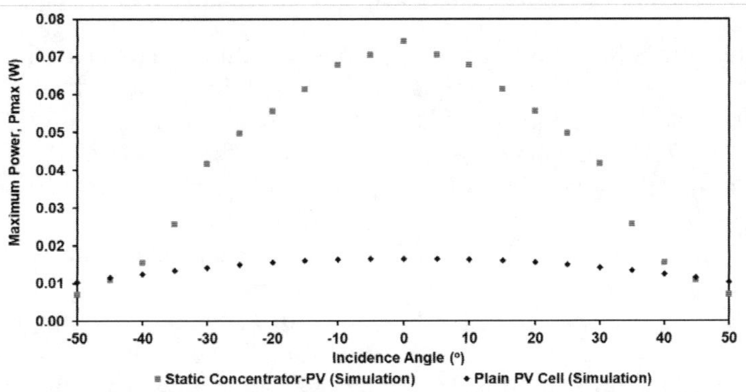

Figure 5. Comparison of the simulated maximum powers for the static concentrator-PV configuration and a plain PV cell at STCs.

Figure 6 displays the optoelectronic gains for the static concentrator-PV configuration at STCs based on simulations. The simulated optoelectronic gain decreased from a peak of 4.1714 at normal incidence to 3.6047 at a ±15 degrees angle of incidence. When the incident angle exceeds ±30 degrees, the optoelectronic gain value drops to lower than 50% of its maximum value. Beyond the design of half-acceptance angle of ±40 degrees, the optoelectronic gain value is less than 1.

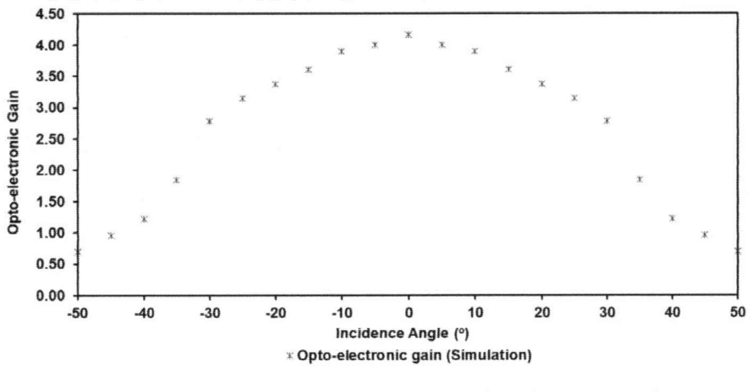

Figure 6. Simulated optoelectronic gains for the static concentrator-PV configuration at STCs.

5. Conclusions

Modelling the performance and predicting the energy generated by any solar PV cell has become increasingly popular. As a result, several researchers have constructed equivalent circuits for PV cells and other modellings to forecast the characteristics and possible photovoltaic cell output power. The performance of a MATLAB algorithm created to model the I-V as well as P-V properties of the static concentrator-PV configuration was discussed in this work. Under STCs, the static concentrator-PV configuration and the plain PV cell are simulated at various incidence angles that ranges from -50 degrees to 50 degrees. The simulations produced an excellent result, allowing the short-circuit current values, open circuit voltage, maximum power, and optoelectronic gain to be calculated. The model's accuracy would be determined by comparing the simulations to the experimental results.

6. References

[1] Qazi S 2017 Mobile Photovoltaic Systems for Disaster Relief and Remote Areas *Standalone Photovoltaic (PV) Systems for Disaster Relief and Remote Areas* (Massachusetts, USA: Elsevier) pp 83–112

[2] Alamoudi A, Saaduddin S M, Munir A B, Muhammad-Sukki F, Abu-Bakar S H, Mohd Yasin S H, Karim R, Bani N A, Mas'ud A A, Ardila-Rey J A, Prabhu R and Sellami N 2019 Using static concentrator technology to achieve global energy goal *Sustainability* **11** 3056:1-22

[3] Li W, Paul M C, Baig H, Siviter J, Montecucco A, Mallick T K and Knox A R 2019 A three-point-based electrical model and its application in a photovoltaic thermal hybrid roof-top system with crosed compound parabolic concentrator *Renew. Energy* **130** 400–15

[4] Mahmoud Y A, Xiao W and Zeineldin H H 2013 A parameterisation approach for enhancing PV model accuracy *IEEE Trans. Ind. Electron.* **60** 5708–16

[5] Mehta H K, Warke H, Kukadiya K and Panchal A K 2019 Accurate Expressions for Single-Diode-Model Solar Cell Parameterization *IEEE J. Photovoltaics* **9** 803–10

[6] Vinod, Kumar R and Singh S K 2018 Solar photovoltaic modeling and simulation: As a renewable energy solution *Energy Reports* **4** 701–12

[7] Ramirez-iniguez R, Muhammad-Sukki F, McMeekin S G and Stewart B G 2018 Optical element (Patent No. US9,910,253,B2) 1–26

[8] Muhammad-Sukki F, Abu-Bakar S H, Ramirez-Iniguez R, McMeekin S G, Stewart B G, Sarmah N, Mallick T K, Munir A B, Mohd Yasin S H and Abdul Rahim R 2014 Mirror symmetrical dielectric totally internally reflecting concentrator for building integrated photovoltaic systems *Appl. Energy* **113** 32–40

[9] O'Gallagher J J 2008 Nonimaging Optics in Solar Energy *Synth. Lect. Energy Environ. Technol. Sci. Soc.* **2** 1–120

[10] Muhammad-Sukki F, Abu-Bakar S H, Ramirez-Iniguez R, McMeekin S G, Stewart B G, Munir A B, Mohd Yasin S H and Abdul Rahim R 2013 Performance analysis of a mirror symmetrical dielectric totally internally reflecting concentrator for building integrated photovoltaic systems *Appl. Energy* **111** 288–99

[11] Bonkoungou D, Koalaga Z and Njomo D 2013 Modelling and simulation of photovoltaic module considering single-diode equivalent circuit model in MATLAB *Int. J. Emerg. Technol. Adv. Eng.* **3** 493–502

[12] Mammo E D, Sellami N and Mallick T K 2013 Performance analysis of a reflective 3D crossed compound parabolic concentrating photovoltaic system for building façade integration *Prog. Photovoltaics Res. Appl.* **21** 1095–103

[13] Welford W T and Winston R 1989 *High Collection Nonimaging Optics* (California, USA: Academic Press)

Acknowledgement

Edinburgh Napier University supported this research through the SEBE SEED-Corn fund, Ministry of Higher Education (MOHE) through Fundamental Research Grant Scheme (FRGS/1/2020/TK0/UTM/02/110) and Universiti Teknologi Malaysia under the Visiting Researcher (Publication) scheme (Q.J090000.21A4.00D20).

Journal of Physics: Conference Series **2053** (2021) 012004 doi:10.1088/1742-6596/2053/1/012004

Synthesis and characterisation of copper vanadate quantum dots / nanostructured zinc oxide

W Q Wee[1], L C Sim[1*], K H Leong[2] and P Saravanan[3]

[1] Department of Chemical Engineering, Lee Kong Chian Faculty of Engineering and Science, Universiti Tunku Abdul Rahman, Jalan Sg Long 9, Bandar Sg Long 43000 Kajang, Selangor, Malaysia
[2] Department of Environmental Engineering, Faculty of Engineering and Green Technology, Universiti Tunku Abdul Rahman, Kampar, Perak, Malaysia
[3] Department of Environmental Science and Engineering, Indian Institute of Technology (ISM), Dhanbad, Dhanbad-826004, Jharkhand, India

*Corresponding author: simlc82@yahoo.com

Abstract. 0D/3D heterojunction photocatalyst, specifically copper vanadate (CuVA) quantum dots (0D) / nanostructured zinc oxide (ZnO) (3D), has been prepared using in-situ reaction method. pH of 8 yielded the optimum nanostructure for the hydrothermal synthesis of ZnO nanorods. Characterisation of heterojunction photocatalysts obtained using Fourier transform infrared (FTIR) spectrometer, scanning electron microscopy (SEM) and energy dispersive X-ray (EDX) analysis indicates that this method is effective for incorporating quantum dots (QDs) onto ZnO. FTIR at wavenumber 3537, 1360 and 734 cm^{-1} showed a decreasing transmission peak as the amount of CuVA increases with respect to ZnO. EDX had showed an increasing trend in copper and vanadium elemental percentage in CuVA-ZnO as the targeted CuVA loading increases. 30 wt% CuVA QDs loaded onto ZnO was found to be the optimal mass loading to degrade sulfadiazine (SDZ), at 17.01% with initial concentration of 10 mgL^{-1}. This work is expected to provide a suitable alternative as a photocatalyst to degrade pollutants, with antibiotics being the focus of this work.

1. Introduction

Quantum dots (QDs) have been studied heavily due to their unique properties of small size (<10 nm) which leads to large surface area, higher electron transfer rate and tuneable optical absorptivity [1-3]. Despite these advantages, QDs are very vulnerable to self-aggregation and high recombination rate of electron-hole pairs [4-6]. To overcome these issues, heterojunction may be employed by loading QDs onto another semiconductor-based material such as titanium dioxide (TiO_2) and zinc oxide (ZnO). The interaction between these semiconductor materials can make QDs more dispersive and stable [7], while suppressing the recombination of electron-hole pairs by interfacial charge separation in the composites [8].

In semiconductor photocatalyst, there exists an energy gap that the electron is needed to overcome to escape from the valence band to conduction band, such gap is called band gap. ZnO has a wide band gap of 3.3 eV [9] while copper vanadate (CuVA) has a narrower band gap at 1.83 eV [10]. Individually, ZnO and CuVA separately are not an efficient photocatalyst due to high recombination rate of electron-hole pairs.

The objective of this work is to synthesise and identify the performance of 0D/3D heterojunctions photocatalyst through the incorporation of CuVA QDs (0D) onto nanostructured ZnO (3D). By incorporating both semiconductors together, a heterojunction photocatalyst is formed that in theory will improving the photocatalytic performance. Heterojunction photocatalyst is promising as the conduction band of semiconductor A will "catch" the excited electron from semiconductor B, thus decrease the recombination rate [11]. As control, the individual semiconductor will undergo the same photocatalytic degradation procedures. To the best of our knowledge, this work is the first to identify the performance of such photocatalyst towards degrading antibiotics, particularly sulfadiazine.

2. Methodology

2.1. Materials

All chemicals were used as received with no further purification, including zinc acetate dihydrate ($Zn(CH_3COO)_2 \cdot 2H_2O$, $\geq$ 98.0%, Sigma-Aldrich), hexamethylenetetramine (HMTA, $\geq$ 99.0%, Sigma-Aldrich), copper (II) nitrate trihydrate ($Cu(NO_3)_2 \cdot 3H_2O$, $\geq$ 99.0%, R&M Chemicals), ammonium metavanadate (NH_4VO_3, $\geq$ 99.0%, Sigma-Aldrich), 2-propanol (C_3H_8O, $\geq$ 99.5%, Sigma-Aldrich), sodium hydroxide (NaOH, $\geq$ 98.0%, Sigma-Aldrich), ammonium hydroxide (NH_4OH, 28.0 – 30.0%, Sigma-Aldrich) and sulfadiazine ($C_{10}H_{10}N_4O_2S$, 99%, Alfa Aesar).

2.2. Synthesis of zinc oxide

Zinc Oxide (ZnO) precursors were synthesised through hydrothermal method [12]. It involved mixing 1.25 mmol of $Zn(CH_3COO)_2 \cdot 2H_2O$ and 0.375 mmol of HMTA in 100 mL of distilled water under stirring for 30 min. The mixture was raised to pH 8 using 0.1 M sodium hydroxide solution. The solution was transferred into a Teflon liner stainless-steel autoclave (100 mL) and heated at 110 °C in oven for 15 h. The white precipitate formed were separated by centrifugation at 10000 RCF and washed with distilled water and 2-propanol to remove any residue precursor. The cleaned white precipitates were then dried at 60 °C overnight. After drying, the precipitate was annealed at 400 °C for 2 h, forming nanostructured ZnO.

2.3. Synthesis of copper vanadate / zinc oxide composites

The composites of CuVA QDs and ZnO were synthesised by using in-situ reaction method [13]. The overall yield of pristine CuVA QDs during preliminary studies were found to be 50%. Using this as basis, the ZnO loading is adjusted to the yield on top of 10 wt% dosage calculation. The ZnO was dispersed into 50 mL of 5 mmol/L $Cu(NO_3)_2 \cdot 3H_2O$ solution and stirred until homogeneous. 50 mL of 5 mmol/L NH_4VO_3 solution were added dropwise into the mixture. As shown in figure 1, the solution slowly turned yellow indicating formation of CuVA QDs. The pH was maintained at 7 using 0.1 M ammonium hydroxide solution throughout the synthesis. After adding all NH_4VO_3, the mixture was ultrasonicated for 40 min to allow the vanadate to crystallise. The obtained precipitates were washed similarly to the washing steps involved in ZnO synthesis. The obtained sample were named 10 wt% CuVA-ZnO.

These procedures were repeated for different weight percentage of CuVA, namely 30 wt%, and 50 wt%. The sample were denoted as x wt% CuVA-ZnO with x being the weight percentage of copper vanadate in the sample. As comparison, pristine CuVA QDs was also synthesised with the above method in the absence of ZnO.

Figure 1. Copper (II) nitrate solutions at (a) beginning of adding NH_4VO_3 and (b) end of adding NH_4VO_3.

2.4. Characterisation
Fourier transform infrared (FTIR) spectra were collected on a Nicolet iS10 FTIR Spectrometer to identify the interaction between CuVA and ZnO. Morphology and elemental distribution of samples were performed by scanning electron microscopy (SEM) and energy dispersive X-ray (EDX) analysis. It was caried out on Hitachi S-3400N variable pressure SEM at 15 kV and 20 kV acceleration voltage, respectively.

2.5. Photocatalytic degradation performance
Sulfadiazine (SDZ) was selected to investigate the photocatalytic performance of prepared catalyst in an aqueous solution under visible light irradiation [14]. The visible light source is a 13 W LED bulb (Philips Standard LED Bulb 13 W E27 6500K). The distance between the bulb and top liquid surface of solution was 10 cm. For each test, 100 mg of catalyst was dispersed in 100 mL solution of 10 mgL^{-1} SDZ. Prior to introducing the catalyst, a first 4 mL aliquot was collected and denoted as time, T = -960 min. To ensure adsorption-desorption equilibrium, the mixture was stirred overnight in the dark prior to the light irradiation for 16 hrs. During light irradiation process, 4 mL of the aqueous solution was collected every 20 min for 2 h and filtered using a 0.22 μm syringe filter to separate the photocatalysts. The concentrations of SDZ, including before and after irradiation were analysed using PG Instrument T60UV-Visible Spectrophotometer at 262 nm within 24 hrs of sample collection. This wavelength was determined to be the maximum wavelength of SDZ through a full-range scanning of 10 mgL^{-1} SDZ as shown in appendix A (figure A1) at 200 – 800 nm using Agilent Cary 100 UV-Visible Spectrophotometer. The photocatalytic degradation performance was calculated by the following equation:

$$\text{Degradation} = \frac{C_i}{C_0} \tag{1}$$

C_0 is the initial concentration of SDZ and C_i the concentration of SDZ at time *i*.

3. Results and Discussion
The white coloured powder of ZnO shows rod-like structure under SEM as shown in figure 2. Preliminary study was done at various reaction pH at 7, 8 and 11 [15]. At pH of 7, it was found that the ZnO possess hexagonal rod-like structure in clumps at approximately 10 μm. At pH 8, the ZnO possess nanorod-like structure at ~2 μm. At pH 11, the ZnO possess inconsistency in shapes and sizes. The SEM images of pristine ZnO at different pH are shown in figure 2. Based on this preliminary study, pH 8 were chosen to be the synthesise parameter to conduct further steps.

Figure 2. SEM image of pristine ZnO with reaction (a) pH 7 at 5 k magnification, (b) pH 8 at 20 k magnification and (c) pH 11 at 25 k magnification.

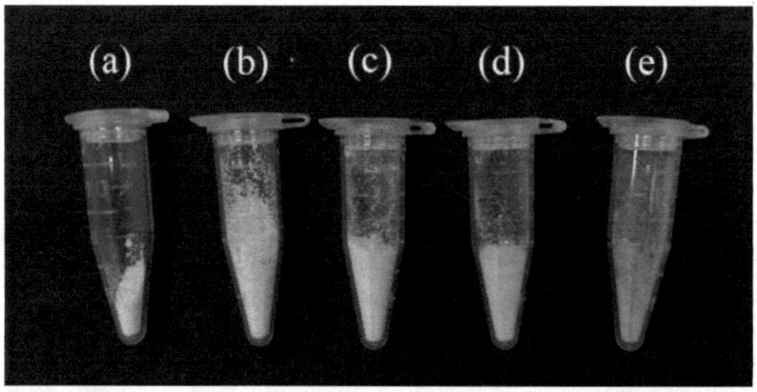

Figure 3. Colour of x wt% CuVA-ZnO. (a) ZnO, (b) 10 %, (c) 30 %, (d) 50 %, (e) CuVA.

Successful introduction of CuVA QDs onto ZnO was confirmed through EDX analysis. The details of EDX analysis are included in appendix B. In-situ reaction method had shown to be an effective method in incorporating CuVA onto ZnO. As shown in appendix B, as the targeted weight percentage increased, the composition of copper and vanadium increased linearly while zinc decreased linearly. The powder also progressively became greener as the weight fraction of CuVA increased (figure 3).

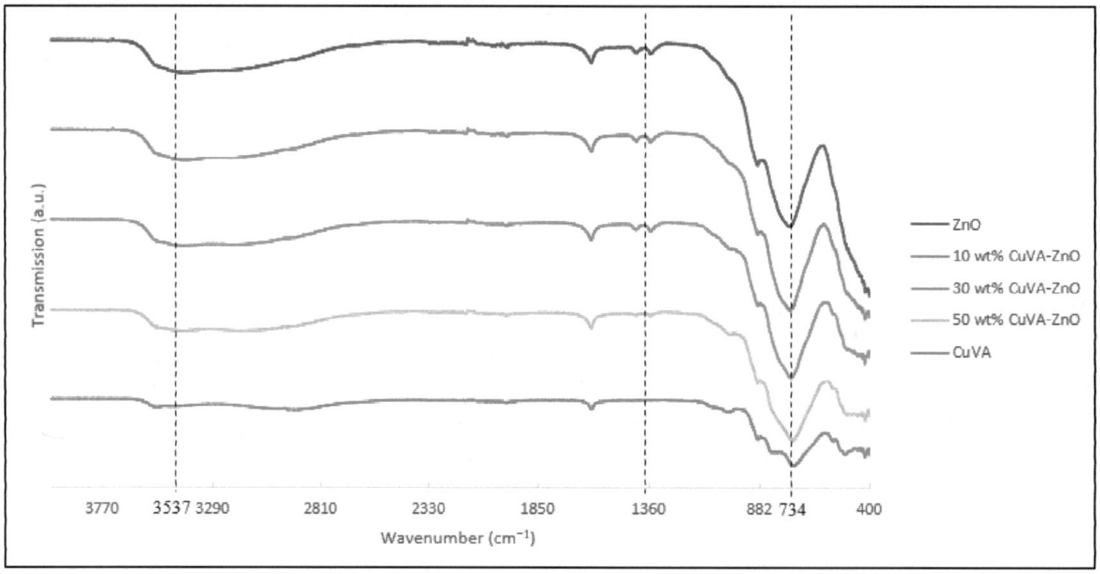

Figure 4. FT-IR Spectra of ZnO, 10, 30, 50 wt% CuVA-ZnO and CuVA.

In the FTIR spectra spectrum (figure 4) of 10, 30 and 50 wt% CuVA-ZnO, the broad band corresponding to ZnO from 2800 cm^{-1} to 3400 cm^{-1} was decreased consistently compared to that of pure ZnO. At 734 cm^{-1}, the peak shows a downward trend in sharpness consistently. The double peak around 1360 cm^{-1} that present in ZnO also became less pronounce as the percentage of CuVA increased. Thus, this confirms that the incorporation of CuVA at their respective percentage is successful.

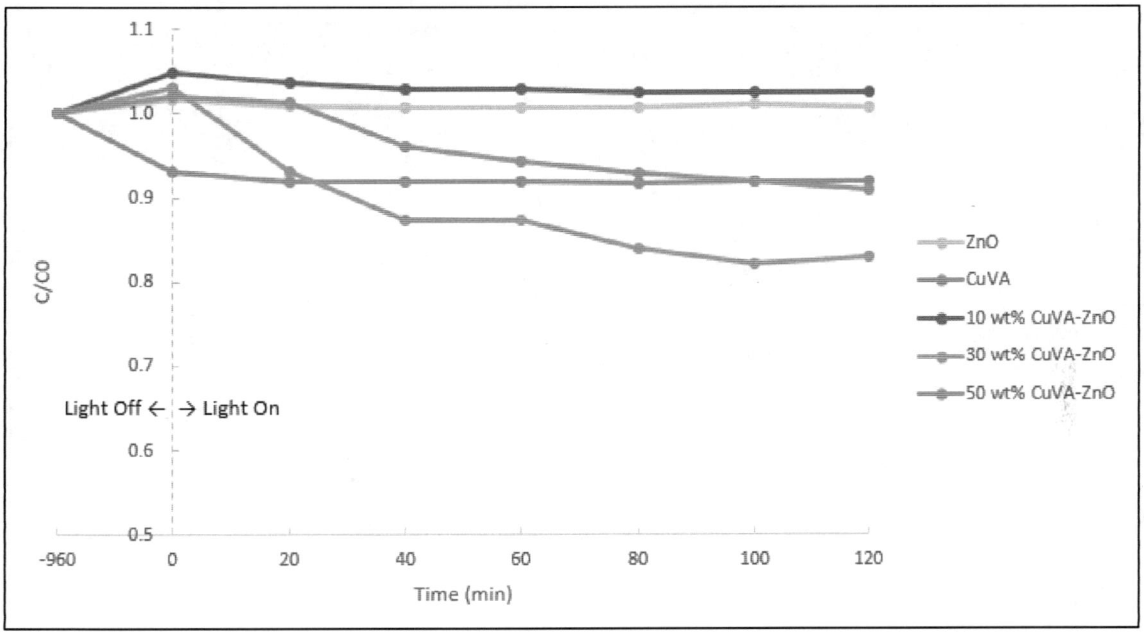

Figure 5. Photocatalytic degradation efficiencies of SDZ over different photocatalysts.

The degradation of 10 mgL^{-1} SDZ under visible light irradiation (figure 5) was monitored over CuVA-ZnO. After irradiation for 120 min, SDZ showed the greatest degradation at 17.01% under the presence of 30 wt% CuVA-ZnO, higher than that of ZnO, CuVA, 10 and 50 wt% CuVA-ZnO by a fair margin. Compared to negligible degradation of 10 wt% CuVA-ZnO over two hours of light time and 50 wt% CuVA-ZnO degrading 9.18%, 30 wt% CuVA were the optimal loading in CuVA-ZnO in terms of degradation of SDZ. Moderate CuVA concentration provides enough photoactive species and at the same time, it does not block the ZnO surface for initiation of oxidation reaction [10]. Apart from CuVA, all the samples showed no observable adsorption of SDZ before light on. This implies that at the absence of light, the catalyst was unable to decrease the concentration of SDZ in any observable manner.

4. Conclusion

The synthesis of ZnO through hydrothermal process and synthesis of different weightage of CuVA-ZnO (10, 30 and 50 wt%) through in-situ reaction has been achieved in this work. In-situ reaction method is particularly attractive because (i) it took relatively short amount of time to prepare and (ii) with ZnO acting as the nucleation point, the QDs were able to directly grow on ZnO surface evenly. Further studies are necessary to identify the catalytic mechanisms of the heterojunction composites as well as the optimal parameter to be used to degrade SDZ to improve the performance (sample loading amount, pH of SDZ solution and SDZ concentration).

Acknowledgement
The research was supported by the Ministry of Higher Education (MoHE) Malaysia through Fundamental Research Grant Scheme project FRGS/1/2019/TK10/UTAR/02/5.

Appendix A

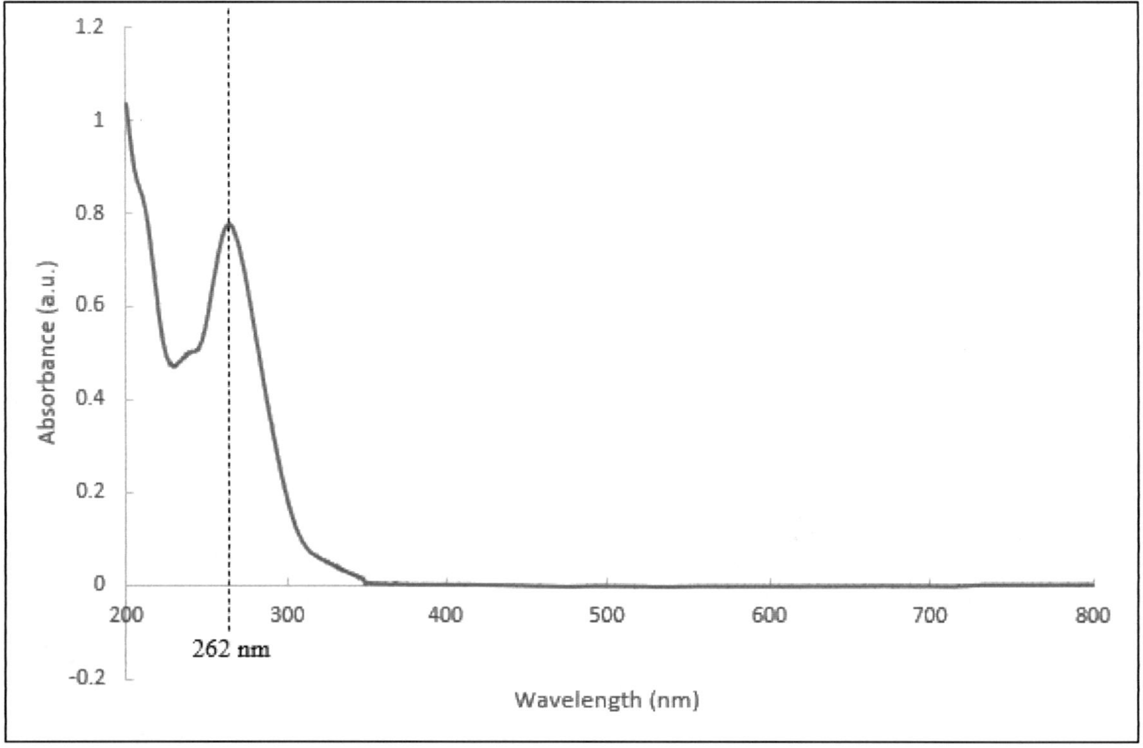

Figure A1. Full-range scan of 10 mgL^{-1} SDZ using double-beam UV-Vis spectrophotometer.

From Figure A1, the maximum local peak of SDZ is found to be 262 nm. Moreover, SDZ does not appear to absorb any visible light (380 – 800 nm).

Appendix B

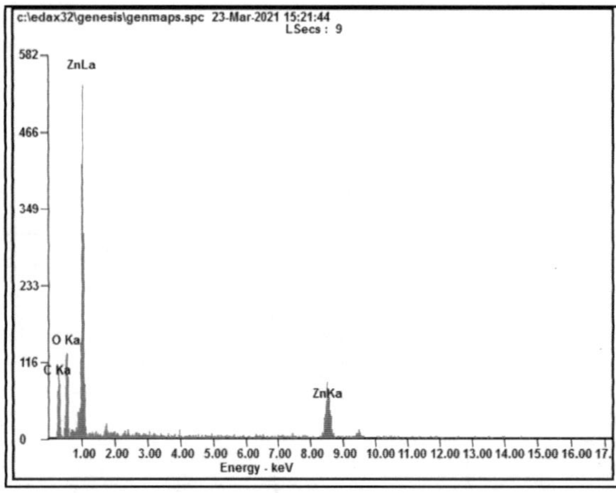

Element	*Wt%*	*At%*
CK	30.63	55.49
OK	20.85	28.36
ZnK	48.52	16.15
Matrix	Correction	ZAF

Figure B1. EDX Spectrum and percentage distribution of ZnO.

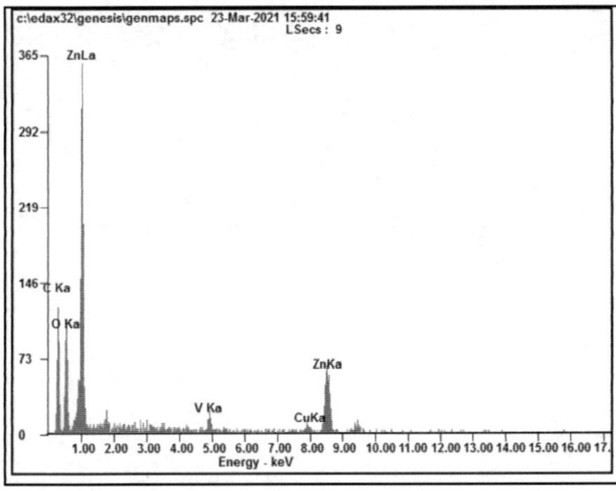

Element	*Wt%*	*At%*
CK	33.10	59.73
OK	17.39	23.56
VK	02.86	01.21
CuK	03.65	01.24
ZnK	43.00	14.26
Matrix	Correction	ZAF

Figure B2. EDX Spectrum and percentage distribution of 10 wt% CuVA-ZnO.

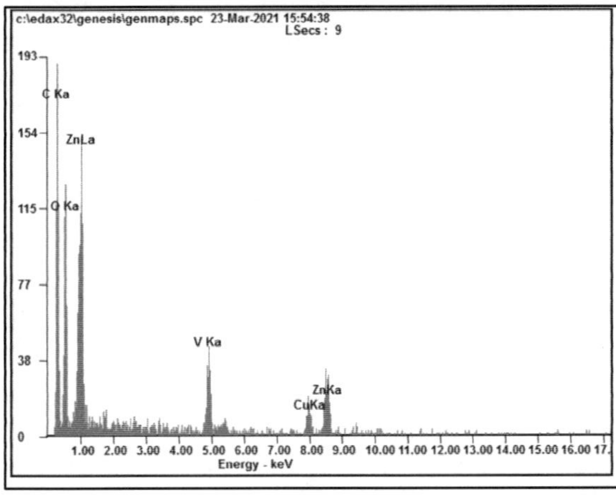

Element	*Wt%*	*At%*
CK	41.34	62.88
OK	23.27	26.58
VK	07.36	02.64
CuK	08.52	02.45
ZnK	19.51	05.45
Matrix	Correction	ZAF

Figure B3. EDX Spectrum and percentage distribution of 30 wt% CuVA-ZnO.

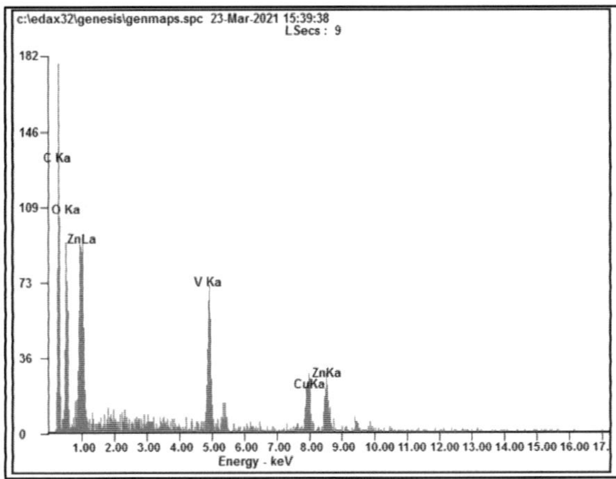

Element	*Wt%*	*At%*
CK	37.37	61.74
OK	19.33	23.97
VK	11.72	04.57
CuK	15.70	04.90
ZnK	15.88	04.82
Matrix	Correction	ZAF

Figure B4. EDX Spectrum and percentage distribution of 50 wt% CuVA-ZnO.

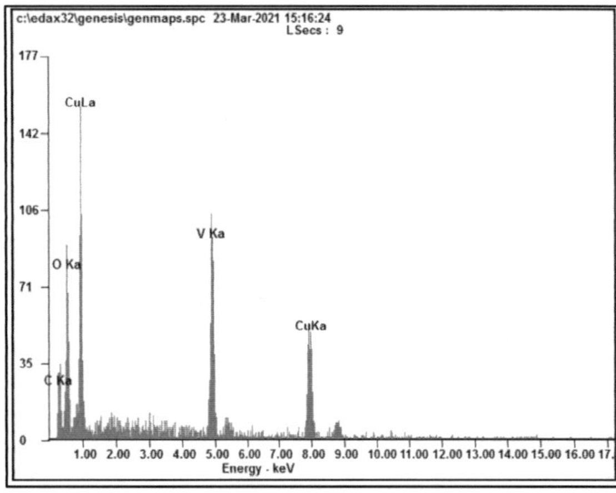

Element	*Wt%*	*At%*
CK	13.52	30.43
OK	24.03	40.62
VK	22.43	11.91
CuK	40.02	17.03
Matrix	Correction	ZAF

Figure B5. EDX Spectrum and percentage distribution of CuVA.

High carbon content in the analysis were due to the sample being placed above a carbon tape during analysis. From figure B1 – B5, it is observed that as the intended mass loading of CuVA increased, the observed percentage of Cu and V increased in a linear matter while presence of Zn also decreased linearly.

References

[1] Liu X, *et al.* 2017 *Adv. Mater.* **22** 1605646
[2] Zhang S, *et al.* 2018 *ACS Nano* **12** 751–8
[3] Zhang Y, Qi F, Zhou X, Sun H, Zhang W, Liu D and Song X 2017 *J. Colloid Interface Sci.* **498** 105–11
[4] Yamashita S, Hamada, M, Nakanishi S, Saito H, Nosaka Y, Wakida S and Biju V 2015 *Angew. Chem., Int. Ed.* **54** 3892–6
[5] Wang R, *et al.* 2018 *Appl. Catal. B* **225** 228–37
[6] Wang J, Wang Z and Zhu Z 2017 *Appl. Catal. B* **204** 577–83
[7] Fu Y, Li Z, Liu Q, Yang X and Tang H 2017 *Chinese J. Catal.* **38** 2160–70
[8] Shi R, Li Z, Yu H, Shang L, Zhou C, Waterhouse G, Wu L and Zhang T 2017 *ChemSusChem* **10** 4650–6.
[9] Kamarulzaman N, Kasim M F and Rusdi R 2015 *Nanoscale Res Lett* **10** 346
[10] Pulipaka S, Boni N and Meduri P 2020 *ACS Appl Energy Mater* **3** 6060–4
[11] Wang H, *et al.* 2014 *Chem Soc Rev* **43** 5234–44
[12] Lin L, Liu T, Zhang Y, Liang X, Sun R, Zeng W and Wang Z 2016 *Ceram. Int.* **42** 3138–44
[13] Ye M, Zhao Z, Hu Z, Liu L, Ji H, Shen Z and Ma T 2017 *Angew. Chem., Int. Ed.* **56** 8407–11
[14] Naciri Y, Bouddouch A, Bakiz B, Taoufyq A, Ezahri M and Benlhachemi A 2020 *Mater. Today Proc.* **22** 48–51
[15] Zhou Q, Lu Z, Wei Z, Xu L, Gui Y and Chen W 2018 *Front. Chem.* **6** 194

Research and development of high efficiency low emission combined cycle power plant arrangements

I I Komarov, O V Zlyvko, A N Vegera, B A Makhmutov, I A Shcherbatov*

Department of Innovative Technologies of High-Tech Industries, National Research University "Moscow Power Engineering Institute", Krasnokazarmennaya street, 14, Moscow, Russian Federation, 111250

*Corresponding Author: ShcherbatovIA@mpei.ru

Abstract. Coal-fired steam turbine thermal power plants produce a large part of electricity. These power plants usually have low efficiency and high carbon dioxide emission. An application of combined cycle power plants with coal gasification equipped with carbon capture and storage systems may increase the efficiency and decrease the harmful emission. This paper describes investigation of the oxidizer type in the integrated gasification combined cycle combustion chamber and its influence upon the energy and environmental performance. The integrated gasification combined cycle and oxy-fuel combustion technology allow the carbon dioxide capture and storage losses 58% smaller than the traditional air combustion one. The IGCC with air combustion without and with carbon dioxide capture and storage has 53.54 and 46.61% and with oxy-fuel combustion has 34.94 and 32.67% net efficiency. Together with this the CO2 emission drops down from 89.9 to 10.6 gm/kWh. The integrated coal gasification combined cycle with air oxidizer has the best net efficiency.

1. Introduction

Nowadays coal-fired steam turbine thermal power plants with efficiency below 38% produce almost 10 000 TWh of the world electricity. For the last two decades the carbon dioxide atmospheric content grows for 2 ppm per year which surely contributes to the global warming [1, 2]. Moreover, the large coal resources and its low price make it an attractive energy source for the electricity production. This defines the importance of coal technology with minimal carbon dioxide emissions.

The carbon dioxide emission may be decreased by application of carbon capture and storage (CCS) systems and by the thermal power plant (TPP) efficiency improvement [3, 4]. The efficiency of electricity production from coal may be increased by transition to the integrated coal gasification combined cycle (IGCC) [5–9].

Integrated coal gasification combined cycles usually include steam-oxygen and steam-air gasification because in the combined cycle all gasification components are available. The syngas produced with the steam-oxygen technology has higher heating value than syngas produced with other methods [10, 11]. It is worth mentioning that the integrated coal gasification combined cycle has rather low efficiency below 45% which is mostly caused by the coal gasification losses. Moreover, the coal-fired combined cycle power plants have relatively low initial parameters which also determines their lower efficiency compared to the advanced power plants [12–15].

Reduction of carbon dioxide emissions at combined-cycle TPPs using syngas in the gas turbine combustion chamber is traditionally ensured by equipping of a carbon capture and storage system, which consumes from 8 to 12% of the generated power plant electric power [16]. The CO_2 may be separated from other gases before or after fuel combustion. Also known is the oxy-fuel combustion technology where fuel burns in the almost pure oxygen without the nitrogen dilution, so the carbon dioxide capture consumes less power. The oxy-fuel combustion allows capture of 99% of carbon dioxide which means the practically zero emission as compared with the carbon dioxide capture before or after combustion in air with the capturing degree below 89% [17–21]. Therefore, the investigation of oxy-fuel combustion integrated gasification combined cycle with CCS is of great interest [22, 23].

This paper reviews possible transition to the coal-fired combined cycle technology with high turbine inlet parameters and the CO_2 emission decrease by oxygen fuel combustion. The analysis compares thermal and environmental performance of IGCC with high working fluid parameters and syngas firing in air or almost pure oxygen.

2. Air and oxy-fuel combustion integrated coal gasification combined cycle

This investigation is devoted to the combined cycle power plant with steam-oxygen coal gasification, carbon capture and storage systems. These power plants specific feature is the oxidizer type supplied to the combustion chamber for the syngas combustion. In the first case the oxidizer is air, in the second case—oxygen.

Figure 1 presents a cycle arrangement of the integrated coal gasification combined cycle with air fuel combustion. The gasification block *3* produces syngas by the steam-oxygen blow method and includes gasifier and ash handling. The air separation unit (ASU) *1* produces oxygen for gasification by air separation into oxygen and nitrogen. It compresses the oxygen up to the gasification pressure in the inter-cooled compressor *2*. Steam is taken from the steam turbine *14*. Then the gas is supplied to the gas turbine combustion chamber *5*. The air compressor *4*, gas turbine *6* and power generator *7* are mounted on the common shaft. Compressor *4* raises the air pressure up to the combustion chamber *5* pressure. Some air is supplied to the gas turbine *6* cooling.

After the combustion products have expanded in the turbine *6* the exhaust gas is directed to the double-pressure drum type heat recovery steam generator (HRSG). In HRSG the heating gas transfers heat to water medium by heat exchanger surfaces from the high-pressure steam superheater *8* to the condensate gas heater *13*. The steam turbine exit steam is condensed in the condenser *16*. Then the condensate pump *19* supplies water to the condensate gas heater *13*. The condenser *16* outlet water temperature is lower than the acceptable inlet temperature of the heater *13*, therefore, a part of heated condensate is supplied to re-circulation by the pump *20*. After heater *13* the water passes the deaerator *15* that is fed with the superheated low-pressure steam. Then the feed water pumps *17* and *18* split the water into two flows with high and low pressures.

After the feed water pump *17* water subsequently passes the vaporizer *12* and low-pressure superheater *11* and enters the steam turbine mixer where it meets the high-pressure steam and expands. Downstream the high-pressure feed water pump *18* water enters the economizer *10* and then, similar to the low-pressure circuit, enters the high-pressure vaporizer *9* and superheater *8*. The high-pressure steam enters the steam turbine *14* where it expands. The turbine is mounted on the generator *7* shaft.

Figure 2 shows the second cycle arrangement which differs from the first one by the fuel combustion in the combustion chamber *5* with the oxygen produced by the ASU *1*. The oxygen is compressed in the inter-cooled compressor *2*. The turbine *6* inlet temperature is reduced by the water injection into the syngas and combustion products. The water is supplied by the pump *21*. The high temperature turbine *6* is cooled with the air compressed by the air compressor *4*. The compressor *4*, gas turbine *6* and power generator *7* are mounted on a single shaft. Other cycle arrangement elements of the figure 2 are equal to the ones of figure 1.

Table 1 presents the main integrated coal gasification combined cycle computer simulation parameters. The coal composition is assumed according to the data [10]: moisture (a.r.) 8.10%, ash 14.19%, carbon 74.04%, hydrogen 4.08%, nitrogen 1.67%, oxygen 7.36%, sulphur 0.65%, volatile matter (dry) 28.51%, chlorine 0.01 %. The 1700°C gas turbine inlet temperature corresponds to the latest developments [9].

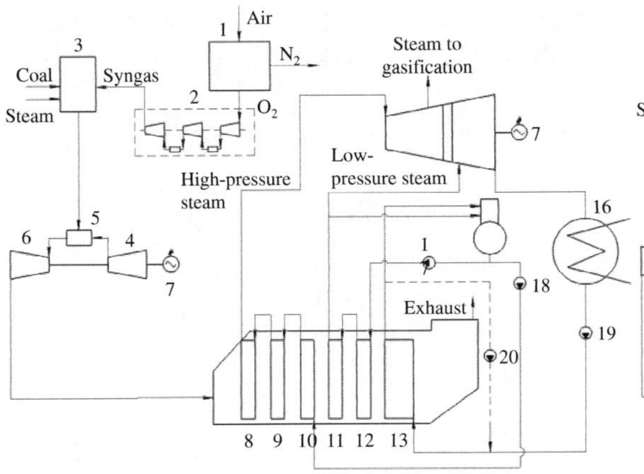

Figure 1. Flow chart of the IGCC with syngas combustion in air.

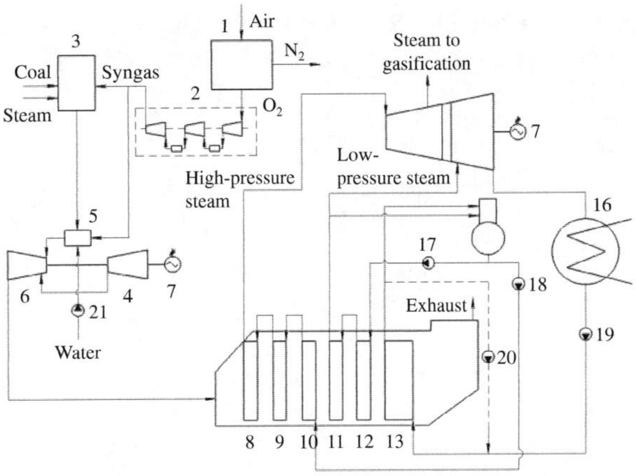

Figure 2. Flow chart of the IGCC with syngas combustion in oxygen.

Table 1. Input data for the integrated coal gasification combined cycle arrangements analysis.

<table>
<tr><th>Parameter</th><th>Air combustion</th><th>Oxygen combustion</th></tr>
<tr><td>Oxygen purity, %</td><td colspan="2">95.6</td></tr>
<tr><td>Gasifier pressure, MPa</td><td colspan="2">3.5</td></tr>
<tr><td>Gas turbine inlet temperature, °C</td><td colspan="2">1700</td></tr>
<tr><td>Coolant specific flow, %</td><td colspan="2">13.6</td></tr>
<tr><td>Gas turbine inlet pressure, MPa</td><td>3</td><td>3.5</td></tr>
<tr><td>Turbine internal efficiency, %</td><td colspan="2">89</td></tr>
<tr><td>Pumps internal efficiency, %</td><td colspan="2">85</td></tr>
<tr><td>Compressors internal efficiency, %</td><td colspan="2">88</td></tr>
<tr><td>Mechanical, electric motor, power generator, heat transport efficiency, %</td><td colspan="2">99</td></tr>
<tr><td>High-/low-pressure steam, MPa</td><td>8.55/0.70</td><td>12.35/0.70</td></tr>
<tr><td>High-/low-pressure superheater minimal temperature difference, °C</td><td>20/20</td><td>120/20</td></tr>
<tr><td>High-/low-pressure vaporizer minimal temperature difference, °C</td><td>50/10</td><td>150/30</td></tr>
<tr><td>Condenser pressure, kPa</td><td colspan="2">4</td></tr>
<tr><td>High-/low-pressure feed water pump outlet MPa</td><td colspan="2">13.0/0.9</td></tr>
<tr><td>Condenser pump outlet pressure, MPa</td><td colspan="2">0.6</td></tr>
<tr><td>Condensate temperature at the condensate gas heater inlet, °C</td><td colspan="2">60</td></tr>
<tr><td>Deaerator pressure, MPa</td><td colspan="2">0.45</td></tr>
</table>

The integrated coal gasification combined cycle arrangement computer simulation involved a set of simulation models with following purposes:

- Air separation unit produces oxygen [24].
- Steam-oxygen blown gasifier transforms coal to syngas [10].
- Combined cycle gas turbine produces electrical power.

Figure 3 presents connections of the mass and energy flow models for the air syngas combustion power plant, gasification and air separation units. The gasification block inputs are the coal composition, humidity and ash. Moreover, at the gasifier inlet the flow of oxygen compressed up to the gasifier pressure and the steam flow taken from the steam turbine bleeding are assumed. The syngas flow with the gasifier outlet composition, temperature, pressure and mass flow enters the combustion chamber of integrated coal gasification combined cycle model. The integrated coal gasification combined cycle power is spent for the internal consumption including the air separation unit compressors drive and the rest is supplied to consumers. In the case of oxygen syngas combustion (figure 3b) the air separation unit outlet oxygen enters the integrated coal gasification combined cycle oxygen compressor.

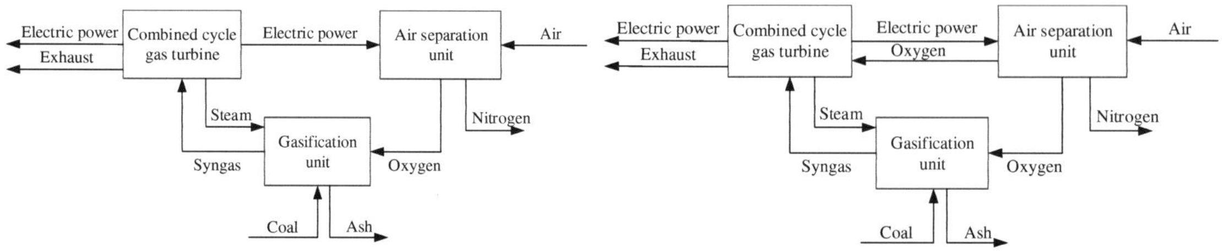

(a) with air syngas combustion (b) with oxygen syngas combustion

Figure 3. Connections between mass and energy flow simulation models of IGCC.

The ASU and gasifier block models are simulated using MS Excel. The air separation unit simulation assumed air as a two-component mixture of nitrogen (79%) and oxygen (21%). The gasifier simulation involved the following assumption:

- The gasifier internal pressure is kept constant.
- Hydrogen, nitrogen and oxygen completely take part in the syngas formation process.
- Ash and syngas leave the gasifier with equal temperature.

The combined cycle gas turbine simulation was carried out using Aspen Plus (figure 4). The computer simulation model of the combined cycle power plant with syngas air combustion includes the following elements (figure 4a): air compressor with air bleeding for cooling, cooled gas turbine, double-pressure heat recovery steam generator with six surface heat exchangers, steam turbine with steam bleeding for gasification, condenser, deaerator, auxiliary equipment with condensate, feed water, recirculation and circulation pumps.

The oxygen syngas combustion combined cycle model (figure 4b) differs from the model in figure 4a by the oxygen compressor that increases the oxygen pressure before the combustion chamber and the additional water pump which supplies water to the combustion chamber.

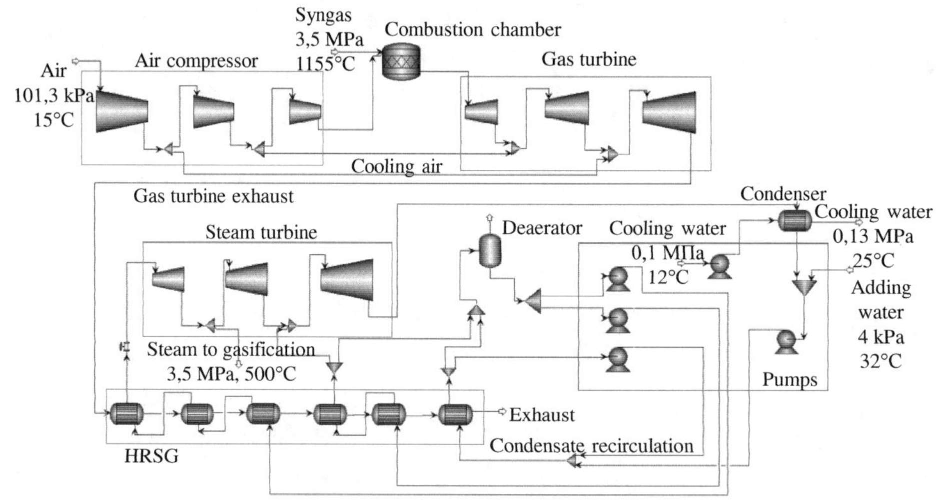

(a) with air combustion

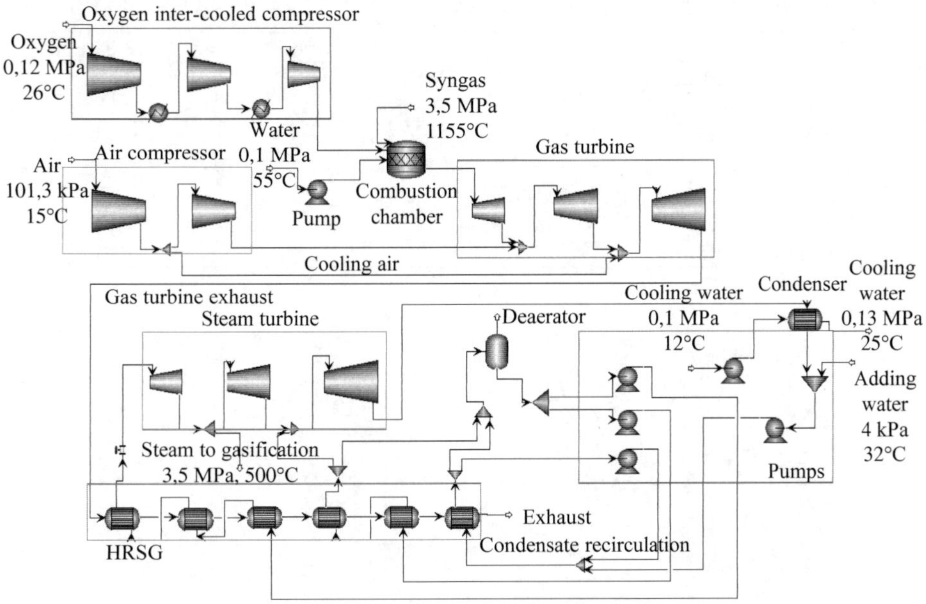

(b) with oxygen combustion

Figure 4. Computer simulation models of IGCC.

The combined cycle gas turbine computer model involves a few assumptions as the following:

- Compressor bleeding mass flow is not dependent on the cooling system efficiency.
- The model does not consider the cooling air injection after each vane or blade row.
- The syngas combustion is stoichiometric.
- Specific internal efficiencies of compressor and gas turbine and the combustor losses are constant.

3. Results

Table 2 summarizes the integrated coal gasification combined cycle main energy and environmental performance obtained from the cycle arrangement analysis. The transition from air to oxygen combustion without the carbon dioxide capture and storage losses reduces the thermal power plant efficiency from 53.54 to 34.94%. This transition combined with the carbon dioxide capture and storage losses reduces the TPP efficiency from 46.61 to 32.67%. The efficiency of simulated air blown IGCC with CCS and gas turbine inlet temperature of 1700°C at 46.61% is in line with currently working Nakoso IGCC LHV efficiency of 42,9% with gas turbine inlet temperature of 1200°C [25].

The transition from air to oxygen combustion in combined cycle power plant with coal gasification leads to a remarkable reduction of the thermal efficiency which is mostly caused by higher heat losses with the HRSG exhaust gas (figure 5). In the case of oxy-fuel combustion this is caused by the turbine inlet temperature reduction by the water injection. Therefore, the gas turbine and heat recovery steam generator exhaust gas for 57% consists of water vapour and its heat content includes the evaporation heat. Thus, the HRSG exhaust with its high heat content is at the lowest temperature acceptable for the heat recovery steam generator because the water condensing on the condensate gas heater surface is not acceptable.

Table 2. Integrated coal gasification combined cycle arrangement analysis.

Parameter	Air combustion		Oxygen combustion	
	Without CCS	With CCS	Without CCS	With CCS
Gross power, MW	226.0		231.5	
Net power, MW	207.3	180.5	171.7	160.6
Auxiliary electricity requirements, MW	18.7	45.5	59.8	70.9
Net efficiency, %	53.54	46.61	34.94	32.67
CO_2 emissions, gm/kWh		89.90		10.63

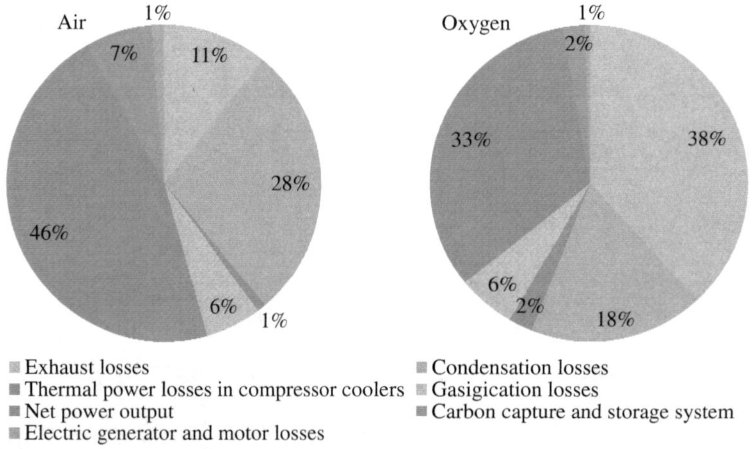

Figure 5. Heat distribution in the integrated coal gasification combined cycle with different oxidizers.

Another important factor influencing the efficiency is the auxiliary electricity requirements. In the oxygen syngas combustion case the auxiliary electricity requirements are 1.56 times higher than in the air combustion one (figure 6) in spite of the two times lower CO_2 capture and storage consumption (figure 7). The 3.4 times higher auxiliary consumption share in gross power is mostly because of the ASU and oxygen compressor consumption. Unlike in the IGCC with air combustion, in the integrated coal gasification combined cycle with oxygen combustion oxygen is produced and compressed not only for the gasification process but also for the syngas combustion.

The oxy-fuel combustion technology undoubtedly has advantages in terms of the high environmental safety. The degrees of carbon dioxide capture with air and oxygen combustion are 89 and 99%. In other words, the integrated coal gasification combined cycle carbon dioxide emission drops down because of the transition from air to oxygen combustion from 89.90 to 10.63 gm/kWh.

The oxy-fuel combustion technology allows for the lowest emissions, so it is of great interest to improve its efficiency. The heat recovery steam generator exhaust gas consists of 57% water vapour which makes it the first target for the thermal efficiency improvement. The exhaust related losses may be reduced by adding surface type condensing heat exchanger in the exhaust gas flow path (figure 8). The heat exchanger condenses water vapour from the exhaust gas and the obtained thermal energy may be transferred to the supply, or district water. In the case of the direct and reverse supply water temperatures assumed at 70 and 40°C the exhaust gas heat recovery degree is 80% and up to 97% of the water vapour will be condensed.

This cycle arrangement allows for additional recovery of 150 MW heat power in the integrated coal gasification combined cycle with carbon capture and storage system but this technical solution is related to a few problems. Specifically, a problem is the stability of oxy-fuel combustor with large water injection and the condensing heat exchanger that would combine the high heat transfer efficiency and high carbonic acid resistance.

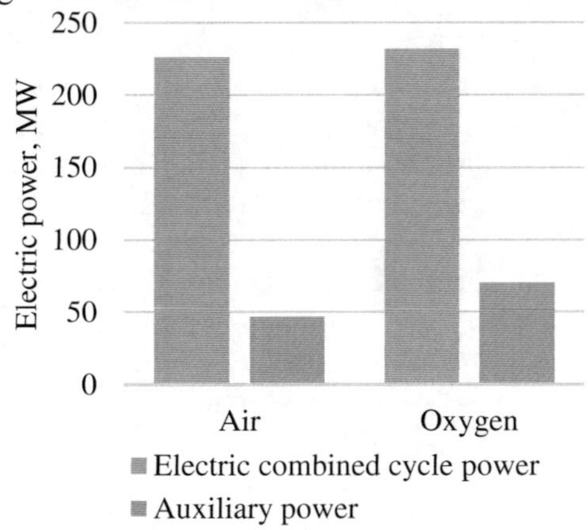

Figure 6. IGCC gross power and auxiliaries electricity requirements with different oxidizers.

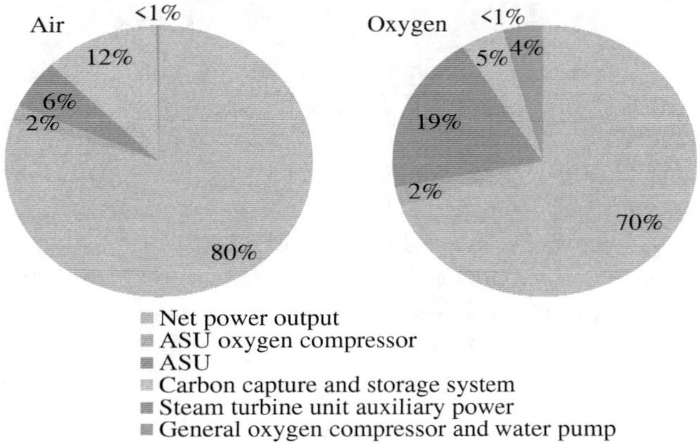

Figure 7. Gross power distribution of IGCC with different oxidizers.

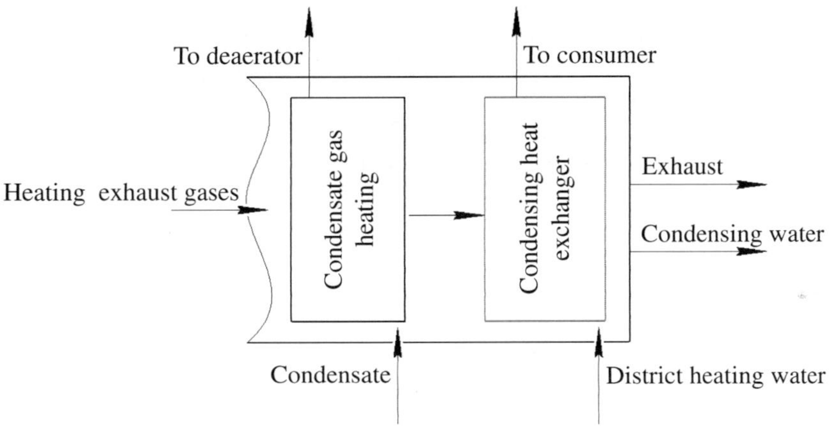

Figure 8. Modification of the IGCC heat recovery steam generator exhaust and oxy-fuel combustion.

4. Conclusions

Cycle arrangements of the combined cycle power plant with steam-oxygen coal gasification and carbon capture and storage systems and their computer simulation models are developed. The integrated coal gasification combined cycle versions use air or oxygen oxidizers. The integrated coal gasification combined cycle with air combustion is traditional and the integrated coal gasification combined cycle with oxygen syngas combustion involves the turbine inlet temperature reduction down to 1700°C by water recirculation and injection into combustion chamber.

Based on the results of thermodynamic analysis of the integrated coal gasification combined cycle arrangements, it was established that:

- The integrated coal gasification combined cycle with air combustion without and with carbon dioxide capture and storage has 53.54 and 46.61% net efficiency and the carbon dioxide emission of 89.90 gm/kWh using carbon capture and storage.

- The integrated coal gasification combined cycle with oxy-fuel combustion without and with carbon dioxide capture and storage has 34.94 and 32.67% net efficiency and the carbon dioxide emission of 10.63 gm/kWh using carbon capture and storage.

In the case of oxygen syngas combustion, the heat losses with the heat recovery steam generator exhaust gas are about 35% which is 4.5 times higher than in the case of air combustion. This effect is caused by the remarkably high water vapour content in the exhaust gas of about 57%. The transition from air to oxygen combustion requires power consumption increase of air separation unit and oxygen compressor from 8 to 21% of the plant power gross power.

The efficiency of the oxy-fuel combustion integrated coal gasification combined cycle may be improved by a condensing heat exchanger that transfers the exhaust gas vapour condensing heat to the delivery water. It allows for additional recovery of 150 MW heat power.

The main problems of the oxy-fuel combustion integrated coal gasification combined cycle design is the combustion stability in oxygen-syngas combustion chamber with water injection and design of a corrosion resistant condensing heat exchanger.

5. References

[1] Oberschelp C, Pfister S, Raptis C E and Hellweg S 2019 *Nature Sustainability* **2(2)** 113–21
[2] Edenhofer O 2015 *Science* **349(6254)** 1286–87
[3] Rogalev A, Komarov I, Kindra V and Zlyvko O 2018 *Entrepreneurship and Sustainability Issues* **6(1)** 429–45
[4] Kianfar E, Pirouzfar V and Sakhaeinia H J 2017 *J. Taiwan Inst. Chem. E.* **80** 954–62
[5] Xia C, Ye B, Jiang J and Shu Y 2020 *J. Clean. Prod.* **271** 122615
[6] Ge H, Zhang H, Guo W, Song T and Shen L 2019 *Fuel* **241** 118–28
[7] Ryzhkov A, Bogatova T and Gordeev S 2018 *Fuel* **214** 63–72
[8] Zaryankin A, Mager A, Rogalev A and Komarov I 2014 *WIT Trans. Ecol. Envir.* **190(1)** 251–60
[9] Ito E, Okada I, Tsukagoshi K, Muyama A and Masada J 2009 *Proc. of the ASME Turbo Expo: Power for Land, Sea, and Air* (Orlando / ASME)
[10] Cormos C-C 2012 *Energy* **42(1)** 434–45
[11] Gnanapragasam N V, Reddy B V and Rosen M A 2010 *Int. J. Hydrogen Energ.* **35(10)** 4933–43
[12] Emun F, Gadalla M, Majozi T and Boer D 2010 *Comput. Chem. Eng.* **34(3)** 331–8
[13] Kanniche M and Bouallou C 2007 *Appl. Therm. Eng.* **27(16)** 2693–702
[14] Lee J C, Lee H H, Joo Y J, Lee C H and Oh M 2014 *Energy* **64** 58–68
[15] Hoffmann B S and Szklo A 2011 *Appl. Energ.* **88(11)** 3917–29
[16] Ye B, Jiang J, Zhou Y, Liu J and Wang K 2019 *J. Clean. Prod.* **222** 476–87
[17] Kanniche M, Gros-Bonnivard R, Jaud P, Valle-Marcos J, Amann J M and Bouallou C. 2010 *Appl. Therm. Eng.* **30(1)** 53–62
[18] Descamps C, Bouallou C and Kanniche M 2008 *Energy* **33(6)** 874–81
[19] Komarov I, Kharlamova D, Makhmutov B, Shabalova S and Kaplanovich I 2020 *E3S Web Conf.* **178** 01027
[20] Allam R, Martin S, Forrest B, Fetvedt J, Lu X, Freed D, Brown Jr G W, Sasaki T, Itoh M and Manning J. 2017 *Enrgy Proced.* **114** 5948–66
[21] Wang Y, Zhao L, Otto A, Robinius M and Stolten D 2017 *Enrgy Proced.* **114** 650–65
[22] Shi B, Xu W, Wu W and Kuo P C 2019 *Energ. Convers. Manage.* **195** 290–301
[23] Rogalev A, Kindra V and Osipov S 2018 *AIP Conf. Proc.* **2047** 020020
[24] Ebrahimi A, Meratizaman M, Reyhani H A, Pourali O and Amidpour M 2015 *Energy* **90** 1298–316
[25] Nunokawa M, Asano T 2014 *Mech. Eng. J. 1(5)* TEP0045-TEP0045

Acknowledgments

This study conducted by Moscow Power Engineering Institute was financially supported by the Ministry of Science and Higher Education of the Russian Federation (project No. FSWF-2020-0020).

Coal-fired steam turbine power plant using oxygen-rich air as oxidizer

I I Komarov, S K Osipov, I A Shcherbatov*, B A Makhmutov and I B Kaplanovich

Department of Innovative Technologies of High-Tech Industries, National Research University "Moscow Power Engineering Institute", 14 Krasnokazarmennaya str., Moscow, 111250, Russian Federation

*Corresponding Author: ShcherbatovIA@mpei.ru

Abstract. Mitigation of harmful and greenhouse gas emissions produced by the coal combustion in thermal power plant is a topic goal. Reduction of the nitrogen oxides, sulfur and ash emissions makes remarkable progress but the carbon dioxide emission still meets considerable difficulties mostly caused by the low greenhouse gas content in the flue gas. A prospective solution to this problem may be the fuel combustion in oxygen-enriched air, which increases the flue gas carbon dioxide content. In this technology, the carbon dioxide content in flue gas is higher and this results in its easier capture. This paper presents the thermodynamic analysis results of a steam turbine power production facility that burns coal in the air with high oxygen content. The computer simulation shows that the oxygen content increase from 21 to 95.6% increases the carbon dioxide content in flue gas by a factor of 3.3 and lowers the power consumption for carbon dioxide capture by 11%. On the other side, the power consumption for pure oxygen production reduces the facility's net efficiency from 28.54 to 21.59%.

1. Introduction

In the last decades, the world community's attention is attracted by the problem of toxic and greenhouse gases emission. The measures of emission mitigation include carbon taxes [1, 2]. Especially attention is attracted to the emission reduction from the coal firing thermal power plant (TPP). The NOx, sulfur and ash emission by TPP reduction is successful [3], but the carbon dioxide one still meets considerable financial and technical difficulties [4]. For example, an application of the carbon dioxide capture technology in steam turbine unit (STU) TPP reduces its total efficiency by 8–12 %.

Three main methods of reducing the carbon dioxide emission by power production steam turbine facilities (figure 1) are the following:

- Pre-combustion capture.
- Post-combustion capture.
- Oxy-fuel combustion capture.

The CO_2 capture before fuel combustion in oxygen requires preliminary coal gasification to transform coal into coal-derived gas, which makes the capture process complicated and expensive [5]. The oxy-fuel combustion method allows high carbon dioxide contents in the flue gas of 80–90%, which simplifies its separation and the harmful emission down to reduction down to 1.5% [6]. The technology shortages are the high combustion temperature of 1100 to 1700°C and the need for a large amount of oxygen production in air separation unit, which hurt the practical application and increase the auxiliary power consumption [7]. Because of that more popular are post-combustion CO_2 capture methods such as chemical absorption, membrane separation, cryogenic separation, and chemical looping combustion.

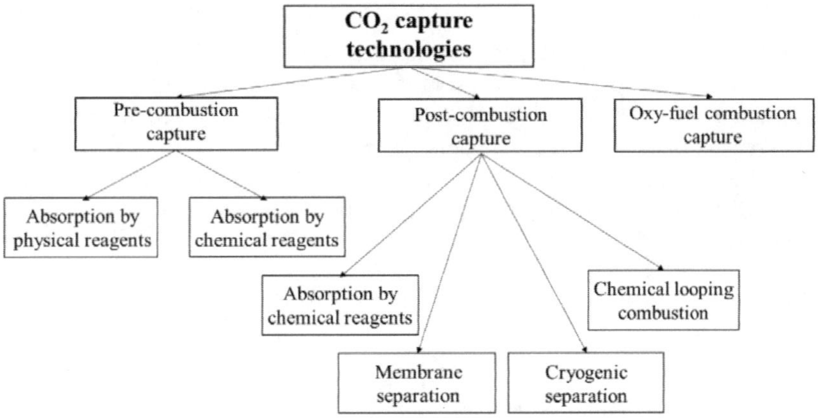

Figure 1. Methods of carbon dioxide capture in TPP.

Papers [8–10] disclose some performance of the post-combustion capture methods (table 1). The chemical absorption with monoethanolamine (MEA) is an efficient approved method of post-combustion capture widely used in Russia and abroad [11]. This method has higher maturity, CO_2 capture degree, and lower power consumption than other capture methods [12]. Also this absorption method captures the carbon dioxide from the larger flue gas flow (which is typical for STU) than the membrane separation technology. The MEA technology power consumption can be reduced by reduction of the nitrogen content in the flue gas.

Table 1. Comparison of the post-combustion CO_2 capture methods.

CO_2 capture methods	Chemical absorption	Membrane separation	Cryogenic separation	Chemical looping combustion
Power consumption, MJ/kgCO_2	4–6	0.5–6	6–10	3.5–5.4
CO_2 capture degree, %	90–98	80–90	>95	~98–99
Technology availability	high	high	mid	low

A possible efficient solution to the carbon dioxide capture problem may be the change to STU with the air enriched with oxygen as the combustion oxidizer. This solution increases the flue gas carbon dioxide content so its separation is simplified. This method requires additional power consumption for the production of highly pure oxygen. Therefore, the issue of reduction of the power consumption for carbon dioxide capture in oxy-fuel STU is still being discussed.

This paper describes the influence of oxygen content in combustion air upon the carbon dioxide capture expenses and the STU efficiency.

2. Coal-fired steam turbine power plant using oxygen-rich air as an oxidizer

The study example is the steam turbine coal-firing unit with the oxygen-enriched air as an oxidizer. For the basis version the K-300-240 power unit (figure 2) operating the condensing steam turbine with steam reheat was chosen. This type of turbine works in more than 20 TPP including the high power Kostromskaya, Konakovskaya, and Iriklinskaya. The efficiency of the TPP with carbon dioxide capture is about 33%. Specific features of the discussed unit are the following:

- Air separation unit (ASU).
- Coal firing boiler with capabilities for coal firing in oxygen-enriched air.

- Flue gas recirculation.
- CO_2 separation from the flue gas flow by chemical absorption.
- CO_2 compression and storage.

The steam turbine unit consists of the condensing turbine K-300-240 LMZ with a steam reheat. The superheated steam enters the power turbine that consists of high, intermediate, and low-pressure turbines. The turbine has eight regenerative bleeding paths directed to three high-pressure feed heaters, a deaerator, and three low-pressure feed heaters. The condensate pump supplies the main part of condensate through the line heater into the low-pressure regeneration line and further into the deaerator. The booster and turbine driven feeding pumps supply the deaerated water to the high-pressure feed heaters and the coal firing steam boiler.

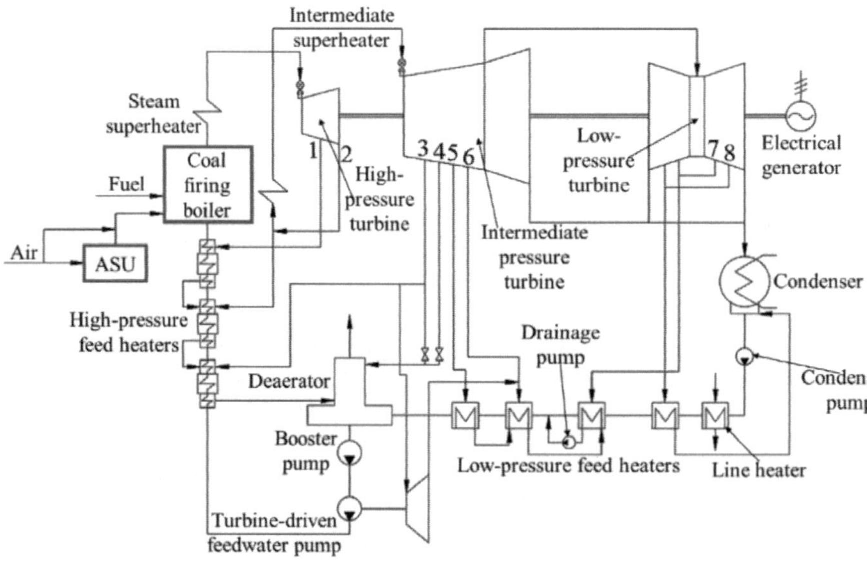

Figure 2. Schematic chart of the steam turbine unit flow.

The boiler consists of furnace walls, steam superheater, economizer, air heater, and the intermediate superheater line (figure 3). Fuel burns in the environment of ASU produced oxygen. The carbon dioxide is separated from the flue gas flow by the chemical absorption MEA technology. Table 2 summarizes the input data for the STU heat flow computer simulation. An example of fuel is the long-flame Kuznetsk coal that forms about 70% of the Russian coal reserves.

A computer simulation model of the STU operating the oxygen-enriched air and the carbon dioxide compression is developed in the Aspen Plus (figure 4) software environment. The CO_2 and H_2O thermo-physical performance are determined with the NIST REFPROP database. The carbon dioxide storage simulation assumes the 10 MPa and 28°C conditions [13].

The boiler combustion process is simulated with the MS Excel code. The simulation assumed the following conventions: constant heat power of the boiler, the adiabatic temperature is equal to the values determined for the air combustion, the adiabatic temperature is reduced by a partly recirculation of flue gas, the exhaust gas temperature is the constant 140°C.

The power consumption for 95.6% pure oxygen production is evaluated by the method presented by paper [14]. The carbon dioxide MEA chemical absorption is simulated by the method in [15]. The carbon dioxide capture degree is assumed at 90%.

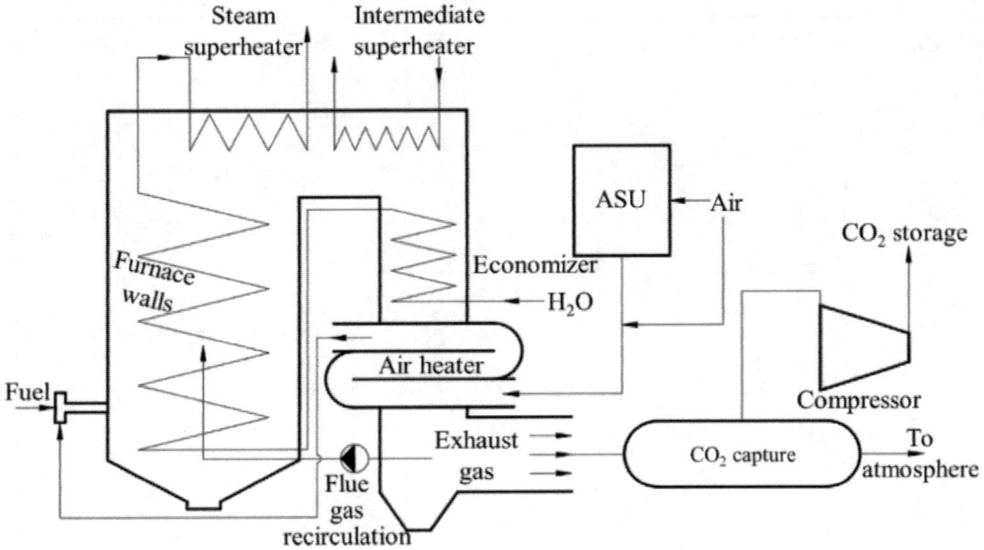

Figure 3. Concept heat flow chart of the coal oxygen combustion boiler with CO_2 capture.

Table 2. Input data for the coal firing STU heat flow analysis [16].

Parameter	Value
STU power, MW	300
Steam initial pressure, MPa	23.5
Steam initial temperature, °C	540
Intermediate superheating pressure, MPa	3.66
Intermediate superheating temperature, °C	540
Condenser pressure, kPa	3.5
Cooling water temperature at the condenser entrance, °C	15
Deaerator pressure, MPa	0.69
Feed water temperature, °C	270
High pressure-turbine internal specific efficiency	0.85
Intermediate-pressure turbine internal specific efficiency	0.91
Low-pressure turbine internal specific efficiency	0.82
Number of regenerative heaters, pcs	7
Approach in the low / high-pressure heaters, °C	5
Heat transition efficiency, %	99
Coal type	long-flame
Kuznetsk coal reserve, billion ton	700
Coal operational humidity, %	11.5
Coal ash content, %	15.9
Coal sulphur content, %	0.4
Coal carbon content, %	56.4
Coal hydrogen content, %	4.0
Coal nitrogen content, %	1.9
Coal oxygen content, %	9.9
Low heating value, MJ/kg	21.9
Volatiles, %	40.5
Grindability index	1.1

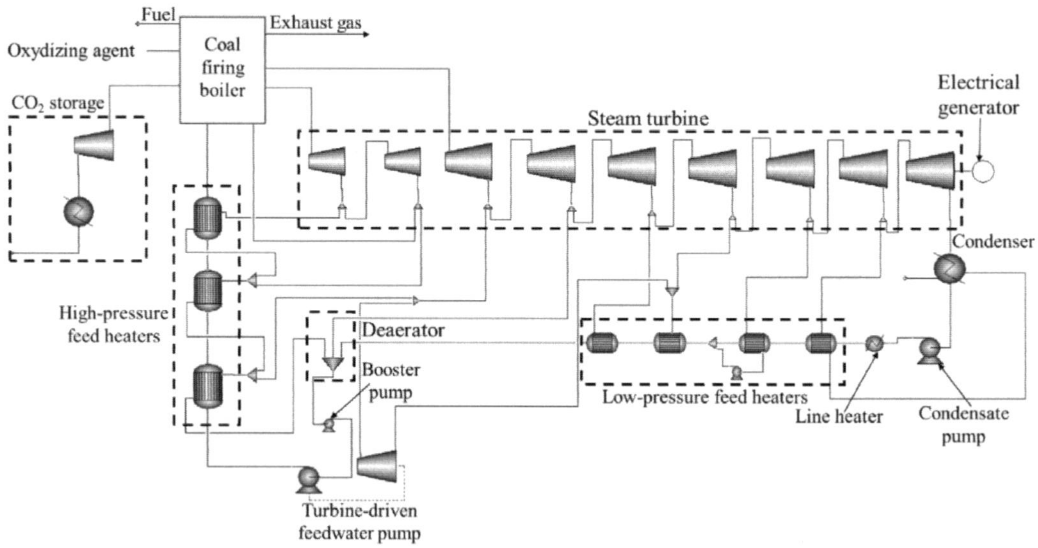

Figure 4. Simulation model of an STU operating oxygen-enriched air and CO_2 compression and storage.

3. Results

Figure 5 presents the oxygen concentration in oxidizer influence upon the carbon dioxide volume content in the combustion products. When the O_2 content grows from 21 to 95.6% the CO_2 content in the combustion products grows 3.3 times. A 1% increase of the oxygen content in air results in a mean 2% of CO_2 content increase. This is because the increase of oxygen reduces the nitrogen content in flue gas and increases the content of oxidizer in reaction with carbon. This result is confirmed by the investigation results in [17, 18], but mentioned papers only considered pure oxygen for combustion or didn't consider the use of oxy-fuel combustion with chemical absorption. The carbon dioxide emissions from TPP without CCS were calculated to be 1100 g/kWh, which is on par with 900 g/kWh reported in literature [19].

Figure 6 shows the oxidizer purity influence upon the boiler efficiency in the form of the flue gas flow and the flue gas heat losses. When the O_2 content grows from 21 to 95.6% the flue gas flow and the heat losses with flue gas drop down for 68.8 and 53.5% respectively. The reason heat losses with the flue gas do not lower as much as the flu gas flow is because of the constant exhaust temperature and increasing heat capacity of flue gas, which is due to the lower nitrogen content. The results show that the determining factor is the smaller amount of the ballast nitrogen N_2 and the smaller total amount of the heated gas.

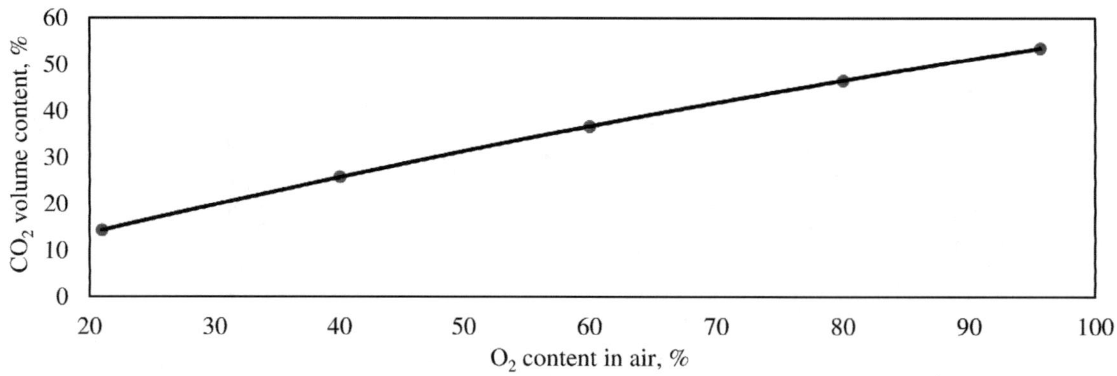

Figure 5. The CO_2 content in exhaust gas vs the O_2 concentration in oxidizer.

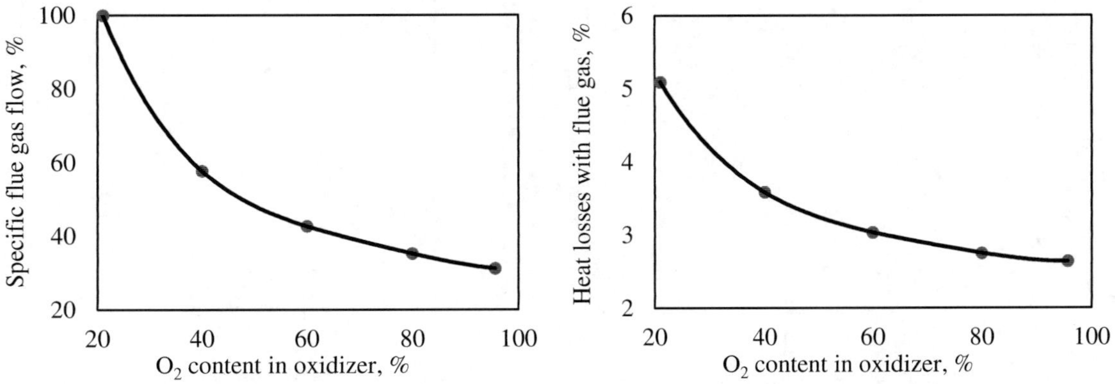

(a) specific flue gas flow vs O_2 content in oxidizer

(b) heat losses with flue gas vs O_2 content in oxidizer

Figure 6. Influence of the O_2 content in oxidizer upon the heat losses with flue gas.

Lower heat losses with flue gas cause lower fuel consumption and increase the boiler gross efficiency (figures 7, 8). An increase in the oxygen content in the oxidizer from 21 to 40% results in a 1.6% reduction of the fuel consumption and the boiler gross efficiency increase of 1.6%. The further oxygen increase from 40 to 60% reduces the fuel consumption by 0.6% and the boiler gross efficiency increase by 0.6%. The further oxygen increases up to 95.6% increases the boiler gross efficiency by 0.4% and reduces the fuel consumption by 0.4%. Thus the oxygen content in the air increase up to 95.6% allows the boiler gross efficiency to increase up to 95.9% which is 3.1% higher than the mean efficiency of currently operating coal firing TPP boilers.

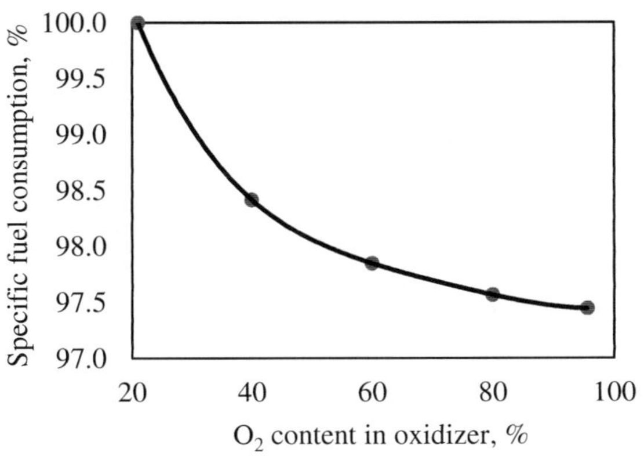

Figure 7. Specific fuel consumption vs O_2 content in the oxidizer.

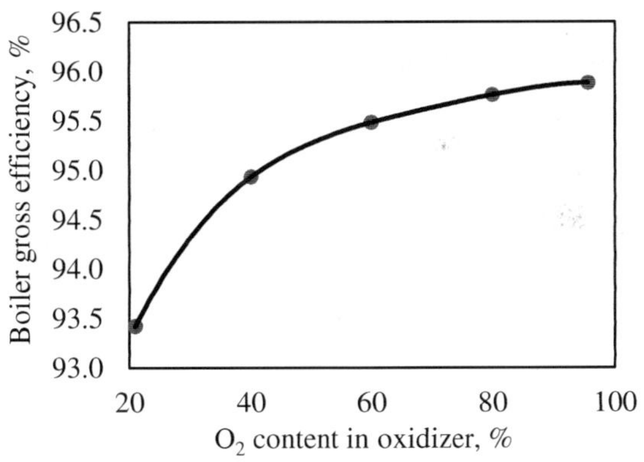

Figure 8. Boiler gross efficiency vs O_2 content in the oxidizer.

Analysis of the coal firing STU auxiliary power consumption shows that the increase of O_2 content requires larger power consumption for its production in ASU (figure 9). An increase of O_2 content in oxidizer up to 95.6% requires the air separation unit power consumption up to 60 MW. The oxygen content increase of 1% requires 1.8% increase in the ASU power consumption.

Fuel combustion in the oxygen-enriched air increases the carbon dioxide content in the boiler flue gas which results in smaller power consumption for the carbon dioxide capture and storage (figure 10). When the O_2 content grows from 21 to 80% the mean power consumption for carbon dioxide capture and storage drops down by 2.5%. The largest drop of the power consumption of about 11% is seen at the oxygen content increase from 21 to 95.6%.

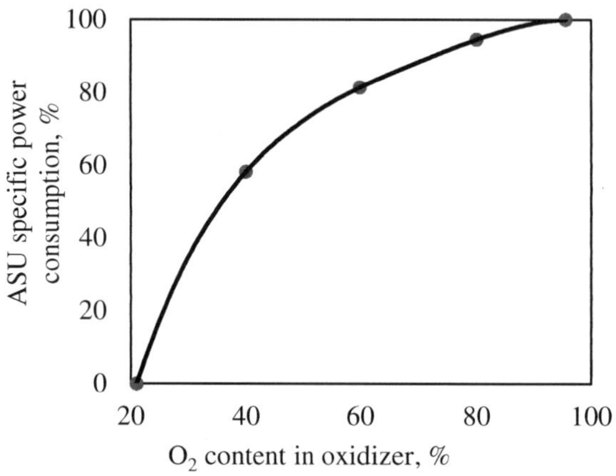

Figure 9. Power consumption for the oxygen production in ASU vs O_2 content in the oxidizer.

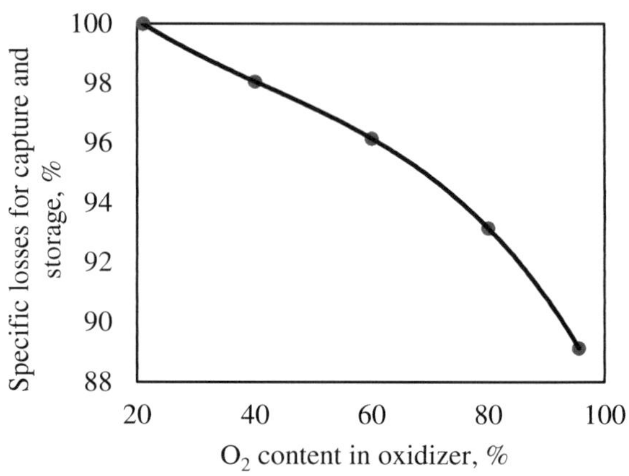

Figure 10. Specific losses for the CO_2 capture and storage vs O_2 content in oxidizer.

Despite the power consumption reduction for capture and storage that follows the oxygen content increase the cycle net efficiency drops down from 28.54 to 21.59% which is mostly caused by the oxygen consumption. At the oxygen content level of 95.6%, ASU power consumption amounts to 20% of the power produced by the steam turbine. The oxygen content change from 21 to 95.6% lowers the carbon dioxide capture and storage system power consumption by only 2.4% of the STU power production.

Finally, the application of oxygen-enriched air with 95.6% oxygen content reduces the TPP efficiency against the standard STU with carbon dioxide capture system by 11.4%.

Through the use of CCS and coal combustion in oxygen enriched air carbon dioxide emissions reduced to 150 g/kWh.

4. Conclusion

Increase of the oxygen content in oxidizer from 21 to 95.6% causes an increase from 21 to 70% of the carbon dioxide content in flue gas and 53.5% reduction of the heat losses with flue gas. Mass flow of coal supplied to the boiler drops down by 2.6% and the boiler gross efficiency grows from 93.4 to 95.9%.

Increase of the oxygen content in oxidizer causes an 11% reduction of the power consumption by carbon dioxide capture due to the higher carbon dioxide content in flue gas. Together with this increases the power consumption of oxygen production in air separation unit, which amounts to 20% of the total power in the thermal power plant operating on the 95.6% oxygen content in the oxidizer.

In total the increase of oxygen content reduces the steam turbine unit net efficiency from 28.54 to 21.59% because of the additional power consumption by the oxygen production. Thus the transition to fuel combustion in the oxygen-enriched air reduces the efficiency of the steam turbine thermal power plant equipped with the carbon dioxide capture and storage system by 11.4%.

Implementation of CCS with oxy-fuel combustion reduced the carbon dioxide emissions from 1100 g/kWh to 150 g/kWh. The use of ASU will allow for higher flame temperature in furnace, smaller air heater surface and carbon capture module. Thus it's not obvious whether the cost of the TPP with CCS and ASU will be higher in comparison with the TPP with air combustion and CCS.

5. References

[1] Elliott J, Foster I, Kortum S, Munson T, Perez Cervantes F and Weisbach D 2010 Trade and carbon taxes *Am. Econ. Rev.* **100(2)** 465–69

[2] Rogalev A, Komarov I, Kindra V and Zlyvko O 2018 Entrepreneurial assessment of sustainable development technologies for power energy sector *Entrepreneurship Sustain. Issues* **6(1)** 429–45

[3] Abdulrasheed A A, Jalil A A, Triwahyono S, Zaini M A A, Gambo Y and Ibrahim M 2018 Surface modification of activated carbon for adsorption of SO2 and NOX: A review of existing and emerging technologies *Renew. Sust. Energ. Rev.* **94** 1067–1085

[4] Rogalev N, Prokhorov V, Rogalev A, Komarov I and Kindra V 2016 Steam boilers' advanced constructive solutions for the ultra-supercritical power plants *Int. J. Appl. Eng. Res.* **11(18)** 9297–306

[5] Sudiro M, Bertucco A, Ruggeri F and Fontana M 2008 Improving process performances in coal gasification for power and synfuel production *Energ. Fuel.* **22(6)** 3894–3901

[6] Kindra V O, Kaplanovich I B, Smirnov M V, Naumov V Y and Zonov A S 2020 Parametric optimization of the semi-closed oxy-fuel combustion combined cycle *J. Phys. Conf. Ser.* **1683(5)** 052028

[7] Rogalev N D, Petrenya Y K, Kindra V O and Rogalev A N 2020 Oxy-fuel power cycles promoting the transition to green and sustainable future in the energy sector *Sustainable Fuel Technologies Handbook* ed S Dutta and C M Hussain (London: Academic Press) chapter 17 pp 513–539

[8] Song C, Liu Q, Deng S Li H and Kitamura Y 2019 Cryogenic-based CO2 capture technologies: State-of-the-art developments and current challenges *Renew. Sust. Energ. Rev.* **101** 265–278

[9] Luis P, Van Gerven T and Van der Bruggen B 2012 Recent developments in membrane-based technologies for CO2 capture *Prog. Energ. Combust.* **38(3)** 419–448

[10] Mukherjee S, Kumar P, Yang A and Fennell P 2015 Energy and exergy analysis of chemical looping combustion technology and comparison with pre-combustion and oxy-fuel combustion technologies for CO2 capture *J. Environ. Chem. Eng.* **3(3)** 2104–2114

[11] Kianfar E, Pirouzfar V and Sakhaeinia H 2017 An experimental study on absorption/stripping CO2 using Mono-ethanol amine hollow fiber membrane contactor *J. Taiwan Inst. Chem. E.* **80** 954–962

[12] Mondal M K, Balsora H K and Varshney P 2012 Progress and trends in CO2 capture/separation technologies: A review *Energy* **46(1)** 431–441

[13] Allam R, Martin S, Forrest B, Fetvedt J, Lu X, Freed D and Manning J 2017 Demonstration of the Allam Cycle: an update on the development status of a high efficiency supercritical carbon dioxide power process employing full carbon capture *Energ. Proced.* **114** 5948–5966

[14] Ebrahimi A, Meratizaman M, Reyhani H A, Pourali O and Amidpour M 2015 Energetic, exergetic and economic assessment of oxygen production from two columns cryogenic air separation unit *Energy* **90** 1298–1316

[15] Kobzar S G, Kovalenko G V, Khalatov A A 2017 computer modeling of the erosion of the convection heat transfer surfaces of boiler TPP 312 *Thermophysics and Thermal Power Engineering* **39(5)** 78-83

[16] Li K, Leigh W, Feron P, Yu H and Tade M 2016 Systematic study of aqueous monoethanolamine (MEA)-based CO2 capture process: Techno-economic assessment of the MEA process and its improvements *Appl. Energ.* **165** 648–659

[17] Wang Y, Zhao L, Otto A, Robinius M and Stolten D 2017 A Review of Post-combustion CO2 Capture Technologies from Coal-fired Power Plants *Energy Procedia* **114** 650–665

[18] Singh D, Croiset E, Douglas P L and Douglas M A 2003 Techno-economic study of CO2 capture from an existing coal-fired power plant: MEA scrubbing vs. O2/CO2 recycle combustion *Energ. Convers. Manage.* **44(19)** 3073–3091

[19] Sethi V K 2018 Low Carbon Technologies (LCT) and Carbon Capture and Sequestration (CCS)-Current Scenario *Res. J. eng. technol. Mang.* **1(01)**

Acknowledgments

This study conducted by Moscow Power Engineering Institute was financially supported by the Ministry of Science and Higher Education of the Russian Federation (project No. FSWF-2020-0020).

Comparative analysis of the efficiency of using hydrogen and steam methane reforming storage at combined cycle gas turbine for cogeneration

I I Komarov, A N Rogalev, D M Kharlamova, V Yu Naumov* and S I Shabalova

Department of Innovative Technologies of High-Tech Industries, National Research University "Moscow Power Engineering Institute", 14 Krasnokazarmennaya str., Moscow, 111250, Russian Federation
*Corresponding Author: NaumovVY@mpei.ru

Abstract. Combined cycle gas turbine cogeneration power plants provide maximal thermal efficiency. The facility thermal load depends upon ambient conditions and it may limit its unloading degree, or control range. Energy accumulation extends the control range by the energy storing during the power excess period and the energy discharge during the high consumption period. This paper considers two accumulation technologies, with the steam methane reforming and with hydrogen storage by using electrolyzer and fuel cell. In steam methane reforming technology a part of heat recovery steam generator steam passes to the synthetic gas production and reduces the electric power production during the low load periods. During the peak load periods, the synthetic gas is burned in combined cycle gas turbine and peak load gas turbine combustors. In hydrogen accumulation technology the excessive power produces hydrogen in an electrolyzer, which is stored and used during the peak load periods in fuel cells for additional electricity production. The simulation was carried out in the Aspen Plus software, the thermo-physical fluid properties were determined using the Peng-Robinson state equation. Application of the steam methane reforming accumulation technology increases the facility contribution margin by 2.6%, the hydrogen accumulation technology reduces this margin by 0.1%.

1. Introduction

The combined cycle cogeneration natural gas burning facilities have maximal thermal efficiency in electricity production [1–3]. In the case of climatic conditions in Russia, district heating is especially important. District heating systems work on the heat from co-generation combined cycle gas turbine (CCGT-CHP) facilities. The heat supply must be kept on the given level but it limits the CCGT-CHP control range. So the CCGT-CHP maneuverability improvement methods are topical.

Nowadays one can see the rising variation of electricity daily prices caused by an increase of the power system load deviation. The night load reduction causes excessive system power and this results in a sharp decline of day-ahead market electricity prices even down to zero [4]. So the increase of power facility maneuverability by the introduction of high capacity accumulation systems is becoming increasingly topical.

Different methods of energy accumulation are used for the extension of the facility control range. Energy may be accumulated by electro-chemical or air accumulators, or in pumped hydroelectric energy storage, etc. [5]. In the case of heat load, the accumulator type may be steam, or hot water, or phase transition, or filling [6].

Hydrogen energy accumulation is a prospective power control technology. In this technology hydrogen or hydrogen-enriched synthetic gas is produced during the power drop periods and is used for electricity and heat production in the peak load periods. The hydrogen production methods are numerous [7] but the main are the hydrogen production in electrolyzers or out of fossil fuel by the thermo-chemical method.

Now the electrolysis method is used for 2% of the hydrogen production because of the high power consumption of about 4 kWh/Nm^3 of H_2 [8]. The electrolysis method employs the decomposition of demineralized water with direct current. The decomposition products of the (1) reaction are 99.98% pure hydrogen and oxygen.

$$2H_2O \rightarrow 2H_2 + O_2. \quad (1)$$

Two direct current carrying electrodes are placed into a conductive fluid. The water electric conductivity is increased with water solvents of electrolytes, the most usual is alkali. In some electrolyzers the electrolyte is replaced with solid non-porous membranes which reduces heat losses. The electrode oxidation is prevented with the precious metal coating but it increases the electrode price [9]. Solid oxide electrolyzers (SOE) are yet experimental hydrogen production devices. Usually, they operate at high temperatures of about 800°C which allows the high-temperature electrolysis. In the SOE case, a part of the electrolysis energy is supplied in the form of heat. The SOE is not yet introduced into the industry because of its shortages which are: long startup to operating temperature, chemical stability, and short life of 10 years [10]. The electrolysis consumes at least as much energy as is produced by hydrogen combustion so its constant operation is very expensive. The electrolysis efficiency varies in a very wide range. The alkaline electrolyzers have efficiency up to 62%, the proton membrane and SOE ones have 67 and 75% respectively [11]. The total efficiency of the storage system may be 25 to 45% [12].

The most efficient hydrogen production method is the catalytic reforming of hydrocarbons. For example, the energy consumption for hydrogen production by the steam methane reforming (SMR) is 2 kWh/Nm^3 of H_2 at the efficiency level of 74–85% but this is not "green" hydrogen because the SMR side product is the carbon dioxide. The steam methane reforming environmental performance may be improved by the carbon dioxide capture and storage which makes the hydrogen environmentally harmless "light blue" [13]. The SMR technology cost is relatively low so it is the most popular hydrogen production technology applied for 76% of the world production. Other methods of hydrogen production by hydrocarbon reforming have lower efficiency of 60–75% for example the auto-thermal reforming [7].

The CCGT in literature are being modeled based on the developed algorithms or by the means of commercial software products. Most notable programs for such simulation are Gatecycle [14], Thermoflow and Aspen Plus [15. 16], which is used the most.

So the review of the financial and production efficiency of the energy accumulation in the gas state is of great interest.

2. Materials & Methods

2.1. CCGT-CHP with steam methane reforming and hydrogen energy storage

Figures 1 and 2 show the energy accumulation by steam methane reforming and by electrolyzer and fuel cell facility. Figure 3 shows the flow chart of GTCC-CHP with SMR facility. In this facility, the high-pressure steam is partly directed to the steam transformer which produces steam for steam methane reforming facility. Methane burns in the air and provides heat for methane reformer; the exit gas is supplied to a heat recovery steam generator (HRSG); heat from SMR facility is removed to the district water. During the peak power period, a part of synthetic gas is supplied to the CCGT-CHP combustor and the other part is supplied to the peak load gas turbine. The steam transformer exit condensate and the peak load gas turbine exhaust gas transmit their heat to the district heating water. The main parameters for this scheme computer simulation are summarized in Table 1.

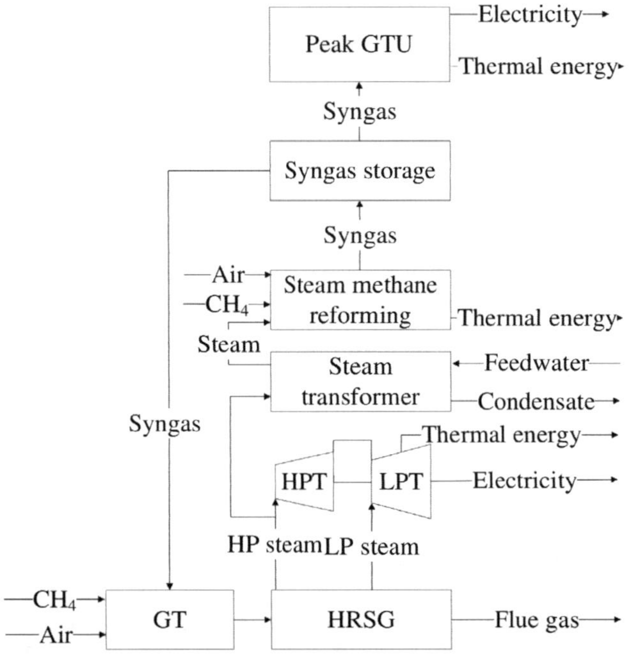

Figure 1. CCGT CHP with a steam transformer and a peak load gas turbine flow chart.

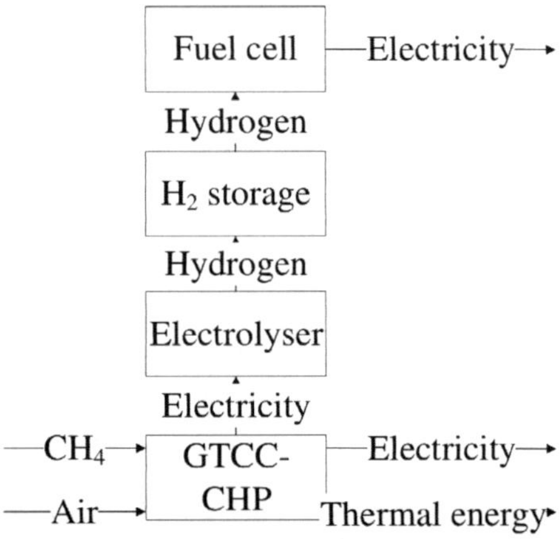

Figure 2. CCGT-CHP with hydrogen accumulation by electrolyzer and fuel cell.

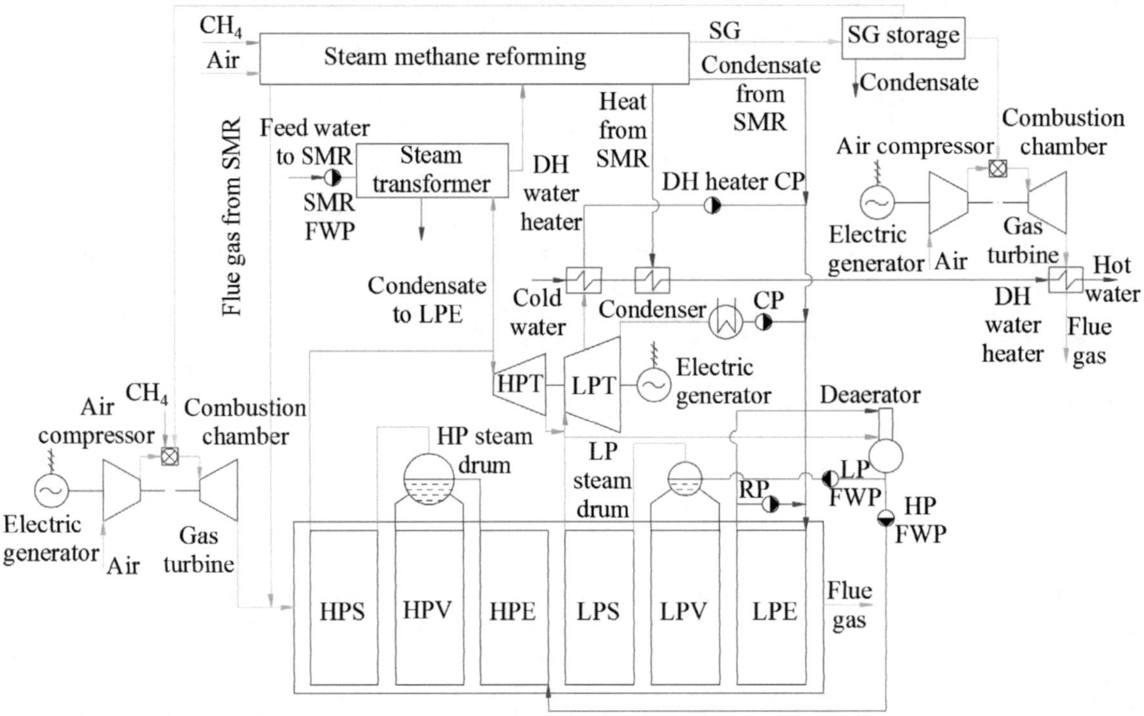

CP – condensate pump, DH – dictrict heating, FWP – feed water pump, GT – gas turbine, HP – high-pressure, HPE – high-pressure economizer, HPS – high-pressure superheater, HPT – high-pressure steam turbine, HPV – high-pressure vaporizer, LP – low-pressure, LPE – low-pressure economizer, LPS – low-pressure superheater, LPT – low-pressure steam turbine, LPV – low-pressure vaporizer, RP – recirculation pump, SG – synthetic gas.

Figure 3. CCGT-CHP with steam methane reforming accumulation, a steam transformer and a peak load gas turbine flow chart.

Table 1. CCGT-CHP with steam methane reforming accumulation analysis input data [14–19].

Parameter	Value
Methane temperature, °C	15
Methane pressure, MPa	0.7
Gas turbine entry temperature, °C	1700
Gas turbine entry pressure, MPa	3
Coolant specific flow, %	13.6
Internal turbine efficiency, %	90
Internal pump efficiency, %	85
Internal compressor efficiency, %	88
Mechanical efficiency, %	99
Electric motor efficiency, %	99
Power generator efficiency, %	99
High-pressure steam pressure, MPa	8
Low-pressure steam pressure, MPa	0.8
District heating steam turbine bleeding pressure, MPa	0.06
Steam under-heating in superheater, °C	20
Minimal under-heating in high-pressure vaporizer, °C	30
Minimal under-heating in low-pressure vaporizer, °C	10
Steam turbine exit pressure, kPa	5
Low-pressure feeding pump exit pressure, MPa	1
High-pressure feeding pump exit pressure, MPa	10
Condensing pump exit pressure, MPa	0.8
District heating water heater condensate pump exit pressure, MPa	0.8
Condensate temperature at the inlet of low-pressure economizer, °C	60
Deaerator pressure, MPa	0.7
Under-heating at low-pressure economizer exit to the saturation temperature, °C	10
Synthetic gas pressure, MPa	2
Steam to methane ratio	5
Reformer exit temperature, °C	700
Reformer inlet heating gas temperature, °C	800
Reformer exit heating gas temperature, °C	400
Synthetic gas temperature at the high temperature conversion reactor inlet, °C	450
Synthetic gas temperature at the high temperature conversion reactor exit, °C	275
Synthetic gas temperature at the steam methane reforming facility exit, °C	80
Under-heating in the steam transformer superheater, °C	20
Pressure of intermediate condenser in steam transformer, MPa	2
Pressure of low-pressure condenser in steam transformer, MPa	0.8
Super-cooling in the steam transformer bleeding, °C	10
Condenser cooling temperature at the steam transformer exit, °C	80
Synthetic gas content in the CCGT-CHP combustor inlet, peak load, %	10
Working fluid temperature at the peak load gas turbine inlet, °C	750
Pressure at the peak load gas turbine inlet, MPa	1.45
Exhaust gas temperature after district heating water heater in peak load gas turbine, °C	90
Natural gas price, RUB/kg	6.48
Natural gas low heating capacity, MJ/kg	49.157
Electricity price during load dropdown, RUB/MWh	700
Electricity price during load peak, RUB/MWh	1700
Exchange rate, RUB/USD	75

During the low load period, the CCGT supplies energy to the electrolyzer that splits water into hydrogen and oxygen. The hydrogen is supplied to its storage and during the peak load period, it powers the fuel cell. Published papers on electrolyzers and fuel cells make base for the following efficiency assumptions: 67% for the transformation of electricity to hydrogen in an electrolyzer with proton-exchange membranes and 60% for the hydrogen transformation to electricity in a solid-polymer hydrogen-air fuel cell [7, 9, 11].

The simulation was carried out in the Aspen Plus software that is good for the thermal and physics-chemical simulation (Figure 4) [20]. The thermo-physical fluid properties were determined using the Peng-Robinson state equation.

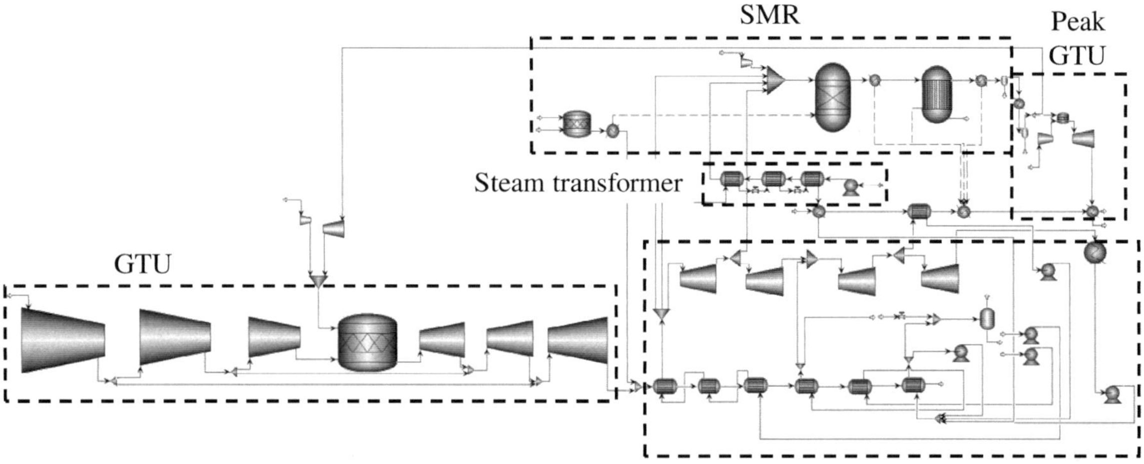

Figure 4. Simulation model scheme of CCGT-CHP with steam methane reforming and peak load gas turbine facility

2.2. Fuel efficiency

In the case of hydrogen accumulation power reduction during the low power periods was caused by electrolyzer operation and corresponded to the power reduction in the facility with SMR accumulation. In this case, the fuel cell produced only electricity.

To take into account the heat of the resulting synthetic gas during the operation of the installation in a drop down load mode, the indicator of the fuel efficiency η_{fuel_eff} was calculated, which takes into account the heat of the syngas, electric and thermal power as a product of the facility:

$$\eta_{fuel\_eff} = \frac{\left(N_{GT} \cdot \eta_{mech} - \dfrac{N_C}{\eta_{mech}}\right) \cdot \eta_{eg} + \left(N_{HPT} + N_{LPT}\right) \cdot \eta_{mech} \cdot \eta_{eg} + Q_{SG\_LHV} \cdot B_{SG\_SMR} + Q_t}{\left(B_{CC\_CCGT} + B_{SMR} + B_{CC\_SMR}\right) \cdot Q_{LHV} + Q_{SG\_LHV} \cdot B_{SG}} \tag{2}$$

$$\frac{-(N_{CCGT\_SC} + N_{SMR\_SC} + N_{SG\_SC} + N_{DH\_CP} + N_{CP} + N_{HP\_FWP} + N_{LP\_FWP} + N_{RP}) / \left(\eta_{mech} \cdot \eta_{em}\right)}{\left(B_{CC\_CCGT} + B_{SMR} + B_{CC\_SMR}\right) \cdot Q_{LHV} + Q_{SG\_LHV} \cdot B_{SG}},$$

where N_{GT} – gas turbine power, MW; N_C – air compressor power, MW; N_{HPT} – high-pressure steam turbine power, MW; N_{LPT} – low-pressure steam turbine power, MW; N_{CCGT_SC} – supercharger power for methane delivery to CCGT, MW; N_{SMR_SC} – supercharger power for methane delivery to SMR, MW; N_{SG_SC} – supercharger power for synthetic gas delivery to CCGT, MW; N_{DH_CP} – district heating water heater condensate pump power, MW; N_{CP} – steam cycle condensate pump power, MW; N_{HP_FWP} – high-pressure feed water pump power, MW; N_{LP_FWP} – low-pressure feed water pump power, MW; N_{RP} – recirculation water pump power, MW; B_{CC_CCGT} – CCGT combustion chamber methane consumption, kg/s; B_{SMR} – methane for reforming consumption, kg/s; B_{CC_SMR} – reformer combustion chamber methane consumption, kg/s; B_{SG_SMR} – synthetic gas generation in steam methane reforming facility, kg/s; B_{SG} – synthetic gas consumption, kg/s; Q_{LHV} – low heating value of methane, MJ/kg; Q_t – heat power, MWt; Q_{SG_LHV} – lower heating value of synthetic gas, MJ/kg; $_{mech}$ – mechanical efficiency, %; η_{eg} – electric generator efficiency, %; η_{em} – electric motor efficiency, %.

2.3. Accumulation financial effect
The accumulation financial effect was assessed with the contribution margin–index, which is the difference between the product selling income and the variable expenses related to production [21]. In the case of cogeneration, the products are heat and thermal energies, and the variable expenses are fuel cost and water preparation expenses, water purchase, treatment chemicals, etc. The thermal energy produced was the same at all loads and the main part of the variable expenses was related to the fuel purchase. Because of that contribution margin index CM may be calculated with the following equation:

$$\mathrm{CM} = t^{\mathrm{peak}} \cdot (P^{\mathrm{peak}}_{\mathrm{elec}} \cdot N^{\mathrm{peak}} - P_{\mathrm{fuel}} \cdot B^{\mathrm{peak}}) + t^{\mathrm{drop}} \cdot (P^{\mathrm{drop}}_{\mathrm{elec}} \cdot N^{\mathrm{drop}} - P_{\mathrm{fuel}} \cdot B^{\mathrm{drop}}), \quad (3)$$

where $P_{elec}{}^{peak}$ – electricity price during peak load period, USD/MWh; $P_{elec}{}^{drop}$ – electricity price during drop down period, USD/MWh; N^{peak} – peak load mode facility supply power, MW; N^{drop} – drop down load mode facility supply power, MW; t^{peak} – facility operation time at peak load mode, s; t^{drop} – facility operation time at drop down load mode, s; P_{fuel} – fuel price, USD/kg; B^{peak} – peak load mode fuel consumption, kg/s; B^{drop} – drop down load mode fuel consumption, kg/s.

3. Results and Discussion
The calculated CCGT-CHP efficiency in condensation mode was 61.6%, which is in accordance with data in literature: 56,11% for gas turbine inlet temperature of 1315°C [16], 58.59% for gas turbine inlet temperature of 1416°C [15].

Figure 5 compares the calculated fuel efficiency of the described above accumulation versions. Both versions reduced the facility fuel efficiency. In the steam methane reforming version, it is caused by the heat bleeding from SMR and the CCGT operation in condensing mode. When the electrolyzer operates the facility fuel efficiency drops down because the efficiency of hydrogen electricity production is smaller than the facility efficiency in the low power with heat supply regime. The peak power fuel efficiency is also lower. In the CCGT-CHP with SMR accumulation it is caused by the low efficiency peak load gas turbine operation and the heat bleeding from the combined cycle steam turbine. In this case, the turbine operates at condensing regime. In the case of electrolyzer and fuel cell, the fuel efficiency is also lower due to the transformation processes energy losses.

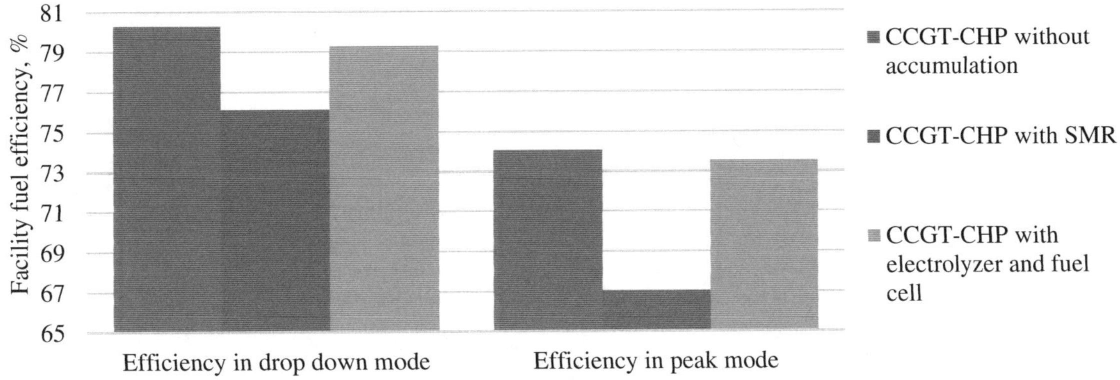

Figure 5. Fuel efficiency at different operation regimes.

Figure 6 shows the CCGT-CHP power with different accumulation systems. The version with SMR accumulation shows the maximal power increase in peak operation which is due to the large amounts of synthetic gas production during the load drop down and the gas combustion in the peak load gas turbine.

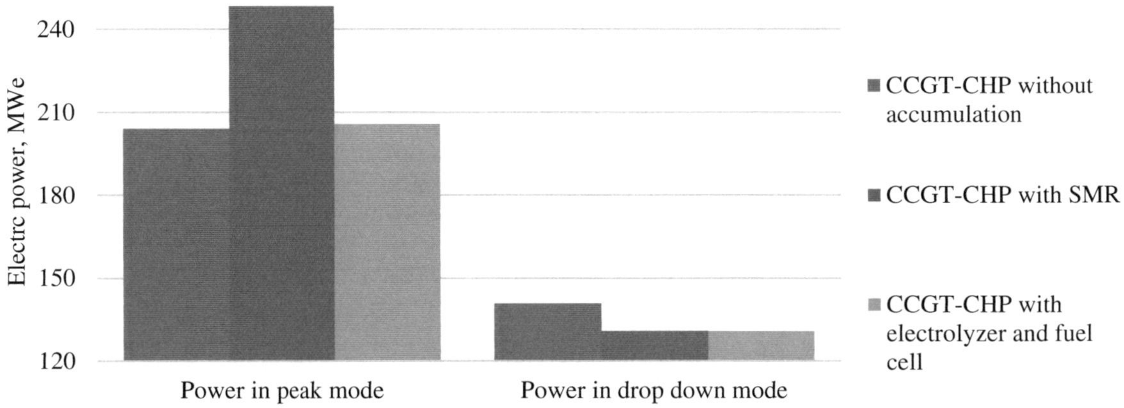

Figure 6. Power of the CCGT-CHP accumulation versions at peak and dropdown loads.

Analysis results summarized in Table 2 show, that the CCGT-CHP transition from the base scheme to SMR accumulation increases the contribution margin, or financial efficiency for 173.6 USD/day, or 2.57%, the transition from electrolyzer and fuel cell provides 196.7 USD/day or 2.9% bigger contribution margin. The hydrogen accumulation with electrolyzer and fuel cell has 0.3% lower contribution margin than the base scheme.

Table 2. CCGT-CHP parameters without and with accumulation.

Parameter	No accumulation CCGT-CHP		CCGT-CHP with SMR		CCGT-CHP with electrolyzer and fuel cell	
	drop	peak	drop	peak	drop	peak
Supply power, MW	140.78	201.82	130.76	248.30	130.76	205.74
Methane massflow, kg/s	4.80	6.85	8.04	6.55	4.80	6.85
Electricity price, USD/MWh	9.33	22.67	9.33	22.67	9.33	22.67
Operation period, h	3	3.07	3	3.07	3	3.07
Sell profit, USD	17999.7		20957.0		17976.8	
Fuel cost, USD	11211.3		13995.0		11211.3	
Contribution margin, the heat supply not considered, USD	6788.4		6962.0		6765.5	

4. Conclusion

Energy accumulation with steam methane reforming allows reduction of the supplied electricity during the low load period for 10 MW or 7.0% and to increase it during the peak load for 44.48 MW or 23.0%. The same supplied power reduction with the electrolyzer operation and the peak power increase with the produced hydrogen used in the fuel cell during the peak load is 3.92 MW or 0.3%.

Accumulation of synthetic gas and hydrogen reduces the facility fuel efficiency. The steam methane reforming accumulation in peak periods reduces the fuel efficiency from 74.1 to 67.0% and at low load from 80.3 to 76.1%. The hydrogen accumulation with electrolyzer and fuel cell reduces the fuel efficiency from 74.1 to 73.5% and from 80.3 to 79.3% at low load.

Accumulation with steam methane reforming increases the daily contribution margin by 173.6 USD or 2.6%. Introduction of hydrogen with electrolyzer and fuel cell reduces the daily contribution margin by 22.9 USD or 0.3%.

References

[1] Zaryankin A, Mager A, Rogalev A and Komarov I 2014 *WIT Trans. Ecol. Envir.* **190(1)** 251–60
[2] Lisin E, Rogalev A, Strielkowski W and Komarov I 2015 *Sustainability* **7(9)** 11378–400
[3] Rogalev N, Prokhorov V, Rogalev A, Komarov I and Kindra V 2016 *Int. J. Appl. Eng. Res.* **11(18)** 9297–306
[4] Hub Indices: Trading System Administrator of Wholesale Electricity Market URL: https://www.atsenergo.ru/results/rsv/hubs/hubs?zone=1
[5] Frate G F, Ferrari L and Desideri U 2021 *Renew. Energ.* **163** 1754–72
[6] Alva G, Lin Y and Fang G 2018 *Energy* **144** 341–78
[7] Shiva Kumar S and Himabindu V 2019 *Mater. Sci. Technol.* **2(3)** 442–54
[8] The Future of Hydrogen 2019 (Paris: International Energy Agency) p 203
[9] Carmo M, Fritz D L, Mergel J and Stolten D 2013 *Int. J. Hydrogen Energ.* **38(12)** 4901–34
[10] Zheng Y, Wang J, Yu B, Zhang W, Chen J, Qiao J and Zhang J 2017 *Chem. Soc. Rev.* **46(5)** 1427–63
[11] Comparison of Fuel Cell Technologies: Office of Energy Efficiency and Renewable Energy URL: https://www.energy.gov/eere/fuelcells/comparison-fuel-cell-technologies
[12] Ceran B 2018 *Energy Policy* **21(3)** 97–110
[13] Mitrova T, Melnikov Yu, Chugunov D and Glagoleva A 2019 *A Hydrogen Economy—A Path towards Low Carbon Development* (Moscow: Skolkovo Moscow School of Management) p 63
[14] Carapellucci R and Giordano L 2019 *Energy* **173** 229–43
[15] Romano MC, Cassotti EN, Chiesa P, Meyer J and Mastin J 2011 *Enrgy. Proced.* **4** 1125–32
[16] Lozza G and Chiesa P 2000 *Proc. of the ASME Turbo Expo: Power for Land, Sea, and Air* (Munich / ASME)
[17] Karpunin A P 2016 *Study of the influence of the parameters of GTU and CCGT on their characteristics based on the technique with detailed taking into account of losses from cooling in a gas turbine* (Moscow: MPEI) p 182
[18] Valamin A Y, Kultyshev A Y, Bilan V N, Goldberg A A, Sakhnin Y A, Shekhter M V, Aguilera H C P, Stepanov M Y, Shibaev T L and Polyaeva Y N 2012 *Therm. Eng.* **59(12)** 883–92
[19] Trukhnii A D 2013 *Combined-Cycle Plants of Power Plants* (Moscow: MPEI) p 648
[20] Rogalev A, Komarov I, Kindra V and Zlyvko O 2018 *Entrepreneurship Sustain. Issues* **6(1)** 429–45
[21] Alivanova S V, Kurennaya V V 2012 *Scientific Journal of KubSAU* **80(06)** 0801206015.

Acknowledgments

This study conducted by Moscow Power Engineering Institute was financially supported by the Ministry of Science and Higher Education of the Russian Federation (project No. FSWF-2020-0020).

The influence of selenium amount added into the graphite box during the selenization of solution deposited CIGSe thin films.

G M Albalawneh[1,*], M M Ramli[2,3], M ZM Zain[4] and Z Sauli[3]

[1]Institute of Nano Electronic Engineering, Universiti Malaysia Perlis, 01000 Kangar, Perlis, Malaysia.
[2]Geopolymer and Green Technology, Center of Excellence (CEGeoGTech), Universiti Malaysia Perlis (UniMAP), Perlis, Malaysia.
[3]Faculty of Electronic Engineering Technology, Universiti Malaysia Perlis (UniMAP), Perlis, Malaysia.
[4]Faculty of Mechanical Engineering Technology, Universiti Malaysia Perlis, Perlis, Malaysia
* Ghadeer.ballawneh@gmail.com

Abstract. Cu(In,Ga)Se2 (CIGSe) semiconductor is an efficient light absorber material for thin-film solar cell technology. The sequential evaporation of precursor solution, followed by the selenization process, is a promising non-vacuum and low-cost approach for CIGSe thin-film fabrication. The main properties of CIGSe thin films are strongly affected by the post-selenization step. Hence, thorough control of selenization parameters is essential for achieving pure crystalline, large grain films needed for high-performance solar cell devices. In this study, the impact of selenium (Se) amount added during the selenization step was evaluated. The structural, morphological, and compositional properties of the selenized thin films were investigated. The CIGSe precursor film was deposited by a spin-coating technique using a thiol/amine-based solution, followed by annealing with different Se amounts (100, 200, and 300 mg) within a partially closed small round graphite container. In all cases, uniform films of 1.2–1.5 µm thickness with a well-defined single chalcopyrite phase were obtained. It was observed that the grain size and Se content increased with increasing Se mass added. Moreover, the sample selenized with 200 mg Se resulted in higher surface coverage, thinner fine-grained layer, and less MoSe2 formation than the excess Se samples.

1. Introduction

Copper indium gallium selenide (CIGSe) chalcopyrite semiconductors have attracted much attention as the solar cell absorber layer with the highest conversion efficiency of over 23% among thin-film solar cells [1–3]. In addition to the high efficiency, the CIGSe thin-film absorbers also have several advantages, such as suitable optical absorption (>105 cm–1), excellent outdoor stability, high defect density, and a tunable bandgap that can be adjusted by varying the Ga content [4,5]. Vacuum deposition techniques such as evaporation or sputtering are commonly used to produce CIGSe solar cells with the highest efficiency. However, conventional vacuum-based techniques result in increased costs and technical obstacles for commercial fabrication [6]. Therefore, solution-based vacuum-free processes are believed to be a scalable, straightforward alternative option that offers an efficient route for a low-cost, large-area, high-throughput, and flexible solar cell production [7,8].

To date, the highest efficiency recorded for a solution-based non-vacuum CIGSe cell has remained around 17.3% prepared by dissolving chalcogenides in hydrazine solution [9]. However, the dangerous nature of hydrazine has restricted this route for large-scale production [10–13]. Thus, alternative free-hydrazine solution routes have been employed for CIGSe absorber layer fabrication [14–19]. Nevertheless, some limitations need to be addressed, such as the preparation process complexity, ink stability, and poor crystallinity of the fabricated film [20–22].

Solution-processed deposition techniques commonly involve two steps. The first stage is the preparation and deposition of homogeneous precursor solution at relatively low temperatures using one of the standard coating methods, such as spin coating, spray coating, or inkjet printing. The second step involves post-selenization at about 450–600 °C to convert the alloy precursor layers into CIGSe absorber layers. Thiol/amine mixtures have recently been used to dissolve several metals and chalcogenide sources effectively to prepare precursor solution and deposit quality CIGSe absorber layers [9,22–24]. Photovoltaic devices based on thiol/amine solutions result in a promising CIGSe absorber properties and device efficiencies up to 13.12% [25–32]. However, some unresolved challenges of this route need to be addressed, such as a) the formation of a bottom layer composed of fine grains, b) a significant elemental reduction in the deposited film, and c) the formation of a thick MoSe2 layer between the Mo and CIGSe layer after selenization, which are all commonly thought to be harmful to solar cell performance [26,29].

Great efforts have been made to improve the solution-based CIGSe absorber quality. Selenization is an essential step to transform coated raw films into good uniformity, high-quality, single-phase CIGSe films with improved crystal size suitable for solar cells[9,22–24]. The selenization process also enlarges grain size, which effectively decreases grain boundary concentration, potentially boosts carrier mobility, reduces the chance of electron-hole recombination, and subsequently increases current generation [33–38]. The influence of selenization parameters, such as temperature ramping rate, annealing temperature, annealing time, and partial pressure of selenium (Se) on the final film, has been widely studied by many researchers [39–41]. However, the effect of Se mass added during the selenization step on the crystallinity and grain growth of the CIGSe absorber layer has not been broadly investigated. In this work, a thiol/amine-based CIGSe precursor solution was spin-coated to deposit a dense CIGSe layer. Selenization was accomplished using Se pellets in a partially closed graphite box whose design was improved compared to the one used in earlier works and those reported in the literature. The Se mass added during the selenization step was varied, and its influence on the morphological, structural, and compositional properties was reported.

2. Experimental methods

2.1 Materials.

Copper(II) acetylacetonate ($Cu(acac)_2$, 99.9%), gallium(III) acetylacetonate ($Ga(acac)_3$, 99.99%), and Se pellets (Se, 99.99%) were purchased from Sigma-Aldrich. Indium(III) acetate ($In(OAc)_3$, 99.9%), 1,2-ethanedithiol ($C_2H_6S_2$, 95%), 1,2-ethylenediamine ($C_2H_8N_2$, 99%), and Se powder (Se, 99.99%) were purchased from Acros (Belgium).

2.2 CIGSe precursor solution preparation

To prepare the CIGSe precursor solution, $Cu(acac)_2$ (481.3 mg, 1.84 mmol), $In(OAc)_3$ (408.7 mg, 1.4 mmol), Ga (220.2 mg, 0.6 mmol), and Se (315.8 mg, 4 mmol) were transferred into a flask. Then, 10 ml of ethylenediamine and 1 ml of ethanedithiol were added. After overnight stirring at 65 °C, a homogeneous dark brown ink was obtained. The resultant ink remained stable for up to three weeks.

2.3 Deposition and selenization of CIGSe films

The CIGSe films were deposited onto 2.5 × 2.5 cm molybdenum (Mo)-coated soda-lime glass (SLG) substrate. The films were produced by spin-coating using the previously prepared ink. The solution was first deposited at a rotation speed of 100 rpm for 5 s, followed by 3,000 rpm for 30 s, with approximately 0.2 ml of the solution was used for each coating step. The samples were annealed using hot plate at 300 °C for 2 min to remove organic additives after each coating cycle. The coating/drying cycles were repeated ten times by targeting 1.2–1.5 µm film thickness. Both deposition and drying steps were conducted in a nitrogen-filled glovebox. The coated films were then selenized inside a tube furnace using Se pellets as a Se vapour source. During selenization, the deposited sample was put in a graphite box along with Se pellets. A small round graphite box was used in this work of about 45 mm diameter and approximately 7 cm^3 volume (with the lid on). The graphite box is shown in figure 1. The graphite box's lid has a 1-mm hole at the centre where excess Se can escape. Initially, the quartz tube was purged with nitrogen and then evacuated to a predetermined pressure of 400 Torr. The selenization heating rate of 16 °C/s was applied to the temperature of 550 °C, which lasted for 30 min and dwelled for 20 min. The quantity of Se necessary to substitute sulphur and supply an adequate Se vapour pressure to recrystallize the coated film was investigated by loading different amounts of Se (100, 200, and 300 mg) into the graphite box during the selenization step.

Figure 1. Digital photograph of used the graphite box.

2.4 Characterization.
The structure of the CIGSe absorber was investigated by X-ray diffraction (XRD) recorded on a Bruker D8 Advance X-Ray diffractometer with a Cu Kα anode (λ = 0.1542 nm) . The surface morphology and grain size of the CIGSe absorber layer were assessed by a JEOL JSM 6460 scanning electron microscope (SEM) using 20 kV acceleration voltage. For verification of the CIGSe layer thickness, SEM cross-sectional images were taken from mechanically cleaved samples and then the images were analyzed with the software MountainsMap from DigitalSurf®. Energy dispersive X-ray spectroscopy (EDX) was performed for elemental analysis, with an opening size of 30 μm^2 and 20 kV voltage.

3. Results and discussion

In this work, the influence of selenization conditions on the final CIGSe film properties was investigated. Spin-coated CIGSe precursor films were selenized with varied Se amounts introduced into the graphite box while keeping all other selenization parameters constant. Optimising the Se amount is expected to enable more control on the Se pressure and diffusion into the precursor film, which leads to a more uniform structural and chemical homogeneity [33–38].

On the other hand, it is generally believed that the graphite box provides enhanced control for the selenization step. It was demonstrated that using a round graphite case instead of a rectangle one resulted in a high-quality CIGSe absorber layer with large grains, and it is more efficient to eliminate the fine-grained layer. The circular graphite box has an improved capability to maintain Se vapour during selenization [42,43]. The graphite box used in this work was designed as a simple, round graphite box, similar to the one used previously [42], but with a smaller size and closer distance between the film surface and the Se source. This design is intended to deliver more Se vapour to the film and suppress the internal Se vapour escaping the box. Besides, a closer space between the sample surface and the Se source would result in uniform and higher Se vapour pressure, resulting in good crystallisation of the CIGSe phase [44].

3.1 Morphological analysis
Surface and thickness characterizations were investigated by SEM images, and then the images were colored and analyzed using MountainsMap software. The cross-section view of the selenized film with different Se amounts, as shown in figure 2, indicates acceptable absorber thickness (1.2–1.5 μm) for solar cell fabrication. Furthermore, all films exhibit a bi-layer structure consisting of a fine-grained layer beneath a large-grained layer. The formation of such a structure is one of the primary issues involving the deposition of solution-based CIGSe film. It is commonly thought that films with thick fine-grain layers are detrimental to solar cell performance [13,26,41]. Therefore, increasing the coarse crystalline layer thickness and eliminating the fine-grained layer introduces a reliable indication of solar cell improvement. For the thin film selenized with 200 mg Se, the cross-section film shows that the absorber layer exhibited good film crystallinity with approximately 700 nm thick large-grained top layer. 300 and 100 mg Se Samples showed a thinner coarse layer compared to the fine-grained layer. The SEM cross-section image in the case of adding 300 mg Se shows a feature of grain accumulation, which can be referred to the $MoSe_2$ formed between the CIGSe layer and the Mo-back contact [30,42].

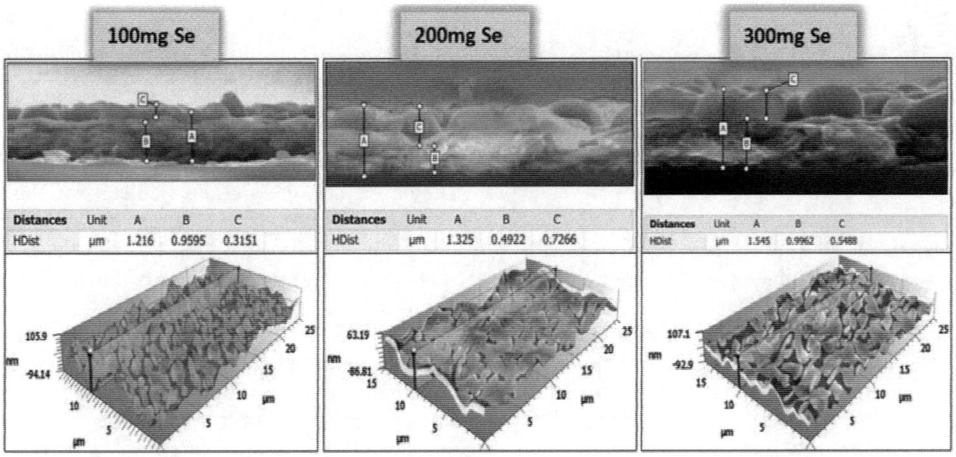

Figure 2. SEM images of cross-sections of the selenized films with 100, 200, and 300 mg of Se added into the graphite box.

From the perspective of grain growth, the higher the Se mass supplied to CIGSe, the larger the resultant grain size, which was accomplished by increasing the Se vapor diffusion. The diameter of some grains exceeded 3 µm when 300 mg of Se was added. The surface coverage, grain mean area and mean diameter are maximum when 200 mg Se was added to the graphite box compared to 100 and 300 mg. a detailed surface analysis for different Se amounts is shown in figure 3.

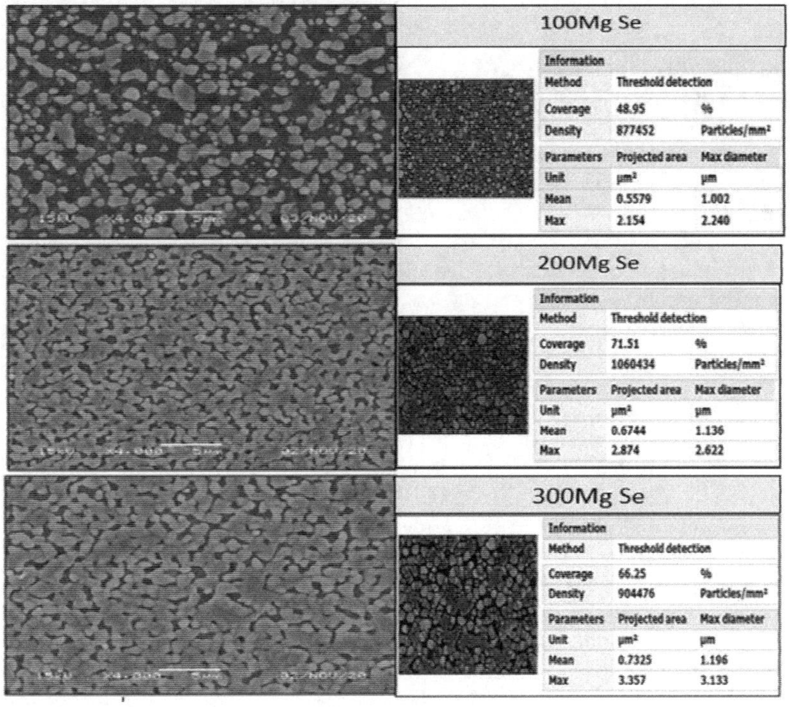

Figure 3. Surface morphology and grain size of the selenized films with 100, 200, and 300mg of Se added during the selenization step.

3.2 Compositional analysis

In CIGSe absorber layer, the [Cu]/[In+Ga] and [Ga]/[In+Ga] ratios have a significant influence on the structural, morphological, optical, and electrical properties. Furthermore, high-efficiency CIGSe absorber layers are obtained with [Cu]/[In+Ga] ratios ranging from 0.8 to 0.9 [45,46]. The Cu-poor film can avoid the formation of undesired secondary phases and aids the formation of a p-type CIGSe absorber [47]. [Ga]/[In+Ga] ratio is the most important parameter that will determine the bandgap energy of CIGSe thin-film. Although the higher Ga ratio could increase bandgap values, the gallium-rich absorber layer shows deteriorated photovoltaic. [Ga]/[In+Ga] =0.3 is the optimum ratio for high-efficiency solar cells [48,49]. Thus, in this work, the targeted film composition for the films was $Cu_{0.92}In_{0.7}Ga_{0.3}Se_2$. The desired ratio for cations was controlled by adding specific ratios in the precursor solutions as described in the experimental section.

EDX was used to determine the composition of the selenized thin film using different Se amounts, and the results are listed in table 1. All samples showed slightly poor Cu composition [Cu/(Ga+In)] ~ 0.8 and [Ga/(Ga+In)] ~ 0.3, which are the preferred values of high-efficiency CIGSe absorber material [2]. In contrast, the Se-to-metals ratio [Se/(In+Ga)] noticed an increase from 1.3 to 2.4. It is expected that increased Se partial pressure, corresponding to a higher amount of Se, allows the selenization to occur more deeply and results in a higher Se content.

Table 1. EDX analysis for the CIGSe absorbers selenized with different selenium amounts (atm%).

	Cu	In	Ga	S	Se	Cu/(In+Ga)	Ga/(In+Ga)	Se/(In+Ga)
100mg Se	24.30	20.30	9.61	4.54	41.23	0.81	0.32	1.38
200mg Se	20.41	16.74	8.17	4.91	49.77	0.82	0.33	2.00
300mg Se	17.87	16.10	6.63	4.53	54.86	0.79	0.29	2.41

3.3 Structural and phase analysis

The XRD spectra of the CIGSe sample deposited on the Mo-SLG substrate and selenized with different Se amounts are displayed in figure 4. All samples have a well-defined chalcopyrite crystal structure and show a preferred orientation of (112) assigned to $CuGa_{0.3}In_{0.7}Se_2$ chemical formula (JCPDS Card File No, 35-1102), as well as Mo, $MoSe_2$ secondary phases. It is remarkable that as the Se amount increases, the intensity of $MoSe_2$ peaks increases. The formation of thick $MoSe_2$ is referred to the direct contact of Se vapour with the Mo layer through many pores in the CIGSe layer [50,51]. Even though the formation of a thin $MoSe_2$ layer is important to assist the back ohmic contact, thicker $MoSe_2$ is significantly detrimental to the absorber layer stability for the final solar cell performance [39–41,52]. Another aspect of the formation of thick $MoSe_2$ at the back contact is that it has a significant impact on the adhesion of CIGSe/(MoSe2)/Mo layers to the substrate [53].

To examine the quality of the CIGSe absorbers selenized with different Se amounts, the full width at half-maximum (FWHM) value of (112) peak was estimated. As listed in the inset data of figure 4, it is clear that the films had the highest peak value of XRD and the narrowest value of FWHM when 300 mg Se was introduced, indicating that the crystallinity of CIGSe absorber improved with the increasing Se amount as the FWHM decreased. However, the Mo/CIGS interface is often of poor quality with a lower Mo(110)/$MoSe_2$(110) peak ratios. Hence, a trade-off between the film bulk and the Mo-contact properties has to be made. Therefore, 300 mg Se is considered too much amount and seems to harm device performance. The formation of a thick $MoSe_2$ layer can also be confirmed by the SEM image shown in figure 2 for 300 mg Se, demonstrating that the CIGSe/Mo interface is presumably considered $MoSe_2$ because it has a totally different morphology compared to that selenized with a lower Se amount.

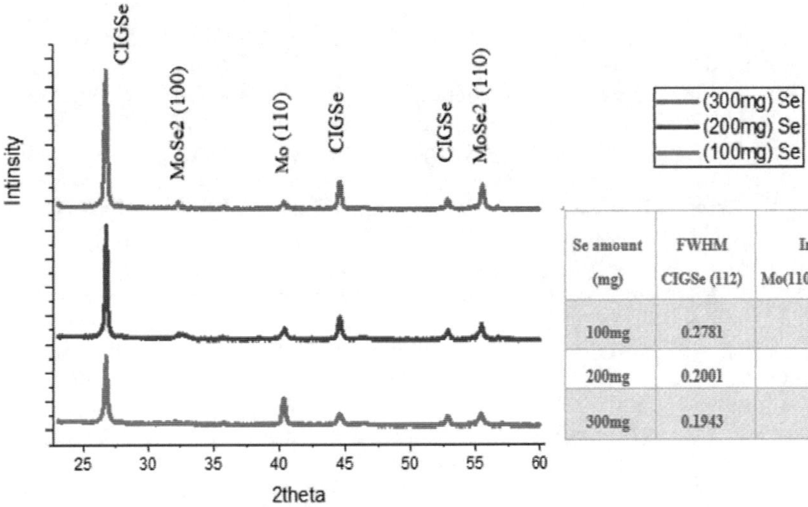

Se amount (mg)	FWHM CIGSe (112)	Intensity $Mo(110)/MoSe_2(110)$
100mg	0.2781	2.12
200mg	0.2001	0.78
300mg	0.1943	0.36

Figure 4. Spectra of CIGSe absorbers selenized with 100, 200, and 300 mg of Se and selenized. The table introduces the (FWHM) values of the leading CIGSe peak and $Mo/MoSe_2$ intensity ratio of the most noticeable peaks.

4. Conclusions

This study focused on optimising the selenization step in a sequential growth process to fabricate a quality CIGSe absorber layer by spin-coating using a thiol/amine-based precursor solution. The selenization step was executed by varying the Se amount. Structural, compositional, and morphological properties of CIGSe thin films were investigated. The findings demonstrated that an increase in the Se amount resulted in enhanced incorporation of Se into the absorber via Se vapour pressure with Se/(In+Ga) ratio ranging from 1.4 to 2.4. In all cases, a well-defined CIGSe phase with a clear bilayer structure was obtained. The sample selenized with 200 mg Se resulted in higher surface coverage and thinner fine-grained layer than the deficient and excess Se samples. The CIGSe films prepared with high Se amounts resulted in larger grains and improved crystallinity. Simultaneously, excess Se induced the formation of a thicker MoSe2 layer between CIGSe and Mo, which plays a deleterious role in the CIGSe absorber layer performance. The influence of Se content during selenization on solar cell performance is not considered in this study. However, it is believed that there is a reasonable effect on the final photovoltaic device performance.

5. References

[1] Feurer T, Reinhard P, Avancini E, Bissig B, Löckinger J, Fuchs P, Carron R, Weiss T P, Perrenoud J, Stutterheim S, Buecheler S and Tiwari A N 2017 Progress in thin film CIGS photovoltaics - Research and development, manufacturing, and applications Prog. Photovoltaics Res. Appl. 25 645–67

[2] Ramanujam J and Singh U P 2017 Copper indium gallium selenide based solar cells – a review Energy Environ. Sci. 10 1306–19

[3] Singh U P and Patra S P 2010 Progress in polycrystalline thin-film Cu(In,Ga)Se2 solar cells Int. J. Photoenergy 2010 1–19

[4] Nakamura M, Yamaguchi K, Kimoto Y, Yasaki Y, Kato T and Sugimoto H 2019 Cd-Free Cu(In,Ga)(Se,S)2 Thin-Film Solar Cell With Record Efficiency of 23.35% IEEE J. Photovoltaics 1–5

[5] Green M A, Dunlop E D, Levi D H, Hohl-Ebinger J, Yoshita M and Ho-Baillie A W Y 2019 Solar cell efficiency tables (version 54) Prog. Photovoltaics Res. Appl. 27 565–75

[6] Powalla M, Jackson P, Witte W, Hariskos D, Paetel S, Tschamber C and Wischmann W 2013 High-efficiency Cu(In,Ga)Se2 cells and modules Sol. Energy Mater. Sol. Cells 119 51–8

[7] Hibberd C J, Chassaing E, Liu W, Mitzi D B, Lincot D and Tiwari A N 2010 Non-vacuum methods for formation of Cu(In, Ga)(Se, S)2thin film photovoltaic absorbers Prog. Photovoltaics Res.

Appl. 18 434–52

[8] Lee D and Yong K 2013 Non-vacuum deposition of CIGS absorber films for low-cost thin film solar cells *Korean J. Chem. Eng.* **30** 1347–58

[9] Zhang T, Yang Y, Liu D, Tse S C, Cao W, Feng Z, Chen S and Qian L 2016 High efficiency solution-processed thin-film Cu(In,Ga)(Se,S) $_2$ solar cells *Energy Environ. Sci.* **9** 3674–81

[10] Katritzky A R 2002 Hydrazine and Its Derivatives: Preparation, Properties, Applications. *J. Am. Chem. Soc.* **124** 6504–6504

[11] Bob B, Lei B, Chung C H, Yang W, Hsu W C, Duan H S, Hou W W J, Li S H and Yang Y 2012 The development of hydrazine-processed Cu(In,Ga)(Se,S) 2 solar cells *Adv. Energy Mater.* **2** 504–22

[12] Schmidt E 2001 *Hydrazine and its Derivatives : Preparation, Properties, Applications* (Wiley-Interscience)

[13] Regmi G, Ashok A, Chawla P, Semalti P, Velumani S, Sharma S N and Castaneda H 2020 Perspectives of chalcopyrite-based CIGSe thin-film solar cell: a review *J. Mater. Sci. Mater. Electron.* **31** 7286–314

[14] Suresh S and Uhl A R 2021 Present Status of Solution-Processing Routes for Cu(In,Ga)(S,Se)2 Solar Cell Absorbers *Adv. Energy Mater.* **11** 2003743

[15] Kim B and Min B K 2018 Strategies toward highly efficient CIGSe thin-film solar cells fabricated by sequential process *Sustain. Energy Fuels* **2** 1671–85

[16] Karunakaran S K, Arumugam G M, Yang W, Ge S, Khan S N, Lin X and Yang G 2019 Recent progress in inkjet-printed solar cells *J. Mater. Chem. A* **7** 13873–902

[17] Uhl A R, Katahara J K and Hillhouse H W 2016 Molecular-ink route to 13.0% efficient low-bandgap CuIn(S,Se) $_2$ and 14.7% efficient Cu(In,Ga)(S,Se) $_2$ solar cells *Energy Environ. Sci.* **9** 130–4

[18] Xie Y, Chen H, Li A, Zhu X, Zhang L, Qin M, Wang Y, Liu Y and Huang F 2014 A facile molecular precursor-based Cu(In,Ga)(S,Se) $_2$ solar cell with 8.6% efficiency *J. Mater. Chem. A* **2** 13237

[19] Wang G, Wang S, Cui Y and Pan D 2012 A Novel and Versatile Strategy to Prepare Metal–Organic Molecular Precursor Solutions and Its Application in Cu(In,Ga)(S,Se) $_2$ Solar Cells *Chem. Mater.* **24** 3993–7

[20] Uhl A R, Fella C, Chirilă A, Kaelin M R, Karvonen L, Weidenkaff A, Borca C N, Grolimund D, Romanyuk Y E and Tiwari A N 2012 Non-vacuum deposition of Cu(In,Ga)Se2 absorber layers from binder free, alcohol solutions *Prog. Photovoltaics Res. Appl.* **20** 526–33

[21] Park S J, Cho Y, Moon S H, Kim J E, Lee D K, Gwak J, Kim J, Kim D W and Min B K 2014 A comparative study of solution-processed low- and high-band-gap chalcopyrite thin-film solar cells *J. Phys. D. Appl. Phys.* **47** 305–43

[22] Kim B and Min B K 2018 Strategies toward highly efficient CIGSe thin-film solar cells fabricated by sequential process *Sustain. Energy Fuels* **2** 1671–85

[23] Todorov T K, Gunawan O, Gokmen T and Mitzi D B 2013 Solution-processed Cu(In,Ga)(S,Se) $_2$ absorber yielding a 15.2% efficient solar cell *Prog. Photovoltaics Res. Appl.* **21** 82–7

[24] Yalçin L and Öztürk R 2013 Performance comparison of c-Si, mc-Si and a-Si thin film PV by PVsyst simulation *J. Optoelectron. Adv. Mater.* **15** 326–34

[25] Zhao D, Tian Q, Zhou Z, Wang G, Meng Y, Kou D, Zhou W, Pan D and Wu S 2015 Solution-deposited pure selenide CIGSe solar cells from elemental Cu, In, Ga, and Se *J. Mater. Chem. A* **3** 19263–7

[26] Zhao D, Fan Q, Tian Q, Zhou Z, Meng Y, Kou D, Zhou W and Wu S 2016 Eliminating fine-grained layers in Cu(In,Ga)(S,Se) $_2$ thin films for solution-processed high efficiency solar cells *J. Mater. Chem. A* **4** 13476–81

[27] Zhao, Xin Lu M, Koeper M J and Agrawal R 2016 Solution-processed sulfur depleted Cu(In, Ga)Se 2 solar cells synthesized from a monoamine–dithiol solvent mixture *J. Mater. Chem. A* **4** 7390–7

[28] Arnou P, van Hest M F A M, Cooper C S, Malkov A V., Walls J M and Bowers J W 2016 Hydrazine-Free Solution-Deposited CuIn(S,Se) $_2$ Solar Cells by Spray Deposition of Metal Chalcogenides *ACS Appl. Mater. Interfaces* **8** 11893–7

[29] Arnou P, Cooper C S, Uličná S, Abbas A, Eeles A, Wright L D, Malkov A V., Walls J M and Bowers J W 2017 Solution processing of CuIn(S,Se) 2 and Cu(In,Ga)(S,Se) 2 thin film solar cells using metal chalcogenide precursors *Thin Solid Films* **633** 76–80

[30] Fan Q, Tian Q, Wang H, Zhao F, Kong J and Wu S 2018 Regulating the starting location of front-gradient enabled highly efficient Cu(In,Ga)Se2 solar cells: Via a facile thiol-amine solution approach *J. Mater. Chem. A* **6** 4095–101

[31] Uličná S, Arnou P, Abbas A, Togay M, Welch L M, Bliss M, Malkov A V., Walls J M and Bowers J W 2019 Deposition and application of a Mo-N back contact diffusion barrier yielding a 12.0% efficiency solution-processed CIGS solar cell using an amine-thiol solvent system *J. Mater. Chem. A* **7** 7042–52

[32] Albalawneh G and Ramli M 2020 Review—Solution Processing of CIGSe Solar Cells Using Simple Thiol-Amine Solvents Mixture: A Review *ECS J. Solid State Sci. Technol.* **9** 061013

[33] Li W, Tan J M R, Leow S W, Lie S, Magdassi S and Wong L H 2018 Recent Progress in Solution-Processed Copper-Chalcogenide Thin-Film Solar Cells *Energy Technol.* **6** 46–59

[34] Khoshsirat N and Yunus N A M 2016 *Nanoelectronics and Materials Development* (INTECH)

[35] Huang W-C, Tseng C, Chang S, Tuan H, Chiang C, Lyu L and Huang M H 2012 Solvothermal synthesis of zincblende and wurtzite CuInS2 nanocrystals and their photovoltaic application. *Langmuir* **28** 8496–501

[36] Rau U, Taretto K and Siebentritt S 2009 Grain boundaries in Cu(In, Ga)(Se, S)2 thin-film solar cells *Appl. Phys. A Mater. Sci. Process.* **96** 221–34

[37] Idris B, Rafik Z, Kamal D, Abdessalam B and Faouzi G 2015 Size and grain-boundary effects on the performance of polycrystalline CIGS-based solar cells *2015 6th International Renewable Energy Congress, IREC 2015* (Institute of Electrical and Electronics Engineers Inc.) pp 1–5

[38] Zhou H, Hsu W C, Duan H S, Bob B, Yang W, Song T Bin, Hsu C J and Yang Y 2013 CZTS nanocrystals: A promising approach for next generation thin film photovoltaics *Energy Environ. Sci.* **6** 2822–38

[39] Zhu X, Zhou Z, Wang Y, Zhang L, Li A and Huang F 2012 Determining factor of MoSe 2 formation in Cu(In,Ga)Se 2 solar Cells *Sol. Energy Mater. Sol. Cells* **101** 57–61

[40] Chen S C, Wang S W, Kuo S Y, Juang J Y, Lee P T, Luo C W, Wu K H and Kuo H C 2017 A Comprehensive Study of One-Step Selenization Process for $Cu(In_{1-x}Ga_x)Se2$ Thin Film Solar Cells *Nanoscale Res. Lett.* **12**

[41] Klinkert T, Theys B, Patriarche G, Jubault M, Donsanti F, Guillemoles J F and Lincot D 2016 New insights into the Mo/Cu(In,Ga)Se2 interface in thin film solar cells: Formation and properties of the MoSe2 interfacial layer *J. Chem. Phys.* **145**

[42] Zhao D, Fan Q, Tian Q, Zhou Z, Meng Y, Kou D, Zhou W and Wu S 2016 Eliminating fine-grained layers in Cu(In,Ga)(S,Se)2 thin films for solution-processed high efficiency solar cells J. Mater. Chem. A 4 13476–81

[43] Yadav B S, Dey S R and Dhage S R 2019 Role of selenium content in selenization of inkjet printed CIGSe 2 thin film solar cell *AIP Conference Proceedings* vol 2082 (American Institute of Physics Inc.) p 050001

[44] Cevher Z 2018 *Optimization of CuInxGa1-xSe2 Solar Cells with Post Selenization How does access to this work benefit you ? Let us know !* (City University of New York (CUNY))

[45] Virtuani A, Lotter E, Powalla M, Rau U, Werner J H and Acciarri M 2006 Influence of Cu content on electronic transport and shunting behavior of Cu(In,Ga)Se 2 solar cells *J. Appl. Phys.* **99** 014906

[46] Shi Y, Jin Z, Li C, An H and Qiu J 2006 Effect of [Cu]/[In] ratio on properties of CuInS 2 thin films prepared by successive ionic layer absorption and reaction method *Appl. Surf. Sci.* **252** 3737–43

[47] Cojocaru-Mirédin O, Choi P, Wuerz R and Raabe D 2012 Exploring the p-n junction region in Cu(In,Ga)Se 2 thin-film solar cells at the nanometer-scale *Appl. Phys. Lett.* **101** 181603

[48] Song S H, Nagaich K, Aydil E S, Feist R, Haley R and Campbell S A 2010 Structure optimization for a high efficiency CIGS solar cell *Conference Record of the IEEE Photovoltaic Specialists Conference* pp 2488–92

[49] Hanna G, Jasenek A, Rau U and Schock H W 2001 Influence of the Ga-content on the bulk defect densities of Cu(In,Ga)Se2 *Thin Solid Films* **387** 71–3

[50] Weinhardt L, Blum M, Bär M, Heske C, Fuchs O, Umbach E, Denlinger J D, Ramanathan K and Noufi R 2007 Chemical properties of the Cu(In,Ga)Se2/Mo/glass interfaces in thin film solar cells *Thin Solid Films* **515** 6119–22

[51] Wada T, Kohara N, Nishiwaki S and Negami T 2001 Characterization of the $Cu(In,Ga)Se_2$/Mo

interface in CIGS solar cells *Thin Solid Films* **387** 118–22
[52] Pouzet J and Bernede J C 1990 $MoSe_2$ thin films synthesized by solid state reactions between Mo and Se thin films *Rev. Phys. Appl.* **25** 807–15
[53] Ahn S, Kim K H, Yun J H and Yoon K H 2009 Effects of selenization conditions on densification of Cu (In,Ga) Se 2 (CIGS) thin films prepared by spray deposition of CIGS nanoparticles *J. Appl. Phys.* **105**

Acknowledgements
The authors acknowledge the Institute of Nano Electronic Engineering and to the School of Manufacturing Engineering, University Malaysia Perlis, for the support. Special thanks to those who contributed to this project directly or indirectly.

Passive design of buildings: A review of configuration features for natural ventilation and daylighting

S-H Loo[1,4], P I Lim[2], B H Lim[3]

[1] Department of Architecture and Sustainable Design, Lee Kong Chian Faculty of Engineering and Science, Universiti Tunku Abdul Rahman (UTAR), 43000 Kajang, Malaysia
[2] Department of Architecture and Sustainable Design, Lee Kong Chian Faculty of Engineering and Science, Universiti Tunku Abdul Rahman (UTAR), 43000 Kajang, Malaysia
[3] Department of Electrical and Electronic Engineering, Lee Kong Chian Faculty of Engineering and Science, Universiti Tunku Abdul Rahman (UTAR), 43000 Kajang, Malaysia

*seanloo88@yahoo.com

Abstract. Passive design involves the utilization of natural forces such as natural ventilation and daylighting without mechanical input of energy, and is a subset of environmentally sustainable design (ESD), which offers solutions for more environmentally friendly buildings. This review is based on the premise that ESD interventions have an inherent cost on building projects. The aim is to review existing research on applications of various building configurations for facilitating the economical application of passive design, in terms of natural ventilation and daylighting. A systematic review of existing research during the previous decade (2010 – 2020) was conducted. In our findings, we compare and categorise concepts within building configuration in terms of their applicability to natural ventilation and daylighting. Our review identified seven categories of building configuration ie. building geometry, interior, context, envelope, fenestration, building plan and voids, with their associated parameters. It was discovered that building fenestration was the most frequently researched followed by building voids. The identified parameters point to the variety, diversity and trends of research in the field. The knowledge can be used in familiarising, assessing and evaluating various aspects of passive design of building configuration towards energy saving and ESD.

1. Introduction

During the 20th century, energy supply and demand witnessed a paradigm shift, precipitated by the energy crisis of the mid-1900s and a succession of United Nations (UN) initiatives in the late-1900s. Thus, world energy issues garnered even higher attention from the dawn of the new millennium. Building sector is known as a major consumer of energy and contributor of carbon emissions, both globally and locally. Energy efficiency is directly related to carbon emissions that cause global warming and climate change.

The negative impacts of urbanization has given rise to the field of environmentally sustainable design (ESD), that provide counterbalance and relief to the adverse effects caused. ESD offers many solutions for more environmentally friendly buildings, concentrating on passive design methods that utilize natural forces such as natural ventilation and daylighting. It needs to be pointed out that the relative weightage of ESD interventions should be considered, in terms of their potentials for economical application in a building project, to bring benefits over heuristic approaches of ESD interventions on building design. In this regard, building configuration has an important role in determining the actual efficacy of ESD interventions.

In particular, passive design methods such as natural ventilation and daylighting, are the more economical and fundamental types of ESD interventions that are related with building configuration. This paper addresses building configurations for passive design pertaining to natural ventilation and daylighting as the focus areas. The state of the art in current research on these focus areas are reviewed with an intention to identify and explore frequently researched variables in the field.

2. Literature Review

This section presents a broad, general introduction to the concepts and research focus of building features for natural ventilation and daylighting.

2.1 Building features

Firstly, factors affecting building energy performance from the internal and external environment, were categorised into three main groups, namely building characteristics, technologies and equipment as well as occupant behaviours [1]. The interaction of factors was stressed, while aspects of economy and cost-effectiveness was acknowledged to be challenging, especially if without contribution from occupant behaviors [1]. The group of factors known as building characteristics were commended for their importance for energy performance, and the major factors were identified as: building orientation; building shape; building insulation (materials and thickness), window to wall ratio (WWR) and window glazing [1]. Also, the role of the building envelope for thermal comfort and energy performance of high-rise buildings in the tropical climate was examined with reference to components of the roof, external wall, glazing and shading, as well as building orientation and formation [2].

2.2 Natural ventilation

Ventilation is an essential criteria for building habitation, for the removal of pollutants and moisture, as well as the convection of heat [3]. Ventilation occurs naturally as airflow in three known ways: wind driven due to pressure differences, buoyancy driven due to thermal differences, or simultaneously in combination [3]. These two principals have resulted in a range of methods for application in buildings, including: single-sided ventilation; cross ventilation; stack ventilation; vents and rotational ventilators; wind towers and wind catchers; as the primary or traditional types. Newer, solar energy assistive driven types are: solar chimneys; Trombe walls; double-skin facades (DSF); whereas there are larger-scale interventions such as atria [3]. Due to the dependence of natural ventilation on outdoor weather and indoor circumstances, mixed methods utilizing partial mechanical input such as fan-assisted systems were also introduced [3]. The viability of combined types of natural ventilation were discussed and the combination of both wind and buoyancy driven mechanisms was found to be prevalent [3]. Specific attention was given to components of the building façade, including openings, for the purpose of natural ventilation particularly in tropical climates [2]. The most effective features and techniques were identified from a range of heat avoidance, cross ventilation and stack ventilation strategies, in order to support perceived benefits of reduced operation costs, thermal comfort and air quality through natural ventilation [2]. Additionally, a more critical perspective was attempted through concepts such as: natural ventilation potential (NVP), thermal comfort; strategies in ventilating buildings; other passive driven strategies; saving energy; and hybridised systems; quality of indoor air; complementary materials for natural ventilation; as well as building use and operation control [4].

2.3 Daylighting

Daylighting of buildings in urbanised areas has been thoroughly analysed in terms of lighting conditions, lighting physics, visual comfort, illumination standards, simulation software, energy performance, and urban considerations [5]. The utilisation of parametric computational methods to enhance daylighting was explored, through design of elements including louvers, skylight, massing and shading, fenestration, windows, façade and photovoltaics [6]. The "thermal-daylighting balance" describes the rudimentary conflict between daylighting for visual comfort (without glare) and the associated heat gain [7]. The concept of daylight and thermal balancing was defined via aspects such as visual comfort, heat energy balance, views outside or visibility, and material properties [7]. These concepts were related to metrics for thermal and daylighting subcategories of comfort and energy, and the popular metrics include daylight autonomy (DA), interior illuminance, lighting energy usage, heating/cooling energy usage and demand [7]. The approaches for evaluating and achieving the daylight and thermal balance involved research factors such as: building geometry; climatic locality; building openings; glazing technologies; shading design; and others [7].

2.4 Summary

Natural ventilation and daylighting are passive design methods that may be utilised through features integrated into buildings. The many possibilities presented raises concern over the economic feasibility of implementation. Critical perspectives on the risks, potentials and outcomes of natural ventilation and daylighting further increase the scope of considerations, and cause the economic reliability of such passive design methods to deserve attention. In fact, buildings are long term investments that cannot be easily changed once completed. Therefore, the effects of building features on natural ventilation and daylighting cannot be ignored or overlooked in view of passive design for building energy performance. Given the wide range of building features that influence either natural ventilation, daylighting, or both, it is important to distinguish features in terms of their economical application.

3. Methodology

This paper is developed based on secondary sources of past and existing research. As the aim of this review is to identify economical strategies for improving the performance of passive design, therefore parameters of the building itself is given attention over specialised technology related items. The review seeks to extract knowledge on "building characteristics" [1] that have an effect on natural ventilation or daylighting, or both – either directly or indirectly through building energy performance, referred to in this paper as the building configuration. The step-by-step systematic review process was adopted :-

Stage 1 - Research papers from the decade of 2010 – 2020 were retrieved from the ScienceDirect database, via search strings of "building AND (configuration OR shape OR geometry) AND (natural ventilation OR daylighting)" accessed from 3rd April to 14th May 2021. A total of 652 search results were retrieved, 649 from journal articles, with 3 from book chapters.

Stage 2 - The research papers were screened for inclusion and exclusion for this review, based on information in the titles and keywords. A total of 93 research papers were shortlisted after screening. As mentioned above, specialised technology related items were excluded, such as features for natural ventilation including double-skin façades (DSF), wind-catchers and wind-towers, solar chimneys, Trombe walls and Venturi roofs and phase change materials (PCM). Similarly, excluded features for daylighting were: light redirecting systems, light tracking (heliostats, diffuse sunlight), light transport (light tubes and pipes), dynamic and kinetic facades.

Stage 3 - Following that, the abstracts and articles were briefly evaluated for extraction of knowledge related to the variables of building configuration for natural ventilation and daylighting. One article was removed due to being different publication of the same study. Identified variables from the final selected 92 research papers were categorised within seven strategies as parameters that relate to various aspects of building configuration.

4. Results

The variables were grouped into seven categories of building configuration based on observed commonalities among the reviewed papers :-

Building geometry	(B1)	-	the spatial characteristics of a building in a mathematically abstracted way, usually taken in a broad scale and generalised perspective.
Building interior	(B2)	-	the internal spaces that are occupied and inhabited by people

Building context	(B3)	-	the external environment based on the property or site surroundings.
Building envelope	(B4)	-	the external shell or fabric of the building, normally associated with surface elements such as walls, roof and floors but also includes integral features such as balconies.
Building fenestration	(B5)	-	the openings and apertures of a building, forming the porous interface of a building between exterior and interior, primarily windows but also includes shading.
Building plan	(B6)	-	the layout and arrangement of circulation and rooms that form the organising structure of the spaces of a building.
Building voids	(B7)	-	the empty spaces within the enclosed boundary of a building's footprint, usually associated with the absence of enclosing surfaces contiguous to the exterior.

The detailed parameters under the seven types of building configuration categories are presented in table 1. The most frequently found parameters (by order of frequency) were related to:- building fenestration (B5) with 47 articles and building voids (B7) with 31 articles, while the rest of the parameters belonged in building geometry (B1) with 4 articles, building interior (B2) with 17 articles, building context (B3) with 13 articles, building envelope (B4) with 11 articles and building plan (B6) with 12 articles. (Figure 1)

Table 1: Building configuration categories and associated parameters for natural ventilation (NV) and daylighting (DL)

Strategy	Parameters	Source
Building geometry (B1)	Flexibility: Perimeter area / Total floor area (NV, DL)	[8]
	Building depth and width (NV, DL)	[59], [55]
	Height profile (NV, DL)	[59], [77]
Building interior (B2)	Room geometry (NV, DL)	[30], [37], [42], [61], [62] , [66], [73], [83], [89]
	Space type division (DL)	[8]
	Unit size (NV)	[92], [9]
	Unit height and orientation (NV)	[92], [9], [11]
	Partitions layout and surface reflectance (DL)	[12], [86]
	Indoor floor planning (NV, DL)	[33]
	Room sizes (NV)	[78]
Building context (B3)	Building orientation (NV, DL)	[15], [2], [62], [63], [78], [82]
	Solid sky angle (SSA) (DL)	[8]
	Building orientation and spacing (NV)	[13], [33], [38]
	Façade self-shading (DL)	[63]
	Night ventilation (NV)	[90]

	Sky view angle: Terracing, setbacks (DL)	[30]
	Urban aspect ratio (UAR) (NV)	[55]
Building envelope (B4)	Thermal mass (NV)	[19], [69]
	Balconies on façade (NV)	[14], [80], [84]
	Balcony with wing wall (NV)	[87]
	Wind deflection (NV)	[38]
	Roof profile (NV)	[35], [40]
	Wall profile (DL)	[64], [74]
Building fenestration (B5)	Window to wall ratio (WWR) (DL)	[8], [10], [29], [45], [49], [57], [61], [62], [63], [67], [75], [76], [79], [85]
	Window to wall ratio (WWR) (NV)	[2], [55]
	Window to floor ratio (WFR) (NV, DL)	[8]
	Window size (NV, DL)	[59], [42], [50], [38]
	Window position (NV)	[50], [38]
	Window orientation (DL)	[30], [42], [57] , [61], [79], [82]
	Window geometry (DL)	[45], [49], [66]
	Window opening type (NV)	[68]
	Window opening area (WOA) (NV)	[55]
	Openings shape and position (NV)	[36]
	Openings configuration (NV)	[15]
	Asymmetric openings (NV)	[35]
	Skylights (DL)	[59], [12], [30], [70]
	Clerestory windows (DL)	[30]
	Light shelf (DL)	[28], [57], [60], [82]
	Shading devices (DL)	[10], [12], [22], [28], [63], [67], [76], [81], [85]
	Cross ventilation (NV)	[19], [35], [36], [46], [48], [52], [54], [73], [77]
	Single-sided ventilation (NV)	[11], [54], [68], [84], [87], [91]
	Displacement ventilation (NV)	[37], [47]
	Stack ventilation: vertical shaft (NV)	[52]
Building plan (B6)	Linear wings (NV)	[92], [9]
	Rectilinear with irregular corners (NV, DL)	[10]
	Circular radial (NV)	[11]

	Central clustered (NV)	[13], [15], [53]
	Linear clustered (NV)	[15], [53]
	Tall oval: height-width ratio (HWR) and height-thickness ratio (HTR) (NV)	[58]
	Corridors: single-sided, double-sided (NV, DL)	[62], [78]
	Distributed volumes (NV)	[38]
	Building typologies (DL)	[83]
Building voids (B7)	Atrium / building circular cross-section (NV)	[16]
	Atrium wall vertical angle (NV)	[16]
	Atrium enclosure type (DL)	[17]
	Atrium roof (NV, DL)	[16], [17], [27]
	Atrium with clerestory windows (DL)	[43]
	Atria proportions (DL)	[17]
	Atrium well Index (DL)	[17], [21], [24]
	Atrium well geometry (NV, DL)	[25], [27], [32]
	Atrium well façade ratio (DL)	[25]
	Atrium internal balconies (DL)	[21], [25]
	Atrium surface reflectance (DL)	[24], [25], [71]
	Atrium building (NV, DL)	[85], [88]
	Atrium with night ventilation/purging (NV)	[26]
	Atrium with buoyancy-driven ventilation (NV)	[31], [32]
	Courtyard openings (DL)	[20], [72]
	Courtyard morphology (DL)	[88]
	Courtyard proportions (NV, DL)	[19], [48], [71], [72]
	Courtyard building (NV, DL)	[10], [20], [34], [85], [77]
	Light-well with horizontal voids (NV)	[39]
	Light-well size, height and orientation (NV, DL)	[12], [18], [56], [65]
	Light-well roof aperture (DL)	[12], [18]
	Light-well wall vertical angle (DL)	[18], [56]
	Air wells (NV)	[23], [2], [51]
	Double-height space (NV)	[41]
	Vertical and horizontal "wind-paths" (NV)	[13]
	Sky view factor (NV)	[19]

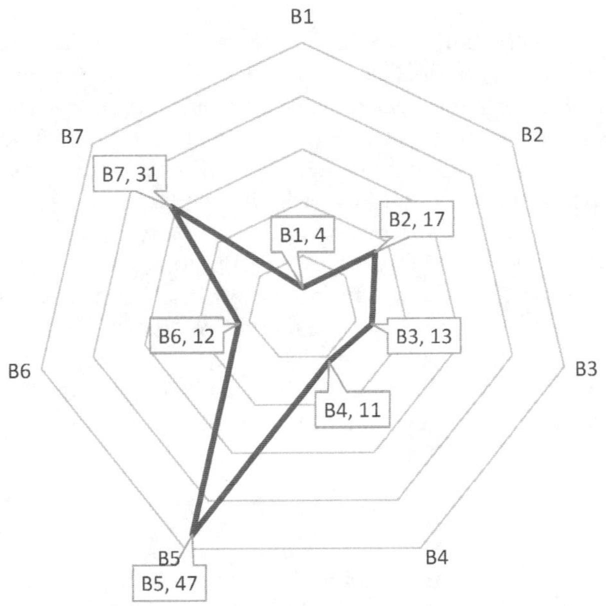

Figure 1: Frequency of parameters in each category based on articles

5. Discussion

Based on the results, there were 37 parameters related to natural ventilation (NV), 25 parameters related to daylighting (DL), and also 16 parameters related to both (NV, DL), as shown in Figure 2. The discussion below provides some explanation on the characteristics of the building configuration categories and their relevance to passive design in terms of natural ventilation and daylighting. The comprehensive range of possibilities are reviewed in order to form an outline of the research domain, in view of the review objective of seeking economical approaches to passive design.

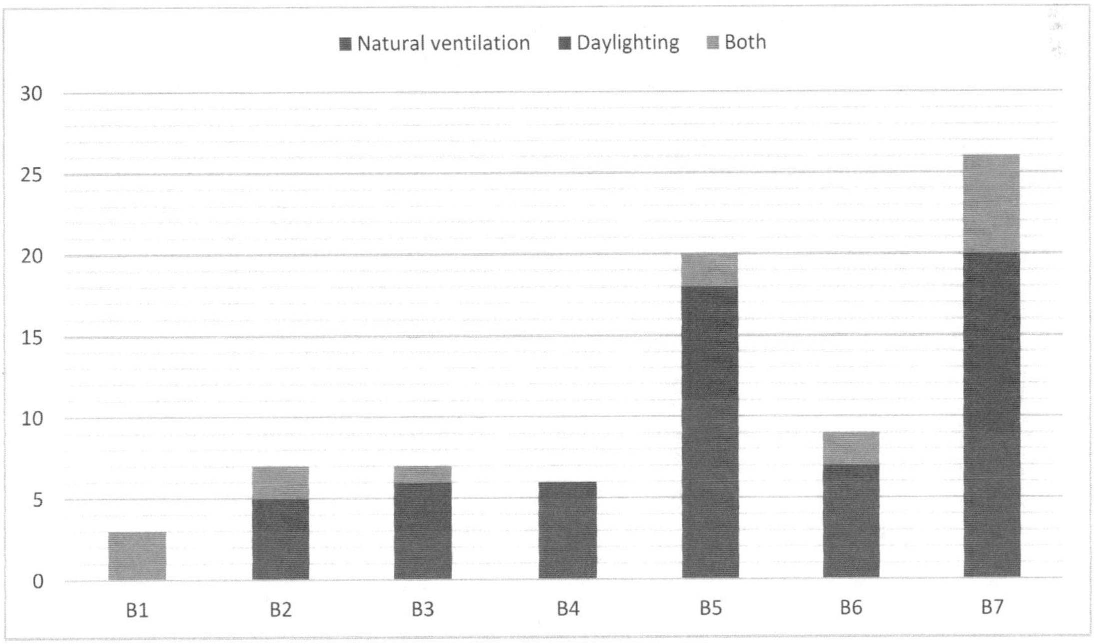

Figure 2: Parameters for natural ventilation, and daylighting and both in each category

5.1 Building geometry (B1)

Building geometry (B1) involves the building's perimeter area, building depth and width, and height profile. Perimeter area is defined as a particular interior zone adjacent to the building façade, and thus able to interface with the exterior environment for on-site renewable resources, such as daylight and airflow [8]. This metric was used to indicate the flexibility of a given floor area for alternatives to active electrical usage and consumption [8]. It indirectly shows the proportion of floor area that is not at the perimeter, ie. Located deeper inside the building, which entails a higher degree of servicing and conditioning [8]. Next, aspect ratio or (building width / depth), was found to have a noticeable effect for daylighting and overall energy use under different climatic conditions, with larger aspect ratio preferred for hotter climates, and vice versa [59]. Similarly, building aspect ratio was among the variables for assessing thermal and visual comfort in buildings with natural ventilation [55]. The height profile, based on the highest point in the building at the roof ridge, had a small negative effect for daylighting, and a larger negative effect for overall energy use, particularly in a colder climate [59]. However, the height profile as defined by relative height of rooms, was found to be suitable for inducing buoyancy-driven stack ventilation [77].

5.2 Building interior (B2)

Building interior (B2) parameters include room geometry parameters as the leading concepts. Room geometry as defined by room depth for a given height is well-established and relied on for its inversely proportional relationship to side-lit daylighting [30], [61], [62]. Furthermore, height and depth of rooms were found to contribute beneficially to single-sided and cross ventilation [73]. A variety of room width-to-depth ratios were also tested for daylighting and thermal performance [42]. However, no noticeable differences between square and rectangular room geometries were present in a basic comparison [66]. Free-form room geometries was explored in relation to their effect on influencing indoor airflows [37]. Many attributes of room geometry were evaluated for daylighting, including room depth and width, floor area, glazing area and glazing to floor and to wall ratios [83]. The vertical inclination angles of walls were found to have a substantial influence on daylighting and thermal performance [89]. Additionally, different room sizes were investigated for the effect on natural ventilation of spaces [78]. Moreover, it was found that the distribution of units in a high-rise residential building, affects the natural ventilation performance of individual units based on their height positions, orientations and sizes [92], [9], [11]. The space type division was parameterized for daylighting as the division of office space into modules differentiated by varied sizes, their horizontal positions in floor plan (corner, middle, etc.) and vertical locations in each storey [8]. The interactions between indoor space planning with improved daylighting performance and natural ventilation were studied in a paper [33]. Finally, the interior partitions layout and surface reflectances, were shown to affect daylighting performance of spaces [12], [86].

5.3 Building context (B3)

Building orientation and spacing were the major parameters compared to the rest. In general, the site orientation of buildings with respect to the N-S-E-W directions etc, and the distance between buildings sharing the same site had effects for natural ventilation [13], [15], [33], [38], [2] as well as for daylighting [62], [82]. For example, the building orientation was varied according to the prevailing wind direction in a study [78]. Also, building orientation was an important factor for shading devices on facades, taking into account daylighting [63]. Façade self-shading, through the use of vertically inclined facade walls, was a factor that was examined for solar control in buildings with daylighting taken into consideration [63]. The solid sky angle (SSA) is calculated from the building height, and shape, height of surrounding buildings, as well as their separated distance and topography [8]. This metric describes the building environment and the sky hemisphere visible from a vertical window, for daylighting application [8]. Also, night ventilation was assessed as a strategy for natural ventilation [90]. An urban aspect ratio (UAR) was applied as a function of building height and the street separation distance to opposite buildings in addition to varying building dimensions, for the purpose of assessing natural ventilation based on thermal and visual comfort [55]. The sky view angle represents the angle subtended by the window and the closest edge of adjacent buildings, determines the relative area of sky visible from a window, and may be improved for daylighting with property setbacks and terraced facades [30].

5.4 Building envelope (B4)

Parameters for building envelope (B4) were balconies on façade, wing wall, wall profile, roof profile, wind deflection, and thermal mass. Four critical features of balconies that affect natural ventilation performance are: availability, location of unit level, depth and orientation of balconies [80]. Balcony geometry was investigated [84] and the airflow effects were compared with buildings without balconies [14]. The addition of wing walls to balconies was studied to improve single-sided ventilation of medium-rise residential units [87]. The use of deflecting devices was mentioned for improving the distribution of airflow at positions lacking ventilation [38]. Wall profile as defined by offsets, or undulations in the building envelope were investigated for their obstruction to daylighting, as exacerbated by balconies [64]. Also, aesthetic façade designs with undulating wall profiles were optimised for daylighting [74]. The roof inclination angle of an isolated space was shown to affect the wind-driven cross ventilation performance of vertically asymmetric wall openings [35]. A variety of roof geometries, such as straight, convex, concave and hybrid types were tested for a building with cross ventilation, using a saw-tooth shaped roof on the leeward side [40]. A paper indicated that the use of thermal mass had a pronounced cooling effect, that could complement the lack of cross ventilation in deep plan, courtyard buildings [19]. Another study showed that thermal mass moderated the internal temperature fluctuations in both summer and winter conditions, and that this could enhance thermal comfort in the latter climatic condition [69].

5.5 Building fenestration (B5)

The most frequent parameter in this category is window to wall ratio (WWR), that is the ratio of glazing area to wall area of the building's façade, or part thereof. The effect of window to wall ratio on energy use and savings in terms of lighting and thermal performance is well-documented. Varied WWR specific to a particular building orientation were shown to influence daylighting and energy use [8], [10], [29], [45], [49], [61], [62], [63], [67], [76], [79]. The importance of sunshading to reduce solar heat gain was considered [76], as well as the inclusion of courtyards [10], [76]. Also, this was balanced with consideration of natural cross-ventilation performance in one paper [55]. Furthermore, a paper also examined window to floor ratio (WFR), measured by the glazed area and floor area, as well as the façade area in relation to floor area, that showed a compounded effect due to compactness on energy use for a given WWR [8]. Moreover, a paper noted the contribution of WWR to effective cross-ventilation in buildings [2]. Parametric variations of façade patterns were varied while maintaining the same WWR, to study the effect on daylighting performance [75]. The next most frequent parameters were window-related variables, cross-ventilation and shading devices. Parameters related to windows were examined for their effect on daylighting and thermal performance, such as their size, geometry, and orientation [59], [30], [42], [45], [49], [61], [66], [79]. Additionally, window performance for natural ventilation was also considered separately from daylighting [38], [50] or as part of the holistic system as based on the window opening area (WOA) [55]. A variety of window opening types were tested for buoyancy-driven, single-sided ventilation performance [59], [68]. The visual comfort in terms of avoiding glare from daylighting was also highlighted [57]. Cross ventilation is a common technique whereby windows or openings are placed on opposing sides of an enclosed space, and this was studied extensively [35], [36], [46], [52], [54], [77]. Cross ventilation was also considered for buildings with courtyards [19], [48], and for its role in buildings with a solar chimney [73]. Shading devices such as static exterior projections and overhangs [10], [22], [28], [67], [76], awnings, internal blinds, roller screens [12], complex geometries [63], and louvered slats [81]. The use of shading devices was also studied for daylighting and thermal performance of courtyard and atrium buildings [85]. Single-sided ventilation involves the use of windows or openings on the wall of only one side of an enclosed space, and was studied for its influence on thermal comfort in buildings [11], [54]. Single-sided ventilation driven by thermal buoyancy was investigated for a variety of window opening types [59], [68]. Also, the separation between two apertures in single-sided ventilation coupled with the angle of incoming wind affected the ventilation rate relative to wind velocity and aperture size. [91] Additionally, single-sided ventilation performance were examined in conjunction with building envelope features, such as balcony geometry [84] and balcony with wing walls [87]. Door openings were included for analysis of natural ventilation in high-rise residential units [15], while wall openings were included for analysis of inter-connected spaces [36]. Displacement ventilation, referring to airflow from lower to higher elevations in an enclosed space, was explored using low windward and high leeward openings [37], and in conjunction with various specialised technologies [47]. Stack ventilation using vertical shaft was included in a study on natural ventilation performance of office buildings [52]. A detailed examination of vertically asymmetrical openings was done in conjunction with inclined roof profile for cross ventilation [35]. Skylights have been shown to be effective for daylighting especially in deeper floor plans, provided with adequate sizing and arrangement [59], [30] or may be combined with light-wells or internal floor voids for multi-storey buildings [12], [70]. Similarly, clerestory windows which are above eye-level glazed areas on walls, are able to improve daylighting performance [30]. The light shelf is an extended projection at openings used to improve daylight distribution into an enclosed space, and was compared with other daylighting systems such as: sun breaker, external louvers and internal blinds or in combination thereof [82]. A building with light shelves was appraised for daylight performance and visual comfort [57]. Furthermore, a light shelf was designed with louvered slats was explored to reduce the cooling load [28], or attached with photovoltaic solar modules to explore power generation [60].

5.6 Building plan (B6)

The parameters in this category are based on the various types of spatial organisations. Buildings with linear wings of corridors projecting radially from a centre were examined for their potential to achieve energy savings through natural ventilation, albeit with careful avoidance of solar heat gains [92], [9]. Rectilinear building plans with projecting corners of varying complexity, and also internal courtyards were simulated for daylighting, natural ventilation and solar control for optimal energy efficiency [10]. A circular radial plan layout for a high-rise residential building, whereby each unit was facing each of eight directions, was analysed for thermal performance through single-sided ventilation, with variegated results [11]. A central clustered plan layout is arranged in a square grid around a centre, and was involved in the investigation of natural ventilation for high-rise residential buildings [13], as well as a variant with concave corners resulting in a cruciform shape [15]. A linear clustered plan layout, spread out along a directional axis, was utilised for investigation on natural ventilation [15]. The central clustered type was compared to the linear clustered type, with the latter essentially having better performance for natural ventilation [53]. A range of generalised building typologies assessed for daylighting, were described as A to Z patterns, included plan layouts that were bent into U, L or O shapes to form outdoor yards and courts [83]. Smaller, distributed building volumes were recommended to enable airflow between each building for natural ventilation [38]. In an investigation on high-rise buildings with oval shaped floor plans, height-thickness ratio (HTR) and height-width ratio (HWR) were examined for wind pressure effects on building surfaces [58]. Finally, plan layouts with single-sided and double-sided corridors were examined for daylighting and thermal performance [62], and also for wind-driven ventilation [78].

5.7 Building voids (B7)

Parameters in this category are commonly related to atria, courtyards, and lightwells. The buoyancy-driven ventilation in atria was investigated for thermal performance in buildings [31], [32]. In contrast, the airflow of wind-induced natural ventilation in a building with a patio, or central courtyard was studied in detail [34]. Night purge ventilation was studied for an atrium to investigate the displacement ventilation of heated air into the cool, still night environment [26]. Various effects of atrium characteristics were examined, such as the vertical angle of atrium walls on natural ventilation and thermal performance [16], and also using the case of a circular atrium cross section with glazed roof instead of opaque roof [16], as well as using rectilinear cross section with various atrium roof types [27]. Comparatively, daylight performance was examined for various types of atrium enclosures, atrium apertures such as skylight and monitors, and well index as defined by geometric proportions of the atrium [17]. Clerestory windows were explored for their effects on daylighting based on the atrium geometry [43]. Similarly, well index was tested for daylighting performance with consideration of internal balconies or corridors in the atrium [21], and the consideration of light reflecting surfaces on the atrium internal walls [24], [71]. Also, light reflectance of internal balcony walls were examined for daylighting, with consideration of well geometry as defined by aspect ratios of the atrium plan and section, compared with atrium façade conditions without balconies and with adjoining room partitions [25]. Additionally, well geometry of atriums affected the natural ventilation and thermal performance of atria through the atrium size, height, width and depth [27], [32]. Finally, atrium buildings were examined for the benefits of atria in terms of combined daylighting and overall thermal performance from natural ventilation [88], as well as compared with courtyard buildings [85].

Courtyard buildings with windows facing inside were compared with conventional buildings [20], or inserted into irregularly shaped buildings [10] to examine daylighting and energy performance. Also, courtyards were combined with solar chimneys for natural ventilation and thermal performance [77]. The effect of courtyard proportions were examined for natural ventilation and thermal performance, with consideration of the courtyard height to width ratio [48] and geometrical relationships to adjoining spaces [19]. Comparatively, daylighting and energy performance was investigated with consideration of the courtyard aspect ratio, as well as the ratio of openings in the courtyard [72]. The contribution of courtyard for daylighting was affected by surface reflectance of internal walls [71]. Three types of courtyard morphology was examined for daylighting and thermal performance, including dual, triad and quadrangle enclosures [88]. The sky view factor (SVF), estimated using equisolid-angle projection lens, was found to be a major determinant of air temperatures in courtyards for natural ventilation [19]. The use of light-wells as floor openings to enable the transmission and distribution of daylight was coupled with skylights [12] or variations thereof [18]. The geometrical characteristics of light-wells were examined for daylighting and energy performance in terms of size only [12], or in combination with height, orientation and wall vertical angle [18], [56]. The size, height and orientation of light-wells are also important for the stack effect application of natural ventilation, in addition to daylighting [65]. The application of horizontal voids connected to light-wells was explored for buoyancy-driven natural ventilation, based on exterior wind directions [39]. Horizontal "wind paths" were explored for a high-rise residential building to enable the passage of airflow for natural ventilation [13]. Air-wells are vertically continuous voids used for buoyancy-driven natural ventilation, such as carrying airflow from individual units in a residential high-rise building [51], and was associated with the operation of vertical airflow channels such as wind towers or wind catchers [23], [2]. Similarly, double-height spaces were explored for buoyancy-driven natural ventilation through the stack effect of warmer air rising and displacement ventilation with incoming cool air [41].

6. Conclusion

This paper focused primarily on critical and strategic decisions concerning architectural and building design, and less on specialised technologies and ongoing innovations at the element-scale. The fact is that buildings are relatively permanent structures and difficult to modify once constructed. While the proliferation of options affords many possibilities, we must not overlook the more fundamental issues of building configuration and their role in providing economical, passive design towards realising the goals of energy efficiency and sustainable development. This review has outlined the range of parameters related to building configuration for passive design. It is acknowledged that the parameters identified were based on the terms and concepts from existing research, which could involve various levels of complexity, considerations and combination of factors. However, the identified parameters point to the variety, diversity and trends of research in the field. Our results should not be taken to indicate the importance of one over others but such knowledge can be used in familiarising, assessing and evaluating various aspects of passive design of building configuration towards energy saving and ESD.

References

[1] Chen S, Zhang G, Xia X, Setunge S and Shi L 2020 *Energy & Buildings* **216** 109944
[2] Aflaki A, Mahyuddin N, Mahmoud Z A and Baharum M R 2015 *Energy and Buildings* **101** 153–162
[3] Zhang H, Yang D, Tam W Y, Tao Y, Zhang G, Setunge S and Shi L 2021 *Renewable and Sustainable Energy Reviews* **141** 110795
[4] Sakiyama, N R M, Carlo J C, Frick J and Garrecht H 2020 *Renewable and Sustainable Energy Reviews* **130** 109933
[5] Nasrollahi N and Shokri E 2016 *Renewable and Sustainable Energy Reviews* **66** 861–874
[6] Eltaweel A and Su Y 2017 *Renewable and Sustainable Energy Reviews* **73** 1086–1103
[7] Yu F, Wennersten R and Leng J 2020 *Building and Environment* 2020 **186** 107330
[8] Ringel G and Capeluto I G 2020 *Sustainable Cities and Society* **61** 102171
[9] Gan J L, Wong H K, Tse K T, Cheng C P, Lo M C and Chan C M 2019 *Journal of Cleaner Production* **231** 1375e1388
[10] Konis K, Gamas A and Kensek K 2016 *Solar Energy* **125** 161–179
[11] Aflaki A, N Mahyuddin and Baharum M R 2016 *Energy and Buildings* **126** 146–158
[12] Bournas I, Abugabbara M, Balcerzak A, Dubois M-C and Javed S 2016 *Journal of Building Engineering* **7** 194–206
[13] Zhou C, Wang Z, Chen Q, Jiang Y and Pei J 2014 *Energy and Buildings* **82** 457–465
[14] Montazeri H and Blocken B 2013 *Building and Environment* **60** 137e149
[15] Gao C F and Lee W L 2011 *Building and Environment* **46** 961e969
[16] Fini A S and Moosavi A 2016 *Energy and Buildings* **121** 265–283
[17] Mohsenin M and Hu J 2015 *Solar Energy* **119** 553–560
[18] Freewan A A Y, Gharaibeh A A and Jamhawi M M 2014 *Sustainable Cities and Society* **13** 32–40
[19] Kubota T, Zakaria M A, Abe S and Hooi C T 2017 *Building and Environment* **112** 115e131
[20] Al-Masri N and Abu-Hijleh B 2012 *Renewable and Sustainable Energy Reviews* **16** 1892–1898
[21] Kim G and Kim J T 2010 *Building and Environment* **45** 304–310
[22] Alzoubi H H and Al-Zoubi A H 2010 *Energy Conversion and Management* **51** 1592–1599
[23] Siew C C, Che-Ani A I, Tawil N M, Abdullah N A G and Mohd-Tahir M 2011 *Procedia Engineering* **20** 363 – 371
[24] Du J and Sharples S 2011 *Solar Energy* **85** 2085–2100
[25] Du J and Sharples S 2011 *Building and Environment* **46** 2142e2152
[26] Lynch P M and Hunt G R 2011 *Building and Environment* **46** 144e155
[27] Hussain S and Oosthuizen P H 2012 *Applied Thermal Engineering* **40** 358e372
[28] Kim G, Lim H S, Lim T S, Schaefer L and Kim J T 2012 *Energy and Buildings* **46**
[29] Goia F, Haase M and Perino M 2013 *Applied Energy* **108** 515–527
[30] Müeller H F O
[31] Hussain S and Oosthuizen P H 2013 *Applied Thermal Engineering* **57** 133e146
[32] Moosavi L, Mahyuddin N, Ghafar N A and Ismail M A 2014 *Renewable and Sustainable Energy Reviews* **34** 654–670
[33] Liu S, Liu J, Yang Q, Pei J, Lai D, Cao X, Chao J and Zhou C 2014 *Energy and Buildings* **68** 686-695,
[34] Driss S, Driss Z and Kammoun I K 2015 *Energy* **90** 917e925
[35] Perén J I, Hooff T, Leite B C C and Blocken B 2015 *Building and Environment* **85** 263e276
[36] Shetabivash H 2015 *Energy and Buildings* **93** 1–15
[37] Stavridou A D 2015 *Frontiers of Architectural Research* **4**, 127–145
[38] Guo W, Liu X and Yuan X 2015 *Procedia Engineering* **121** 573 – 581
[39] Farea T G, Ossen D R, Alkaff S and Kotani H 2015 *Energy and Buildings* **86** 502–513
[40] Peren J I, Hooff T, Ramponi R, Blocken B and Leite B C C 2015 *J. Wind Eng. Ind. Aerodyn.* **145** 102–114
[41] Belpolitia V, Calzolarib M, Davolic P, Altand H and Nassif R 2020 *Procedia Manufacturing* **44** 302–309

[42] Vanhoutteghem L, Skarning G C J, Hviid C A and Svendsen S 2015 *Energy and Buildings* **102** 149–156
[43] Ghasemi M, Noroozi M, Kazemzadeh M and Roshan M 2015 *Journal of Building Engineering* **3** **39–47**
[44] Futrell B J, Ozelkan E C and Brentrup D 2015 *Building and Environment* **92** 591e602
[45] Harmati N and Magyar Z 2015 *Energy Procedia* **78** 2458 – 2463
[46] Ding C, Lam K P and Wong N H 2016 *Procedia Engineering* **169** 264 – 271
[47] Mateus N M, Simões G N, Lúcio C and Graça G C 2016 *Energy and Buildings* **133** 185–196
[48] Micallef D, Buhagiar V and Borg S P 2016 *Energy and Buildings* **133** 658–669
[49] Harmathy N, Magyar Z and Folić R 2016 *Energy* **114** 302e317
[50] Elshafei G, Negm A, Bady M, Suzuki M and Ibrahim M G 2017 *Energy and Buildings* **141** 321–332
[51] Muhsin F, Yusoff W F M, Mohamed M F and Sapian A R 2017 *Energy and Buildings* **144** 1–16
[52] Nomura M and Hiyama K 2017 *Renewable and Sustainable Energy Reviews* **74** 746–754
[53] Weng J, Ge J and Chen S 2017 *Procedia Engineering* **205** 3353–3358
[54] Ma X Y, Peng Y, Zhao F Y, Liu C W and Mei S J 2017 *Procedia Engineering* **205** 3797–3803
[55] Costanzo V and Donn M 2017 *Energy and Buildings* **140** 210–223
[56] Ahadi A A, Saghafi M R and Tahbaz M 2017 *Solar Energy* **155** 679–697
[57] Lim G H, Hirning M B, Keumala N and Ghafar N A 2017 *Energy and Buildings* **141** 175–185
[58] Zhao D X and He B J 2017 *Journal of Building Engineering* **12** 219–228
[59] Fang Y and Cho S 2019 *Solar Energy* **191** 7–18
[60] Warrier G A and Raphael B 2019 *Building and Environment* **159** 106161
[61] Pellegrino A, Cammarano S, Lo Verso V R M and Corrado V 2017 *Building and Environment* **113** 151e162
[62] Zhang A, Bokel R, Dobbelsteen A, Sun Y, Huang Q and Zhang Q 2017 *Energy and Buildings* **139** 371–384
[63] Valladares-Rendón L G, Schmid G and Lo S L 2017 *Energy and Buildings* **140** 458–479
[64] Dabe T J and Adane V S 2018 *Building and Environment* **140** 173–183
[65] Ahadi A, Saghafi M R and Tahbaz M 2018 *Journal of Building Engineering* **18** 220–244
[66] Rubeis T, Nardi I, Muttillo M, Ranieri S and Ambrosini D 2018 *Energy Procedia* **148** 1090–1097
[67] Toutou A, Fikry M and Mohamed W 2018 *Alexandria Engineering Journal* **57**, 3595–3608
[68] Wang J, Zhang T, Wang S and Battaglia F 2018 *Energy & Buildings* **168** 147–164
[69] Kumar S, Singh M K, Mathur A, Mathur S and Mathur J 2018 *Journal of Building Engineering* **20** 569–584
[70] El-Abd W, Kamel B, Afify M and Dorra M 2018 *Solar Energy* **170** 358–368
[71] Acosta I, Varela C, Molina J F, Navarro J and Sendra J J 2018 *Applied Energy* **211** 1216–1228
[72] Guedouh M S, Zemmouri N, Hanafi A and Qaoud R 2019 *Energy Procedia* **157** 435–442
[73] Cuce E, Sher F, Sadiq H, Cuce P M, Guclu T and Besir A B 2019 *Sustainable Energy Technologies and Assessments* **36** 100540
[74] Yi Y K 2019 *Building and Environment* **156** 178–190
[75] Rashwan A, Gizawi L E and Sheta S 2019 *Alexandria Engineering Journal* **58**, 551–557
[76] Xue P, Li Q, Xie J, Zhao M and Liu J 2019 *Applied Energy* **233–234** 62–70
[77] Manu S, Brager G, Rawal R, Geronazzo A and Kumar D 2019 *Building and Environment* **148** 136–156
[78] Yang L, Liu X, Qian F and Du S 2019 *Journal of Cleaner Production* **209** 886e902
[79] Ashrafian T and Moazzen N 2019 *Sustainable Cities and Society* **47** 101483
[80] Izadyar N, Miller W, Rismanchi B and Garcia-Hansen V 2020 *Building and Environment* **170** 106613
[81] Huo H, Xu W, Li A, Cui G, Wu Y and Liu C 2020 *Building and Environment* **180** 10692
[82] Reffat R M and Ahmad R M 2020 *Solar Energy* **209** 258–277
[83] Bournas I 2020 *Building and Environment* **185** 107276
[84] Izadyar N, Miller W, Rismanchi B and Garcia-Hansen V 2020 *Building and Environment* **177** 106847

[85] Asfour O S 2020 *Journal of Building Engineering* **30** 101299
[86] Abdollahzadeh N, Tahsildoost M and Zomorodian Z S 2020 *Journal of Building Engineering* **29** 101171
[87] Ghadikolaei F M, Ossen D R and Mohamed M F 2020 *Journal of Building Engineering* **31** 101316
[88] Zhu L, Wang B and Sun Y 2020 *Building and Environment* **176** 106841
[89] Bakmohammadi P and Noorzai E 2020 *Energy Reports* **6** 1590–1607
[90] Kyritsi E and Michael A 2020 *Building and Environment* **175** 106384
[91] Daish N C, Carrilho da Graça G, Linden P F and Banks D 2016 Building and Environment 108 122e134
[92] Weerasuriya A U, Zhang X, Gan J L and Tan Y 2019 *Building and Environment* **153** 218–232

Journal of Physics: Conference Series **2053** (2021) 012010 doi:10.1088/1742-6596/2053/1/012010

Atmospheric trajectory analysis of Cesium-137 from proposed nuclear power plant site in Bangka Island, Indonesia

Rafieza Rosli[1], Nor Afifah Basri[1,2], Nurlyana Omar[3]

[1]School of Chemical & Energy Engineering, Universiti Teknologi Malaysia, Malaysia
[2]Advanced Nuclear Engineering Research Group, Universiti Teknologi Malaysia, Malaysia
[3]Physics Department, Faculty of Science, Universiti Teknologi Malaysia, Malaysia

*norafifah@utm.my

Abstract. Environmental risk assessment for nuclear power plant (NPP) operation is important to monitor the level of radionuclide concentration in the atmosphere and to determine the health risks associated with potential external radiation exposure. A NPP site was proposed in Bangka Island, Indonesia, 554 km from Johor Bahru, Malaysia. In the event of postulated nuclear emergency, the radionuclide dispersion from Bangka Island can possibly arrive in Malaysia's atmosphere due to its close distance with Malaysian border. The purpose of this study is to assess the trajectories of Cs-137 radionuclide towards Malaysia from Bangka Island NPP site. A simulation was conducted using Hybrid Single Particle Lagrangian Integrated Trajectory Model (HYSPLIT) to describe the movement of radionuclides in the atmosphere from Bangka Island towards Malaysia. The result shows the trajectories and direction of both radionuclides are very much affected by the monsoon season, as well as the meteorological characteristics such as wind speed and direction. In the simulated event of radioactive release in 2019, the trajectories can be seen moving directly towards Malaysia in May, June, July, August, September and October. It is concluded that if Indonesia were to proceed with the NPP construction in Bangka Island, necessary preparation such as emergency plan and risks mitigation should be strategized early in Malaysia to safeguard the public and the environment.

1. Introduction

Radioactive contamination, commonly known as radiological contamination, is the deposition of radioactive substances on a surface or inside a space, where their existence is accidental or undesirable. Radioactive contamination in the environment depends on the characteristics of radioactive pollutant, the level of emissions or radiation exposure, and its pathways through the environment [1]. Radionuclides, whether natural or man-made are present in the environment and in human body. It travels through air, water and through the food chain, and at the same time emits radiation through radioactive decay process. Radioactive material movement in air is influenced by the weather. Climate behaviours such as air currents, wind speed, rain and snow highly influence the movements of radionuclide in air, which eventually effect the area and size of contaminated area from radionuclide distribution [2].

The Fukushima Daiichi nuclear accident in Japan has highlighted the importance of dispersion forecasting system to evaluate the movement of radioactive materials in air originated from nuclear facilities like nuclear power plants (NPP). The dispersion forecasting system should be established not only in the homeowner country, but also at the neighbouring countries with the purpose to evaluate possible risks and to provide detailed emergency plan for possible nuclear emergency events. An early analysis made by the authorities will provide extra measures and ample preparation to the affected countries to protect the public and the environment from unnecessary exposure during nuclear event.

In the Southeast Asia region, Indonesia is the only country that have consistently pursuing the plan to include nuclear power in its national energy mix. As part of the nuclear power infrastructure development plan, research and studies were conducted to identify possible risks and hazard originated from NPP construction operation in Indonesia. There are a number of proposed sites for NPP construction in Indonesia, and the closest location to Malaysia is the Sebagin site in Bangka Belitung province. The probabilistic risk assessments, especially the events related to radioactive release to the environment have been actively conducted by many researchers in Indonesia [3,4]. For example, a study by Pande Made Udiyani and Surip Widodo in 2012 entitled 'The Determination of Atmospheric Dispersion Coefficient for PWR Reactor Accident Analysis in Indonesia Site' was conducted to determine the value of the dispersion coefficient for the selected potential site in order to determine the values of accepted environmental doses for nuclear power plants in Indonesia [5]. Notwithstanding the fact that atmospheric dispersion risk assessments were actively conducted in Indonesia, the importance of risk assessments of radioactive dispersion in the areas beyond the international boundaries of the NPP location cannot be neglected.

This study is conducted to observe the movements of radioactive pollutants toward Malaysia from NPP site in Indonesia during nuclear accident scenario. This study serves as a cornerstone and a valuable source of reference and information for further study to be continued in order to ensure readiness to face nuclear fallout in Malaysia vicinity. The data obtained from this study can also be used and improvised by local authorities in order to plan for emergency procedures in the event of nuclear fallout in the unforeseen future.

2. Study Area

The trajectory observation area in this study is the NPP site in Bangka Island Indonesia as the source of origin, and Malaysia as the affected area. Pulau Bangka is an island located on the east of Sumatera, with a population of approximately 1 million. It is the 9th largest island in Indonesia and the main part of the Bangka-Belitung province. The inverted S-shaped of Bangka Island is roughly 11340 km^2 which generally occupied by low, rounded hills about 50 meters above sea level and separated by broad and open valley [5]. Bangka Island is fairly far from the Indonesia-Malaysia border, and the nearest city is Johor Bahru, about 554 km of direct distance measurement.

3. Methodology

This study utilizes the atmospheric dispersion modelling to simulate the movement pattern of radioactive particles in atmosphere using Hybrid Single Particle Lagrangian Integrated Trajectory System (HYSPLIT) software. HYSPLIT is a computer model developed by the National Oceanic and Atmospheric Administration (NOAA) of USA to measure trajectories of air parcels and the concentration of air pollutants. The model calculation method is a combination between the Lagrangian technique and Eulerian paradigm, where a rotating reference frame for advection and diffusion measurements of air parcels trajectories from their original position, is combined with three-dimensional grid as a reference frame for measuring pollutant air concentrations [6]. Meteorological data input in HYSPLIT is 'NCEP/NCAR Reanalysis' database, extracted from open access global meteorological data provided by National Centers for Environmental Prediction (NCEP) and the National Center for Atmospheric Research (NCAR) of USA [6].

The trajectory simulation was run for a daily emission over a year with reference to meteorological data of 2019. To get a clear observation of atmospheric movement in extreme nuclear emission scenario, the emission input is set equivalent to the estimated release of Caesium during Fukushima accident in 2011 [7,8]. The emission height was set to 60 meters above ground level as the emission origin.

As additional analysis of the radiation impact, another simulation is conducted to find the concentration of radionuclide and the effective dose rate (EDR) in Johor Bahru. In the simulation, the

EDR due to radioactive releases was calculated from the concentration data. For EDR, Caesium (Cs-137) were chosen as the radioactive pollutant input. Cs-137 has long half-life of 30 years and emits gamma radiation. Since Cs-137 stays longer in the environment, it can travel a long distance across the globe. Long term exposure to Caesium can cause stochastic health effects such as cancer or tumour [9]. Table 1 shows the radionuclide input data in HYSPLIT.

Table 1. Radionuclides Input Data.

Properties	^{137}Cs
Phases	Particle
Half-life (days)	10,960
Annual release in normal operation (GBq)	9.4×10^{-1}
Emission rate from Fukushima accident (Bq hr^{-1})	1.0×10^{15}

4. Results and Discussion

The simulation for the radionuclide's trajectory pattern were done using HYSPLIT software. The trajectories were plotted in a regional map for 3 consecutive days to observe the direction and distance of radionuclide movement through the atmosphere. Figure 1 shows the trajectory pattern projected for each month in 2019. The blue, green and red lines indicate the trajectories for the first 3 days (72 hours). The dots on the trajectory lines represents the location and the heights of the observed particle every 24 hours from the emission origin. The dots are used as reference to estimate the time taken by the radioactive particles to travel across the international boundary after the initial release.

Since the emission point is set at 60 meter above the ground, which is slightly above the average height of the land above sea level in Bangka Island, the movement is generally not obstructed by higher topography or tall building in the area. This setting allows the particle movement to reach the furthest distance without consideration of possible obstruction along the dispersion pathways.

4.1 Trajectory Analysis

Generally, the particle movement follows the wind current on different level of altitudes. The wind current depends on the climate characteristics of certain climate seasons. Therefore, general movement of the radioactive particle can be estimated using historical wind data. This scenario is clearly depicted in Figure 1, where the direction of radioactive particles trajectories is different for each month. The radioactive particle is moving towards Malaysia from May to November, which is the period of Southwest Monsoon season according to Malaysian climate. During this period, majority of the wind direction blows toward the north, northeast and northwest of Bangka Island, where Malaysia is located [10]. It can be concluded that the Southeast Monsoon can be classified as a risk period of radioactive dispersion in Malaysia, where the radioactive particles are certainly moves toward Malaysia's atmosphere from Indonesia.

Even though the direction of particle movement can be predicted using seasonal climate characteristics, it is only applicable for general prediction for a very large area. For a smaller area and shorter period of time in the season, the wind direction is different every time and day. With reference on Figure 1, when the emission occurs on 1st January, the particle moves towards the southeast from the emission point. However, on 2nd January, the particle moves toward the east for about 24 hours, before changing direction towards the south on the second day, then towards southeast the 3 days after. This scenario shows that on smaller time scale, the wind direction is unpredictable, which may cause a distinct margin of error in the estimation of possible affected area using the average wind direction.

The graph under the trajectory map shows the movement of the particle in different altitude above the ground level. Depending on the wind current, precipitation process and the surface topography, the

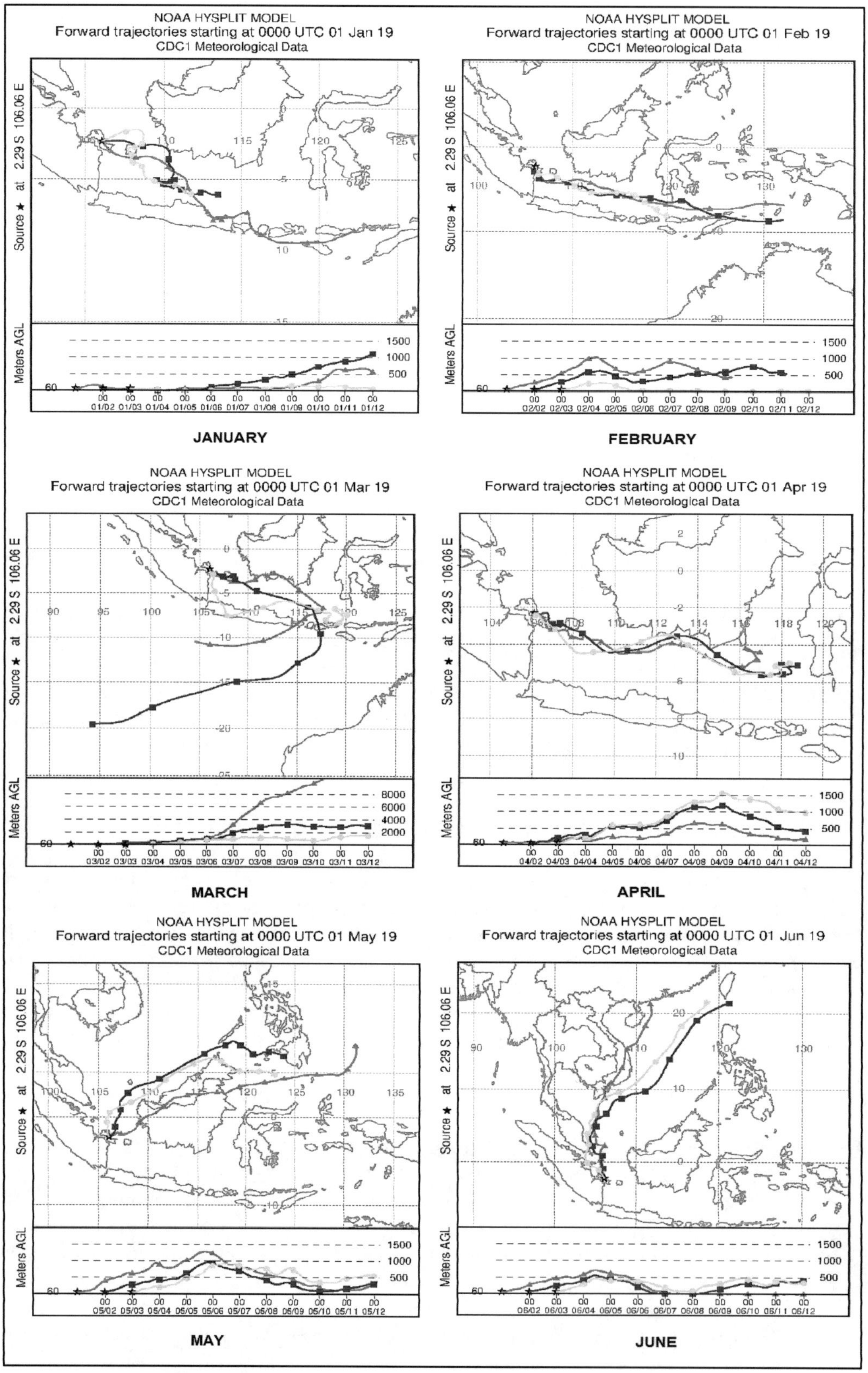

Figure 1. Monthly radionuclide forward trajectories (continue..)

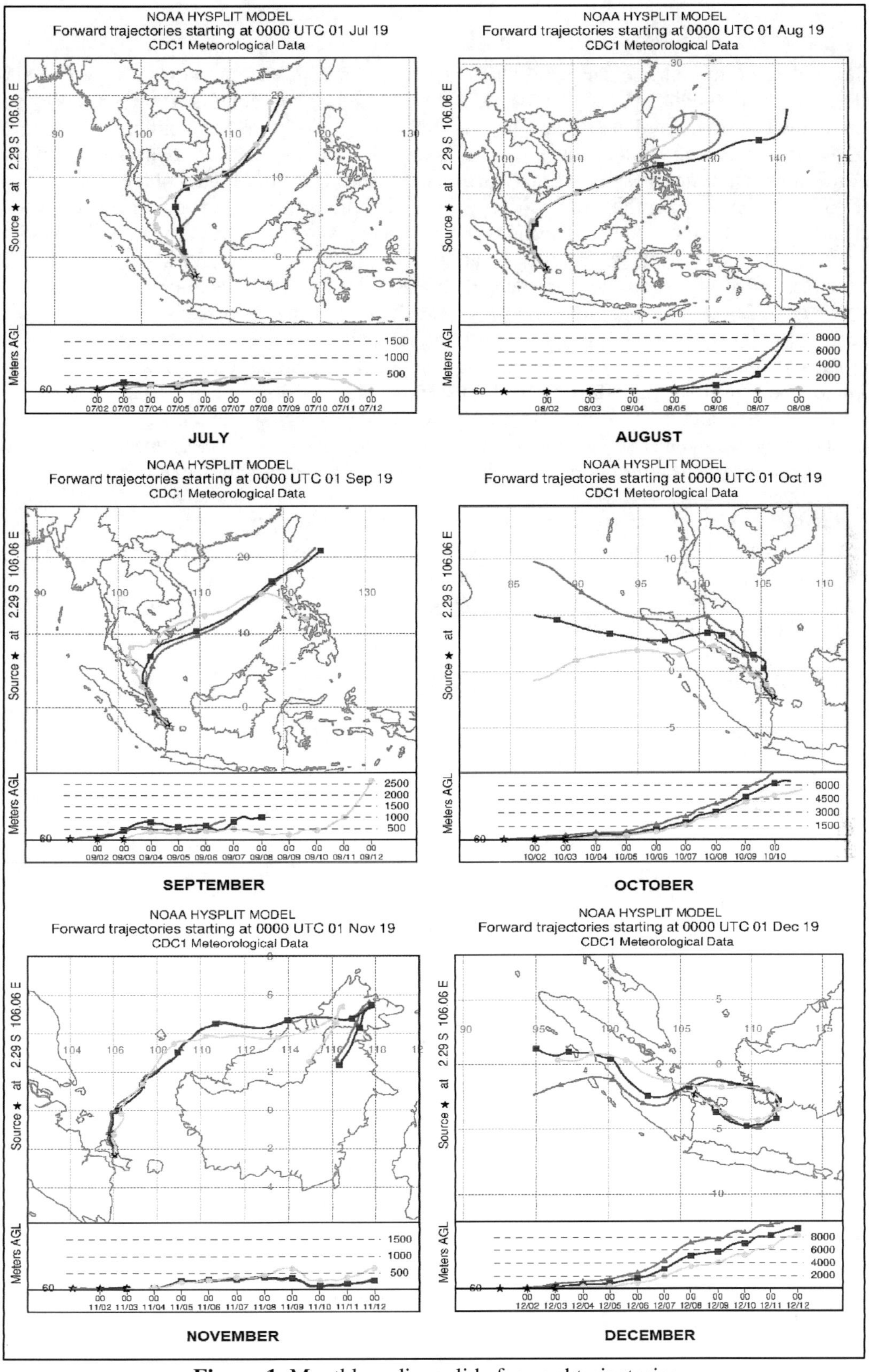

Figure 1. Monthly radionuclide forward trajectories

particles will move to higher or lower altitudes. From the graphs, particle movement above 1000 m altitude can travel farther in 24 hours compared to lower altitudes. This condition corresponds to general assumption that wind speed increase at higher altitudes when there are no significant obstacles [11]. This trajectory scenario shows that normal or accidental release from the NPP site in Bangka Island will arrive in Malaysian atmosphere any time and fully influenced by the meteorological factors. If the amount of emission is large, the risks will be significantly higher. However, the time and location of particle deposition from the atmosphere cannot be estimated from the graph. With only the wind speed and direction data, the distance and the location of particle deposition is not easily predictable. Additional climate data such as the seasonal and climate behaviour like precipitation, cyclone, wet and dry season is needed in the estimation of particle deposition from the atmospheric pathways.

The trajectory analysis is the simplest analysis to predict the movement of radioactive particle in air during nuclear emergency. It can be used as initial risk assessment right after the onset of a nuclear accident to predict the potential affected area from radionuclide dispersion in air. Emergency evacuation or sheltering order can be executed immediately, hence reducing the exposure time from the radionuclide deposition to the effected population. However, the analysis is not capable to estimate size of affected area and the health risks due to radioactive deposition or radiation exposure.

4.2 EDR data

The concentration data and external effective dose rate were obtained in the second simulation. A concentration reading point is set at Johor Bahru coordinate to determine the concentration of Cs-137 and to calculate the EDR in Johor Bahru area following 6 hours emission from NPP in Bangka Island. The EDR is calculated using Equation 1 below,

$$\mathrm{EDR\ (mSv/yr)} = \Sigma \mathrm{DCF}\ [(\mathrm{mSv/yr})/(\mathrm{kBq/m^3})] \times \mathrm{C\ (kBq/m^3)} \times 0.2 \qquad (1)$$

where DCF is the dose conversion factor of the radionuclides [12], C is the activity concentrations of Cs-137 and 0.2 is the occupancy factor.

Table 2 shows the average monthly concentration of Cs-137 and the effective dose rate in Johor Bahru from normal operation scenario and hypothetical accident scenario. The concentration values are different each month because it is influenced by the meteorological characteristics in the dispersion pathways in the simulation. The highest average concentration for normal operation is in April 2019 with 3.35×10^{-14} kBq/m^3, while for the hypothetical accident the highest concentration is in April with 0.31×10^{2} kBq/m^3. During April and May is the period where the Southeast Monsoon is changing to Southwest Monsoon. During the monsoon change, higher amount of rainfall and unstable wind behaviour will occur in Johor Bahru, which caused higher deposition and concentration of Cs-137 in the area.

During normal operation, the average effective dose rate of Cs-137 detected in Johor Bahru is very low due to the controlled emission in the NPP safety requirements. Malaysia Atomic Energy Licensing Board (AELB) guideline stated that for radiation safety assessment, the control limit for external radiation dose rate is 0.5 μSv per hour, and the annual effective dose limit for public is 1 mSv per year [13]. The dose value may not cause health risks to the population in Johor Bahru in a short term, but long period observation should be conducted to determine the probable stochastic effect from it.

Meanwhile, for the hypothetical nuclear accident, the effective dose is significantly higher, way above the 1 mSv annual limit for the public, with maximum reading of 397 mSv/yr in April. From the monthly data, the annual average dose is 1.53×10^{2} mSv for 2019. This value is obtained in a condition where a person is exposed to radiation without shielding and sheltering. Hence, the value did not represent the real total dose received by the person, but only represents the estimated maximum dose rate received from detected concentration in the atmosphere.

Table 2. Average Concentration and Average Effective dose of Cs-137

Month	Average Concentration (kBq/m^3)		Average Effective Dose Rate (mSv/yr)	
	Normal Operation	Hypothetical Accident	Normal Operation	Hypothetical Accident
January	1.35×10^{-14}	0.11×10^{2}	1.45×10^{-15}	1.73×10^{2}
February	2.05×10^{-14}	0.17×10^{2}	2.92×10^{-15}	1.27×10^{2}
March	1.90×10^{-14}	0.17×10^{2}	2.48×10^{-15}	1.54×10^{2}
April	3.35×10^{-14}	0.31×10^{2}	4.22×10^{-15}	3.97×10^{2}
May	3.15×10^{-14}	0.14×10^{2}	4.00×10^{-15}	1.08×10^{2}
June	8.50×10^{-15}	0.08×10^{2}	1.17×10^{-15}	1.24×10^{2}
July	8.50×10^{-15}	0.08×10^{2}	1.32×10^{-15}	1.13×10^{2}
August	6.00×10^{-15}	0.06×10^{2}	9.87×10^{-16}	5.25×10^{1}
September	1.35×10^{-14}	0.11×10^{2}	2.27×10^{-15}	1.26×10^{2}
October	2.85×10^{-14}	0.13×10^{2}	2.83×10^{-15}	7.39×10^{1}
November	3.05×10^{-14}	0.29×10^{2}	2.68×10^{-15}	1.64×10^{2}
December	1.30×10^{-14}	0.13×10^{2}	1.44×10^{-15}	2.21×10^{2}

The EDR calculated in this study is not enough to estimate the absorbed and effective dose for Malaysia. The data obtained is only a simple prediction dose for Johor Bahru, which cannot represent the dose for Malaysia all together. Further analysis is needed to estimate the dose received by the public in different scenario and to predict the possible health effect from the exposure of radiation. Long term effects to the environment should be included in the analysis to establish a full-scale risk assessment for Malaysia from the postulated nuclear accident in Indonesia.

5. Conclusion

From the simulation, it is concluded that dispersion from NPP site on Bangka Island can reach Malaysia as early as 2 days during nuclear emergency equivalent to Fukushima accident in 2011. The Malaysian government should establish an emergency plan for nuclear accident scenario as a preparation to protect the public from unnecessary exposure in such scenario. Collaborative studies should be conducted with Indonesia to identify the potential risks of atmospheric dispersion from Indonesia to Malaysia.

6. References

[1] Aliyu, A. S., Ramli, A. T., & Saleh, M. A. 2014 Environmental impact assessment of a new nuclear power plant (NPP) based on atmospheric dispersion modeling. *Stochastic Environmental Research and Risk Assessment*, *28*(7), 1897–1911. https://doi.org/10.1007/s00477-014-0856-9

[2] Yoshikane, T. *et al.* Long-distance transport of radioactive plume by nocturnal local winds. *Sci. Rep.* **6**, 36584; doi: 10.1038/srep36584 (2016)

[3] Udiyani, P. M., & Husnayani, I. (2017). *Analysis of Radiation Safety in the Nuclear Power Plant (NPP) Site in Normal Operation Condition, Sebagin Site Study.*

[4] Udiyani, P. M., Kuntjoro, S., & Geni, R. S. (2010). *Safety and Sitting Assessment for NPPs Deployment in Indonesia.* (1), 5–9.

[5] Udiyani P. M. and Widodo S. (2012). *The determination of atmospheric dispersion coefficient for PWR reactor accident analysis in indonesia site (Penentuan Koefisien Dispersi Atmosferik Untuk Analisis Kecelakaan Reaktor Pwr Di Indonesia).* Jurnal Teknologi Reaktor Nuklear, 14(2), 121-132. BATAN, Indonesia. Article no. 402/AU2/P2MI-LIPI/04/2012.

[6] Rolph, G., Stein, A., & Stunder, B. (2017). Real-time Environmental Applications and Display system: READY. *Environmental Modelling and Software*, *95*, 210–228.

[7] Leon, J. D., Jaffe, D. A., Kaspar, J., Knecht, A., Miller, M. L., Robertson, R. G. H., & Schubert, A. G. (2011). Arrival time and magnitude of airborne fission products from the Fukushima, Japan, reactor incident as measured in Seattle, WA, USA. Journal of Environmental Radioactivity, 102(11), 1032–1038. https://doi.org/10.1016/j.jenvrad.2011.06.005

[8] Shigekazu Hirao, Hiromi Yamazawa & Takuya Nagae (2013) Estimation of release rate of iodine-131 and cesium-137 from the Fukushima Daiichi nuclear power plant, Journal of Nuclear Science and Technology, 50:2, 139-147, DOI: 10.1080/00223131.2013.757454

[9] Walker, S. (2002). EPA Facts About Cesium-137. 99(July), 2.

[10] Gentilli, J., & Smith, P. J. (2012). *Malaysian-Australian monsoon | meteorology | Britannica*. Retrieved from https://www.britannica.com/science/Malaysian-Australian-monsoon.

[11] Bañuelos-Ruedas, F., Ángeles Camacho, C., & Rios-Marcuello, S. (2011). Methodologies used in the extrapolation of wind speed data at different heights and its impact in the wind energy resource assessment in a region. In Gastón O. Suvire (ed.), Wind Farm - Technical Regulations, Potential Estimation and Siting Assessment. IntechOpen.

[12] Keith Eckerman, Wolbarst, A. B., & Richardson, A. C. B. (1988). Federal Guidance Report No. 11. , United States Environmental Protection Agency.

[13] Atomic Energy Licensing Board (2016). Code of practice on radiation protection relating to technically enhanced naturally occurring radioactive material (TENORM) in oil and gas facilities. LEM/TEK.58 Sem.1. 22 December 2016.

Acknowledgments

This study was funded by Universiti Teknologi Malaysia under Research University Grant (GUP No.: Q.J130000.2651.16J77).

Journal of Physics: Conference Series **2053** (2021) 012011 doi:10.1088/1742-6596/2053/1/012011

Assessing biofouling in Ocean Thermal Energy Conversion (OTEC) power plant – A review

Mohd Zaki Zainal Abidin[1]*, Miradatul Najwa Muhd Rodhi[1], Fazlena Hamzah[1] and Nurul Aimi Ghazali[2]

[1] Biocatalysis & Biobased Material Technology Research Group, School of Chemical Engineering, College of Engineering, Universiti Teknologi Mara UiTM, 40450 Shah Alam, Selangor, MALAYSIA
[2] School of Chemical Engineering, College of Engineering, Universiti Teknologi Mara UiTM, 40450 Shah Alam, Selangor, MALAYSIA

*zakizainal@uitm.edu.my

Abstract. Ocean Thermal Energy Conversion (OTEC) harnesses thermal energy stored at different seawater depths via power generation from a thermodynamic closed-loop cyclical system. Apart from its consistent energy generation, it could be diversified into other side industries, making OTEC an attractive and sustainable source of renewable energy. However, the process that utilises seawater as its main fluid is exposed to biofouling deposition due to unwanted growth and accumulation of biological elements on any contact surfaces, potentially affecting its efficiency and damaging equipment in the process. Considering that biofouling is an inevitable condition that may not be eliminated, a comprehensive study for assessing potential biofouling growth and deposition mechanism is a crucial step for strategizing effective biofouling management in a commercial and large-scale OTEC power plant facility. This review paper focuses on evaluating suitable biofouling assessment techniques specifically for a large-scale OTEC power plant facility. This is achieved by evaluating previous and proposed biofouling assessment techniques relevant to OTEC systems by focusing on their implementation under a realistic OTEC setup. The initial study indicated that the potential of biofouling deposition may be unavoidable in some sections in all OTEC models, despite biofouling-free design consideration. Previous OTEC biofouling studies were evaluated with reported physical and biological assessment approaches indicated the need to further improve these techniques especially in continuous and non-destructive methods. Therefore, several biofouling monitoring systems reported from other water treatment industries were considered for the OTEC systems, with findings indicated the importance of considering important OTEC operational parameters for feasible and robust biofouling monitoring systems. Two major parameters which are seawater intake flow rate and temperature variation at different seawater intake levels were evaluated under OTEC operational evaluation by considering examples of practices conducted in cooling water systems in the power plant industry. A realistic biofouling monitoring setup for mimicking continuous changes in biofouling deposition is required, in this case by side-connecting an operated OTEC power plant facility with a pilot plant setup or a side sampler. This step allows the application of proposed biofouling monitoring techniques under a realistic and uninterrupted biofouling deposition setup.

1. Introduction

Ocean Thermal Energy Conversion (OTEC) is a renewable source of energy that utilises the thermal gradient of the seawater at different sea levels by pumping warm water supply from the sea surface to generate steam that runs a turbine, while concurrently the cold water supply from deep seawater is used to cool down the outlet stream from the turbine for re-condensation [1]. Based on this concept, an optimum OTEC system could be achieved at the minimum sea temperature gradient of 20°C within the sea depth of 1000 m [2]. By considering the fact that the oceans absorb an enormous amount of solar thermal for about 4000 times than the presently consumed by humans [3], OTEC has wide potential to be exploited as a renewable source of energy. Apart from its capability as a continuous and base-load renewable energy generation [4], OTEC has the capability for producing energy up to 3 TW [5], where this number is estimated could meet twice of electricity demand worldwide. Unlike other sources of renewable energy such as solar and wind energy, the applicability of OTEC could be diversified to other industries such as deep seawater (DSW) and desalinated water generation, air conditioning system, aquaculture industries, hydrogen gas production, mineral extraction (i.e. lithium) from seawater and for cosmetic and high-grade food production [6].

In an example of a closed-loop OTEC system, the system operates when warm sea water obtained at the surface level is channelled to an evaporator to vaporize a working fluid in a closed-loop thermodynamic heat engine system, operated using highly-vaporized fluids such as ammonia or R-134a [1]. The evaporator converts the working fluid from liquid to high-pressure vapour flow where this vaporized stream is later used to run a gas turbine for power generation. The stream coming out from the turbine is then cold down by a condenser using deep-sea cold water intake, where the condensed working fluid is pumped and circulated back into the evaporator. The thermodynamic efficiency for about 3-5% could be achieved in an optimum OTEC system, subjected to the optimal temperature difference between hot and cold water temperature [7]. However, this efficiency might decrease by about 1-2% as a result of fouling that could take place inside the heat exchangers (evaporator or condenser) [8].

Huge interest for an alternative source of energy after the world major oil crisis in early 1970 saw large investment for about $260 million by the US federal government in OTEC research with an aim for generating 10,000 MW energy from OTEC by 1999 [9]. The following years (the 1980s-2000s) saw advancement and improvement in OTEC technology with the largest commercial-scale OTEC power production is 100 kW by the Makai OTEC power plant in Hawaii [10]. Large intake of seawater is proportionate to the capacity of OTEC power generation, where a hypothetical 100 MW capacity of OTEC is projected to use enormous amount of seawater for about 400 m^3/sec of warm feed water and 200 m^3/sec of cold feedwater [11]. In terms of Malaysian oceanic profile, the northwest and southeast of Sabah have been identified to be a suitable location for future OTEC development in Malaysia due to the availability of deep seawater greater than 1000 m with the average surface seawater temperature reported at 27°C and approximate temperature of 4°C at the depth of 600 m [12]. This location is projected could produce approximately up to 10 MW of OTEC commercial plant [12].

When considering a system that uses water as the main processing fluid, untreated and raw water contains various physical-chemical properties which directly affects any physical surface that is in contact with water. One such example is on fouling where it could occur as a result of deposition of matters that could disrupt the optimum flow of water in a processing system. This fouling problem could be contributed by various factors, whether scaling or crystalline fouling, organic fouling due to the deposition of organic matter, particle and colloidal fouling as a result of silt or clay deposition or microorganism fouling [13, 14]. Among all these types of fouling, biofouling is considered to be inimitable problem that is influenced by the properties of intake water to a system and optimum conditions, which eventually allow distinctive growth and deposition of microbes on a concerned material surface. Biofouling is the result of undesired deposition of all forms of biological elements on immersed materials [15]. A typical biofouling growth and deposition involve the microbial adhesion on the exposed surface, where the accumulation of microorganism forms aggregates biofilm and the colonization of macrofouling species on the surface of the biofilm layer which is enriched with nutrients and protected from biocides and toxins [16]. The prolonged colonization process allows the attachment and growth of invertebrate larvae and the spores of algae on the submerged surface [15]. Among these macro biofouling species are such as barnacles, mussels, sponges, sea anemones and hydroids, bryozoans, serpulids, algae and others [15]. The final stage of biofouling occurs when the matured biofoulants are accumulated on the surface and in the flowing stream, these biofoulants could be detached from the substrate. Figure 1 illustrates schematic stages involved in biofouling growth and deposition on the exposed surface.

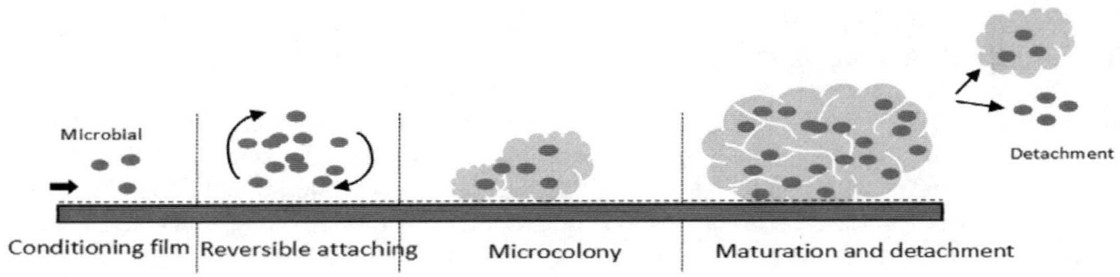

Figure 1. Stages involve in biofouling growth and deposition on an exposed surface [14]

Biofouling poses a major problem in this facility due to its potential deposition on the surface of heat exchangers, reducing the efficiency of heat transfer between seawater and working fluid streams. For example, the presence of a thin biofilm of 25-50 micrometer could reduce the performance of OTEC by 50% [8]. In addition, biofouling deposition particularly for severe macrofouling depositions along the internal wall of contact surface may increase pressure drop inside the heat exchanger and seawater pipelines, as well as impairing the performance of associated auxiliary systems. Apart from that, large deposition of macrofouling which is classified as large-size and distinctive invertebrate larvae and the spores of algae [15] may potentially restrict water pipeline, increasing additional weight to a floating OTEC facility and in the worst condition, might cause serious damage to the processing facilities [17]. The presence of biofouling is also associated with microbially induced corrosion (MIC) where biofilm adhesion on a surface changes the electrochemical interface properties thus inducing corrosion [18].

The earliest study of biofouling in Malaysia was done by Cheah and Chua [19] who recorded more than 34 species in floating net-cages structures such as oysters, algae tunicates and mussels. Relating this outcome with the OTEC potential site in Sabah, Affandy, et al. [20] and [21] conducted two relevant studies on the development of macrofouling at the shallow coastal area of Karambunai Bay. The studies found out that thirteen different biofouling species (sessile and mobile) were identified on submerged PVC plates (2 m x 8 m) throughout their observation in 180 days. These biofouling communities had a significant relationship with environmental factors, in this case, average seawater temperature was recorded within 29.9°C to 30.9°C and low net current speed from 5 cm/s and 7 cm/s.

There are two major focuses of the biofouling management system which are biofouling assessment and detection activities as well as biofouling controlling measures, where in this report, focus will be on adopting continuous and online biofouling monitoring systems. A reliable and robust biofouling detection system provides crucial information of biofouling characteristic profiles either biological or physico-chemical aspects and most importantly, the degree of biofouling deposition at several important sections in the OTEC facility. For example, in the design and engineering aspect, the usage of oversized heat exchangers which intended to provide a large heat transfer area under the anticipation of biofouling deposition may be re-evaluated with prior understanding of the expected severity of biofouling. In addition, continuous assessment of biofouling deposition serves as a guideline in optimum biofouling controlling measures. In this case, the usage of overdose chemical biocides using high-dosage chlorine could be controlled for effective inhibit biofouling deposition [8] as a result of improvement in continuous low-dose chlorination monitoring activity. Despite the importance of acquiring a biofouling monitoring system, current practices in the water-related industries are based on visual inspections and a long period of assessment, albeit that the biofouling growth and deposition are physically undetectable in normal conditions and rapidly changing over a period of time.

Considering the requirement of minimum temperature difference at 20°C and its small efficiency of about 3 to 5 percent [7], feasible maintenance and operating practices in OTEC power plant facilities specifically for managing any biofouling deposition are essential to ensure its consistent potential power generation on a commercial scale. This is crucial for the preparation of operating commercial scale of OTEC facility where any small change in its operation could eventually affect the whole OTEC efficiency. In the following report, focus will be on the biofouling assessment methods and techniques which accommodates commercial and large-scale operation of the OTEC power plant facility. The manuscript consists of a few sections: 1) Overview on several OTEC models and design consideration for potential biofouling deposition 2) Previous reports on biofouling assessment in OTEC system; 3) Potential application of advanced analytical instruments and sensors for detecting biofouling growth and deposition in operated OTEC facilities; 3) Operating aspect for biofouling assessment and 4) Setup for biofouling assessment activities.

2. OTEC model systems and biofouling design consideration

As the demand for renewable energy is expected to increase due to the concern of global warming, it indirectly opens up the opportunity for a feasible and large-scale OTEC power generation, pending the reduction in its overall operational costs [22]. The high operating cost of OTEC is estimated at $40 million per annum [22], contributed partly by the huge pumping cost for transporting large flow rate and low-pressure head of cold deep seawater [23], which consequently contributes to OTEC overall thermal inefficiency [22]. By minimizing biofouling deposition in intake pipelines, indirectly improves the optimal flow rate of seawater intake due to minimal drag force caused by hard-shell biofouling deposit, which ultimately improves OTEC performance. Therefore, the understanding of potential biofouling deposition in OTEC design helps in determining suitable biofouling assessment programs, where this could be achieved by considering various OTEC models and their general process description, as well as ecological factors, specifically in the Malaysian marine environment.

Generally, there are three different OTEC models available on both commercial and laboratory scales: closed systems, open systems and hybrid systems. The closed OTEC systems generate energy by using a closed-loop refrigerant cycle which normally contains ammonia that is vaporized through warm surface seawater before it is passed to a turbine for power generation before cooling it back to liquid refrigerant using cold seawater intake. Different configurations of heat exchangers and turbines for the closed-OTEC system have been proposed for achieving high efficiencies such as first-model of Rankine-based cycle [24], Kalina cycle [25] and Uehara cycle [7]. On the other hand, open-cycle OTEC systems use seawater as a working fluid to generate power in OTEC facilities via rapid flashing of the warm seawater under a low-pressure vacuum chamber [26]. As for hybrid-OTEC systems, the systems which combine both a high-efficiency closed-loop system and an open-OTEC system allow both power generation as well as other side-product production such as high-throughput potable water production [27]. Etemadi, et al. [28] reported that the hybrid cycle uses both seawater and working fluid such as ammonia, fluorinated carbons or hydrocarbons in hydraulic and vapour turbines. The phase change of the water/ammonia vapour in this process turns a turbine producing energy. Even though the selected working fluids used in the process possess qualities due to their excellent transport properties, easy to obtain and low cost but they have a few setbacks due to their flammability and toxicity which might result in both safety and environmental issues. Moreover, the pressure of the liquid used strongly affects the size of the OTEC system, as the high pressure decreases the size of both the turbine and the heat exchangers, which inversely requires thicker wall thickness [29].

The Malaysia hybrid OTEC is the latest OTEC system that has function as power generation and desalinated water production systems [12]. This hybrid OTEC system indirectly uses surface seawater, converts it into vapour in the heat exchanger to produce electricity and at the same time, freshwater can be produced through distillation. A simplified process flow diagram for the hybrid OTEC is illustrated in figure 2. In this system, a vacuum chamber will quickly evaporate warm seawater. The water steam will cause a working fluid to reach its boiling point. Then, electricity is generated by expanding the refrigerant in the turbine, followed by the vaporized fluid condensing inside a heat exchanger, thus generating desalinated water [30]. According to Magesh [31], the hybrid OTEC system is capable to generate nearly 2.28 million litres of desalinated water every day for every megawatt of power.

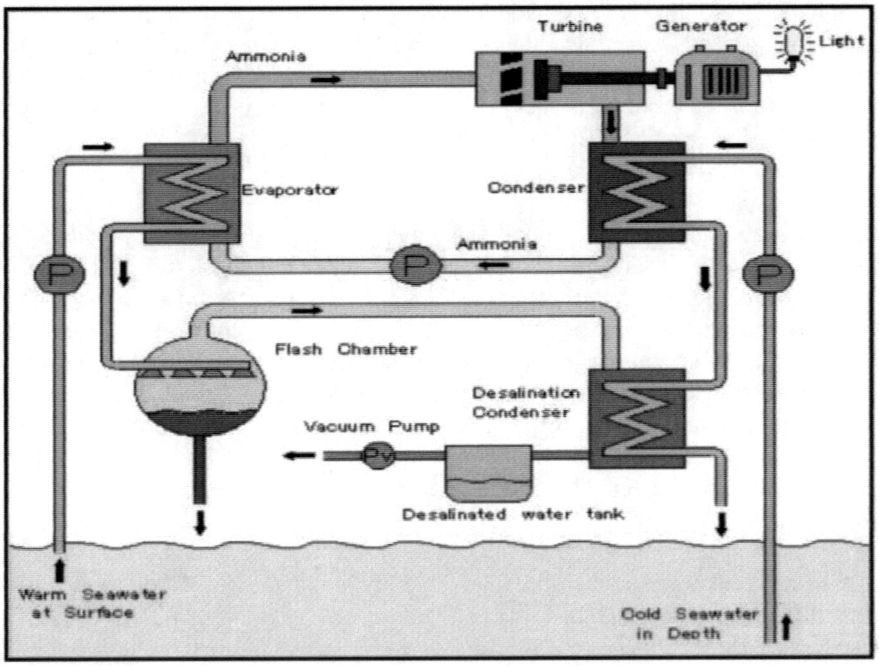

Figure 2. Process layout for OTEC Hybrid cycle with desalination process [32]

Depending on the design of OTEC systems, potential biofoulants that deposit on the surface of heat exchange is inevitably affecting overall efficiency especially an OTEC model that adopts direct-contact seawater heat exchangers for condensers and evaporators. However, a recent study conducted in 2013 reported that a free-biofouling design has been considered since the 1980s, in this study, a brazed aluminium heat exchanger was incorporated in this feature, and currently being used in the Makai OTEC power plant [33]. This biofouling preventive consideration is made possible particularly in a hybrid OTEC system where the heat exchanging process uses vaporized seawater stream generated from a vacuum chamber, thus decreases serious consequences of biofouling in OTEC heat exchangers. This new design feature eliminates the concern of the threat of biofilm in heat exchangers. However, even with this new feature of heat exchangers used in OTEC system, deposition of macrofouling may still unavoided particularly in intake pipeline system for warm surface seawater. In addition, processing units such as strainers, pumps, holding tanks and pipeline fittings are among a few types of equipment in the OTEC power plant that are still exposed to potential biofouling deposition.

Based on the above explanation, it is crucial for OTEC operators to evaluate biofouling in any type of OTEC system used as the exposure to any potential biofouling is almost inevitable. This could be done by exploring any reported techniques used for assessing biofouling specifically for OTEC systems which will be highlighted in the next section.

3. Previous reports on OTEC biofouling assessment

The concern in biofouling study in OTEC is comparative to the interest and advancement of OTEC over a period of years with most of the reports on the assessment of biofouling in OTEC facilities have been initiated since the late 1970s to mid-1980s [8, 15, 34, 35, 36]. These reported techniques were implemented in actual OTEC facilities or the mock heat exchanger for representing close to the actual condition. For example, Berger and Berger [8] conducted their study in Seacoast Test Facilities at Keaholes, Hawaii. Likewise, Panchal and Knudsen [37] conducted a similar analysis at the same facility specifically for determining potential biofouling in tropical water. In contrast, Aftring and Taylor [36] explained in detail their biochemical assessment of biofouling formed on the mock heat exchanger to mimic the actual condition of OTEC condition. The assessment conducted in fields provides realistic representation on biofouling growth and deposition, thus acquiring suitable biofouling evaluation techniques in their setup. Two broad categories of biofouling assessment techniques that were reported under previous OTEC setups are presented here, namely physical and biological and biochemical approaches. The following subsections explain the concept of these techniques, their advantages and their limitations in the current OTEC setup.

3.1 Physical assessment

This method focuses on detecting and analyzing any physical changes that take place due to the presence of biofouling growth and deposition on exposed surfaces. The first method uses a heat transfer monitoring (HTM) device for determining the thickness of micro fouling that is appeared as a thin layer of biofilm on the surface of the pipeline wall. Based on the concept of heat transfer measurement, the thickness of biofilm that acts as heat transfer resistance is measured by measuring the temperature difference between the heated section of fluid inside the pipeline and the reference temperature, in this case at ambient temperature [8, 38, 39]. The apparatus consists of two thick-walled sections of copper cylinders, one section that acts as ambient temperature measurement and another section is a heated wall surrounded by a nichrome heater. These cylinders are inserted inside a section of the pipeline whereupon continuous seawater flow, the presence of biofilm is detected through changes of measured temperature difference.

This method is applicable when the flow velocity is within 2 to 8.4 ft/sec with high sensitivity of temperature difference is about 0.001°F [38]. Berger and Berger [8] reported that the application of HTM by observing the change in fouling resistance (R_f), recorded within R_f of 8 to 10×10^5 °C/m^2·W for about a three-year continuous observation period. However, this device is designed for assessing biofilm thickness in the pipeline and may not be suitable for tubing in a heat exchanger [15]. Moreover, the thickness of biofilm is practically accessible within a short span of the pipeline due to the limited length of the device and may be able to describe accurately biofouling deposition due to its spatial deposition profile on the surface [40].

3.2 Biological assessment

Scanning electron microscopy (SEM) is used to analyze biofouling growth and deposition profile on the contacted surface, providing crucial information of the type of organisms or materials involved, a rough estimation of biofilm thickness and the nature of its growth [15]. Before this visual inspection, the accumulated biofouling sample on a set of attached coupons to the surface of a heat exchanger was extracted by scrapping manually by cutting out a section of the heat exchanger tube [8] or using a special tool to scrape any deposit on the surface of heat exchanger [41]. These coupons were subjected to accumulate biofouling deposition over a period of time, in the report mention by Berger and Berger [8], it was done within six months up to one year before the samples were extracted. The extracted samples formed a biofilm layer comprised of entrapped diatoms and portions of differentiated algae and other organisms. Apart from determining the species present on extracted substrates, this technique evaluates the effectiveness of mechanical cleaning techniques by observing any residual that remains after the physical cleaning process using either sponge rubber balls or bristle "test tube" brushing methods [8, 15] as well as after chlorination approach for retarding microfouling growth [8]. Despite these efforts, the presence of biofilm was still observed after three-year of observation and preventive period [8]. The applicability of SEM is still relevant even in the presence of OTEC setup, however, the destructive method for extracting sample may provide unpresentable biofouling deposition profile on substrates.

Besides the semi-quantitative measurement of SEM, the availability of various biochemical assessments provide a comprehensive biological and chemical characterization of the extracted sample. There are a few biochemical assessments that have been reported under the scope of OTEC systems [15, 36, 42]: adenosine triphosphate (ATP) total organic carbon (TOC), carbohydrate analyses, chlorophyll contents, total dissolved iron content, combustible organic matters and many others methods. Mitchell and Benson [15] and Aftring and Taylor [36] are among the earliest studies of biochemical assessments conducted specifically for OTEC in the 1980s. Among those measurements was the ATP test which detects the concentration of a growing microorganism as ATP is a molecule that could be found in any living cells [15]. Apart from ATP, the TOC measurement that indicates the total nitrogen content and protein that made up the composition of biofilm was also reported in these reports [16]. The measurement of carbohydrates and chlorophyll reported by Aftring and Taylor [36] indicates the presence of organic compounds that promote the bacterial growth of biofilm [43]. Acknowledging the importance of various biochemical assessment which provide quantitative measurement of biofouling sample, the setback of this approach is due to the absence of standard measurement gives discrepancy and inconsistency in reporting severity of biofouling [43]. Moreover, these measurements are only accessible under extensive laboratory procedures and may not be effective in describing constantly-change biofouling growth and deposition profile

The establishment of both physical and biological assessments form the fundamental techniques in assessing biofouling problems, particularly in the OTEC system. However, substantive information that describes biofouling deposition under the actual field parameters is crucial for describing accurately the severity of this deposition, as summarized in figure 3. Factors such as the flow velocity, the physical design of the system, the variation of seawater temperature and the exposure time with seawater are some of the important criteria that need to be considered under the engineering perspective. Considering the several reported biofouling assessment techniques, an improved and reliable biofouling assessment approach is needed, in this case using a continuous, non-destructive and online biofouling detection system. The following sections describe the application of the various analytical instrument in assessing biofouling.

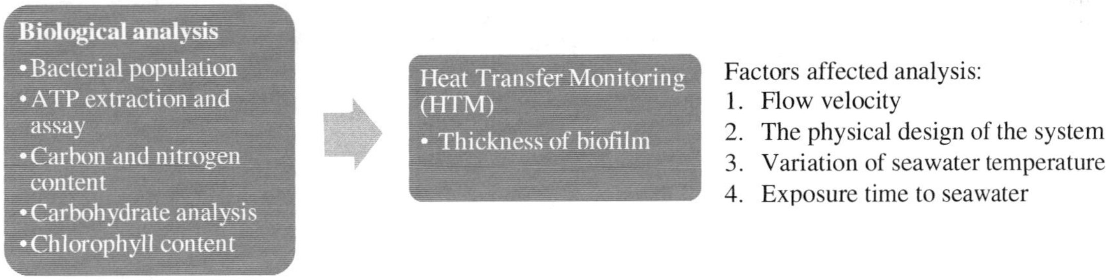

Figure 3. The overview of biofouling assessment techniques conducted in OTEC facilities

4. Potential biofouling assessment techniques

A flexible and effective biofouling monitoring method is required with a focus on several qualities of the continuous, non-invasive, online biofouling detection system, in this case, is based on several studies conducted in membrane distillation (MD) technology used in desalination and water treatment. Several manuscripts have extensively reviewed these methods and techniques under the scope of MD application [13, 14, 16, 44], with most of them are relevant and applicable to the OTEC technical setup. Acknowledging the importance of various applications of analytical instruments in a laboratory setup, this report focuses on identifying suitable monitoring assessments that are suitable under field application. A list of suggested analytical instruments and techniques are presented in table 1, suggesting advantages and setbacks for their implementation under OTEC process conditions.

Although most of these instruments are superior in term of their accuracy and detection in the system, most of them are only applicable in laboratory setup where ex-situ assessment is adopted. Despite its convenience and environmental-controlled laboratory setup, imitating actual conditions for the growth and deposition of biofouling is a challenging task due to complex and constantly-changing environmental factors which are almost non-viable to be replicated under laboratory setup. Even when the scale-up procedure or the modelling and simulation could be implemented, the absence or limited data of biofouling assessment from actual field conditions may overestimate the problem. For example, the variation of nutrient and temperature with respect to different depths of seawater and variation of biofouling deposition with respect to a rapid and constantly changing flow rate of seawater and the size of the deposited section may give measurement far from the actual condition. Several reports documented the usage of analytical instruments in the actual field such [16, 45], but their feasibility to withstand constantly changing environment are yet to be accepted by the industry.

The limitation for implementing suitable biofouling detection system is resolved with the well-addressed parameters that directly control the performance of actual OTEC systems. Therefore, the following section explains factors that contributes to the observation of biofouling in a real OTEC power plant facility.

Table 1. Techniques used in the evaluation of complex biofouling phenomenon

Technique	Concept	Techniques	Advantages
Microscopic and optical techniques	Visual-aided technique to magnify the image of deposited biofouling.	1. Optical coherence tomography [46] 2. Fiber optical sensor (FOS) [47] 3. Differential turbidity measurement (DTM) [48]	1. Ability to detect thin biofouling on a surface Practical techniques with improved features
Spectroscopic	Based on the absorption, scattering or emission characteristic of a matter subjected to the emitted light	1. Raman spectroscopy [49] 2. Fourier Infrared Transform (FTIR) 3. Nuclear Magnetic Resonance (NMR) [50] Photoacoustic spectroscopy	1. High accuracy with an almost negligible amount of required sample 2. A non-invasive and non-destructive method Ability to measure biofilm inflow velocity profile (NMR)
Physical assessment	Changes in physical characteristics in operating conditions with respect to the presence of biofouling	1. Pressure drop detector Heat transfer monitoring system [8]	1. Practical and readily implemented in fields 2. Cheaper Ability to withstand an extreme environment

Electrical techniques	Using electrical techniques to detect biofouling	1. Electrochemical sensors: (a) chemical potential measurement, (b) microelectrode for measuring the chemical concentration 2. Piezoelectric [44] Electrical impedance spectroscopy [51]	1. Time-dependent measurement Ability to detect chemical profiles on the interface.
Biological/ chemical detecting techniques	The on-line monitoring system that adopted biological or chemical sensors for biofouling detector	1. Measurement of metabolic products at biofilm layer [52] 2. Low-concentration diffusible molecules (LCDMS) [53] Electrochemical sensors that detect changes in the interfacial chemistry due to biofilm activities.	Advanced technique that requires simple detectors condition

5. Assessing biofouling in OTEC power plant facilities

Assessing biofouling in an OTEC facility requires an adequate understanding of operating parameters that are subjected to OTEC process specifications. This requires continuous monitoring since any parameter changes that take place inside the system require effective remedies and minimize excessive usage of harmful chemicals for the preventive approaches. As the basis of this study is mostly based on operational aspects of cooling water systems in power plant industries [17], the primary aim of the biofouling monitoring system is to ensure minimum deposition on crucial sections of the OTEC system, in this case at the surface of the heat exchanger as well as seawater intake pipelines. The understanding of these potential factors allows the OTEC plant operator to assess any potential location which likely to form biofouling and possibly for installing online monitoring detectors. There are almost no reports discussing specifically biofouling problem in operational point-of-view. However, this assessment could be done by correlating with a few cases related biofouling operational assessment in cooling water process system that is used power plant industries. This is mainly due to similarity in term of process specification which involves cooling or heating processes using heat exchangers and large usage of natural water obtained from oceans, rivers or lakes [17].

Macrofouling organisms are expected in operational biofouling observation as these organisms are more noticeable in terms of the size and the potential impact on a subjected facility than the microfouling. The degree of fouling is expected to be higher as the apparent seawater usage in the system than a system that uses freshwater sources. The settlement rate of biofouling could take a few hours and up to about 400 hours [17]. However, other parameters may influence the biofouling in biofouling growth and deposition, classified into chemical, physical and biological parameters [54] as shown in table 2. Two main parameters are highlighted in section which are flow rate of seawater intake and temperature of seawater intake as both are the two major parameters that have significant impact to the operability of the OTEC power plant [35].

Table 2. Parameters involved in experimental setup for biofouling assessment [54]

Chemical Parameters	Physical Parameters	Biological Parameters
Substrate type	Temperature	Microorganism type
Substrate concentration	Fluid shear stress	Culture type (mixed or pure)
pH	Heat flux	Suspended cell concentration
Inorganic ions	Surface composition	Antagonist organism
Dissolved oxygen	Surface texture	
Microbial inhibitors	Fluid residence time	

5.1 Flow rate of intake seawater

The rapid flow of water provides sufficient oxygen and nutrients for the growth of macrofoulants, thus propagating colonization of macrofoulants. In contrast, the sea water flow which is under restricted flow condition and well below its critical velocity profile may accomodate fouling inside the processing system [35], thus it is difficult to correlate the relationship the velocity to the biofouling growth and formation. There is possibility of the hindrance of biofouling settlement at rapid flow due to the high shear rate of water than the shear rate of the settlement. However, a typical flow rate of water at 1.4-1.8 ms^{-1} across the heat exchanger and 2-3 ms^{-1} through inlet pipelines does not prevent the settlement of biofouling [17].

The observation of biofouling in operating condition is often observed inside a heat exchanger as well as in sea water intake pipelines. The accumulated biofouling over a period of time increases thickness of biofouling up to 30 cm from the wall [17], causing blockage to the internal section of a direct-contact sea water heat exchanger, as shown in figure 4. Under such condition, the presence of biofouling in the heat exchanger could be detected by assessing the reduction of heat transfer, where the fouling acts as heat transfer resistance as well fluid frictional resistance [55]. Conversely, biofouling accumulation in pipelines could be done by measuring pressure loss in the flowing system [54]. The presence of biofouling detector may not guarantee full assessment of biofouling as the detector might not be able to distinguish other elements of fouling such as deposition of inorganic compounds such as scaling and corrosion. Therefore, Jenkins [35] has suggested to relate the relationship between shear effect due to velocity with the effect of corrosion. The placement of biofouling detectors could be strategically positioned in biofouling-prone sections by considering mechanical and environmental design factors related to OTEC power plant facility.

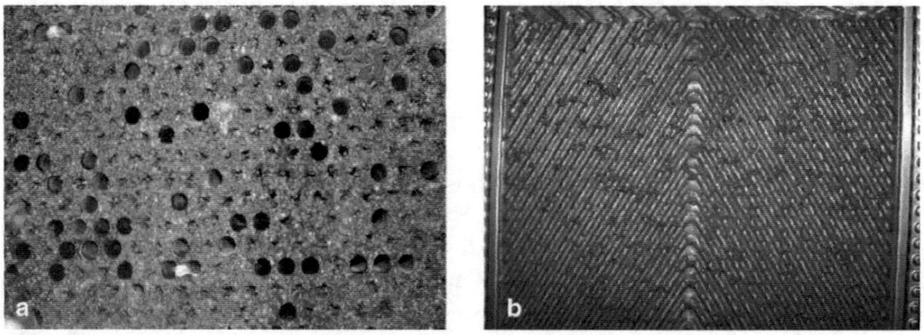

Figure 4. Examples of blockage of biofouling and corrosion in a tubular heat exchanger (**a**) and plate-fin heat exchanger (**b**) in a power plant station [17]

5.2 Variation of temperature at different sea water level

The variation of the temperature profile in an OTEC operating system depends primarily to the depth of seawater intake, where a uniform warm surface seawater temperature at 28°C is typically achieved within 35 to 100 m before it reduces gradually to 4.4°C at the sea depth of 800 to 1000 m [56]. The potential of biofouling growth and accumulation is likely to be higher in tropical seawater environment [17] where constant temperature throughout the year provide steady development of biofouling community than the seasonal-change seawater area [20]. Under OTEC process specification, it is expected high potential biofouling growth and accumulation could take place in the warm surface seawater intake than the deep cold seawater intake [15, 35]. The warm surface seawater exposed to continuous sunlight favours the growth of biofouling, with different potential biofouling that exists at different sea levels [15]. A sea water temperature ranged between 20°C to 50°C is considered to be the optimum temperature for microbial growth with the ability of the microorganism to survive in a wide range of temperatures [37].

Factors such as the location of the OTEC power plant to the location of both warm and cold seawater intake, the size and length of the pipeline as well the amount of heat loss to the surrounding (particularly for cold seawater intake) relative to the length of pipelines indirectly affect the temperature profile of transported seawater, thus subjecting OTEC plant facility with different severity of biofouling deposition. For example, potential biofouling is expected to be higher in onshore facilities than offshore facilities due to the high concentration of organisms near to onshore area [56]. Additional processing units such as desalination and depressurization units which involved along with the main OTEC system may have a significant impact on the flow rate and temperature profile of the system.

6. Setting up environment for biofouling assessment

As much as biofouling detection techniques are concerned, it is essential to evaluate the biofouling under a realistic setup where biofouling could form. Implementing directly these detection techniques in an operating OTEC system is almost impossible due to strict and delicate process requirements, albeit similar problem is faced in laboratory-scale setup due to limitations in imitating the same actual setup and parameters for replicating biofouling deposition. Despite their limitation compared to the actual scale, the laboratory scale data provides a practical solution for assessing the effectiveness of several biocides and other biofouling controlling measures before it could be implemented in actual operating conditions. This could be achieved when considering several requirements to grow the biofoulants in a suitable medium, nutrients and parameters (i.e. temperature, oxygen, salinity, pH level and others), as well as the representation of shear stress of the biofoulant on the contact surface [54]. This could be represented on various scales, ranging from a bench-scale flow cell known as Robbin device [54], a medium-scale flow loop consists of a controlled-process pipeline setup [57] and a dummy heat exchanger setup that is connected to an operating facility [36].

It is favourable to have an in-situ monitoring device which can measure realistic operating condition over a period of time without interruption. One of such options for this is by using a side stream where the water from the OTEC facility is passed through pipelines that represent the same design characteristics (i.e. roughness of the surface, size and materials) and operating parameters [17]. Current practise in monitoring the growth of fouling organisms inside cooling circuits at the power station is by using biofouling monitor which consists of a chamber with removable test panels. For example, the usage of an aquarium-like box as a side-stream sampler such as Bio-box and KEMA Biofouling Monitor® (presented in figure 5) is used to mimic the flow of industrial pipeline condition and strategically located at biofouling-prone section such as intake pipelines [58].

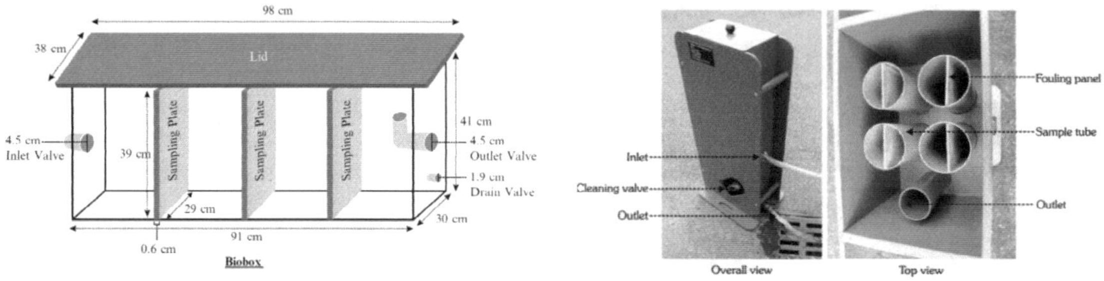

Figure 5. Illustration of (a) Biobox and (b) KEMA Biofouling Monitor [58]

Another option is to build a pilot plant scale process system that could explore all possible parameters where the biofouling deposition could occur close to the actual process conditions [54]. Despite additional technical consideration and cost, these two setups are feasible for evaluating potential biofouling deposition and the effectiveness of using biocides in the close representation of actual conditions.

7. Conclusions

Previous biofouling assessment techniques conducted in OTEC systems showed their reliability in monitoring the realistic condition of biofouling deposition, contributed by their application in an actual OTEC system. However these methods are lack continuous and non-destructive assessment, thus minimizing evaluation on progressive evaluation in biofouling deposition. Therefore, advanced biofouling detection techniques from other water-related industries could be employed in the OTEC system, with emphasis on its feasibility for its implementation in actual OTEC process setup. In this case, the operational understanding in biofouling assessment for OTEC power plant facility is crucial for designing practical and realistic implementation of the proposed biofouling detection techniques, in this case, based on practices done in the cooling system of power plant industries. Two main parameters which are flow rate and the seawater temperature variation are considered in this study, along with factors contributed by physical design and environmental factors concerning the OTEC power plant facility. However, a thorough evaluation of biofouling assessment could only be done under realistic and continuous biofouling growth and deposition setup. In this study, laboratory setup was found to be a practical approach in assessing the capability of analytical instruments inconvenient and small-scale setup, but limited in representing realistic biofouling deposition. Therefore, the application of a side-stream sampler or pilot plant facility that is connected to an OTEC facility provides the opportunity for evaluating realistic biofouling conditions without affecting the OTEC system.

8. References

[1] R. Pelc and R. M. Fujita, "Renewable energy from the ocean," *Marine Policy,* vol. **26**, no. 6, pp. 471-479, 2002.
[2] M. L. Syamsuddin, A. Attamimi, A. P. Nugraha, S. Gibran, A. Q. Afifah, and N. Oriana, "OTEC potential in the Indonesian seas," *Energy Procedia,* vol. **65**, pp. 215-222, 2015.
[3] L. A. Vega, "Ocean thermal energy conversion primer," *Marine Technology Society Journal,* vol. **36**, no. 4, pp. 25-35, 2002.
[4] C. C. Mitigation, "IPCC special report on renewable energy sources and climate change mitigation," *Renewable Energy,* vol. **20**, no. 11, 2011.
[5] G. C. Nihous, "A preliminary assessment of ocean thermal energy conversion resources," 2007.

[6] H. Kobayashi, S. Jitsuhara, and H. Uehara, "The present status and features of OTEC and recent aspects of thermal energy conversion technologies," *National Maritime Research Institute, Japan,[Online], Available: http://www. nmri. go. jp/main/cooperation/ujnr/24ujnr_paper_jpn/Ko bayashi. pdf,* 2001
[7] H. Uehara and Y. Ikegami, "Optimization of a closed-cycle OTEC system," 1990.
[8] L. R. Berger and J. A. Berger, "Countermeasures to microbiofouling in simulated ocean thermal energy conversion heat exchangers with surface and deep ocean waters in Hawaii," *Applied and environmental microbiology,* vol. **51**, no. 6, pp. 1186-1198, 1986.
[9] J. Daly, "Hawaii About to Crack Ocean Thermal Energy Conversion Roadblocks?."
[10] H. Knight, "Oceans of power: the resurrection of ocean thermal," *New Scientist,* vol. **221**, no. 2958, pp. 48-51, 2014.
[11] S. Banerjee, L. Duckers, and R. Blanchard, "A case study of a hypothetical 100 MW OTEC plant analyzing the prospects of OTEC technology," *IN: Dessne, P. and Golmen, L.(eds). OTEC Matters,* vol. **1**, no. 98, p. 129, 2015.
[12] S. T Thirugnana, A. B. Jaafar, T. Yasunaga, T. Nakaoka, Y. Ikegami, and S. Su, "Estimation of Ocean Thermal Energy Conversion Resources in the East of Malaysia," *Journal of Marine Science and Engineering,* vol. **9**, no. 1, p. 22, 2021.
[13] R. A. Al-Juboori and T. Yusaf, "Biofouling in RO system: mechanisms, monitoring and controlling," *Desalination,* vol. **302**, pp. 1-23, 2012.
[14] H. Maddah and A. Chogle, "Biofouling in reverse osmosis: phenomena, monitoring, controlling and remediation," *Applied Water Science,* vol. **7**, no. 6, pp. 2637-2651, 2017.
[15] R. Mitchell and P. Benson, "Micro and macrofouling in the OTEC program," *An overview ANUOTEC. BLM-11, Argonne, National Labs,* 1980.
[16] H.-C. Flemming, "Microbial biofouling: unsolved problems, insufficient approaches, and possible solutions," in *Biofilm highlights*: Springer, 2011, pp. 81-109.
[17] H.-C. Flemming, P. S. Murthy, R. Venkatesan, and K. Cooksey, *Marine and industrial biofouling*. Springer, 2009.
[18] H. A. Videla and W. G. Characklis, "Biofouling and microbially influenced corrosion," *International Biodeterioration & Biodegradation,* vol. **29**, no. 3-4, pp. 195-212, 1992; C. C. de Carvalho, "Marine biofilms: a successful microbial strategy with economic implications," *Frontiers in marine science,* vol. **5**, p. 126, 2018.
[19] S. Cheah and T. Chua, "A preliminary study of the tropical marine fouling organisms on floating net cages [in Malaysia]," *Malayan Nature Journal (Malaysia),* 1979.
[20] M. Affandy, J. Madin, K. Jakobsen, and M. Auluck, "Development and succession of sessile macrofouling organisms on the artificial structure in the Shallow Coastal Waters of Sabah, Malaysia," in *IOP Conference Series: Earth and Environmental Science*, 2019, vol. **236**, no. 1: IOP Publishing, p. 012046.
[21] M. Affandy, J. Madin, K. Jakobsen, and M. Auluck, "Macrofouling Development on Artificial Structure at Karambunai Bay, Sabah Malaysia," in *Journal of Physics: Conference Series*, 2019, vol. **1358**, no. 1: IOP Publishing, p. 012011.
[22] L. A. Vega, "Ocean thermal energy conversion," *Encyclopedia of Energy Technology and the Environment,* vol. **3**, pp. 2104-2119, 1995.
[23] C. Bernardoni, M. Binotti, and A. Giostri, "Techno-economic analysis of closed OTEC cycles for power generation," *Renewable energy,* vol. **132**, pp. 1018-1033, 2019.
[24] F. Chen, L. Liu, J. Peng, Y. Ge, H. Wu, and W. Liu, "Theoretical and experimental research on the thermal performance of ocean thermal energy conversion system using the rankine cycle mode," *Energy,* vol. **183**, pp. 497-503, 2019.
[25] X. Zhang, M. He, and Y. Zhang, "A review of research on the Kalina cycle," *Renewable and sustainable energy reviews,* vol. **16**, no. 7, pp. 5309-5318, 2012.
[26] H. Link and B. Parsons, "Potential of proposed open-cycle OTEC experiments to achieve net power," in *OCEANS'86*, 1986: IEEE, pp. 207-212. ; B. Parsons, D. Bharathan, and J. Althof,

"Open cycle OTEC thermal-hydraulic systems analysis and parametric studies," in *OCEANS 1984*, 1984: IEEE, pp. 370-377.

[27] A. Thomas and D. L. Hillis, "First production of potable water by OTEC and its potential applications," in *OCEANS'88.'A Partnership of Marine Interests'. Proceedings*, 1988: IEEE, pp. 1045-1048.

[28] A. Etemadi, A. Emdadi, O. AsefAfshar, and Y. Emami, "Electricity generation by the ocean thermal energy," *Energy Procedia,* vol. **12**, pp. 936-943, 2011.

[29] M. Syed, G. Nihous, and L. Vega, "Use of cold seawater for air conditioning," in *OCEANS 91: ocean technologies and opportunities in the Pacific for the 90's, October 1-3, 1991, Honolulu HI*, 1991.

[30] J. Herrera, S. Sierra, and A. Ibeas, "Ocean Thermal Energy Conversion and Other Uses of Deep Sea Water: A Review," *Journal of Marine Science and Engineering,* vol. **9**, no. 4, p. 356, 2021.

[31] R. Magesh, "OTEC technology–a world of clean energy and water," in *Proceedings of the World Congress on Engineering*, 2010, vol. **2**: WCE London, UK, pp. 1-6.

[32] Y. Ikegami, "Activity and Future Prospect status of Ocean Thermal Energy Conversion-for sustainable energy and water resource," in *Preceedings of the 13th Asian Congress of Fluid Mechanics*, 2010, pp. 1183-1186.

[33] A. Pacific and C. Energy, "OTEC Heat Exchangers: Performance and Power Output for Historical and Modern Designs," 2013.

[34] J. Fetkovich, G. Grannemann, L. Mahalingam, D. Meier, and F. Munchmeyer, "Studies of biofouling in ocean thermal energy conversion plants," Carnegie-Mellon Univ., Pittsburgh, PA (USA), 1977.

[35] J. F. Jenkins, "Corrosion and Biofouling of OTEC System Surfaces-Design Factors," Civil Engineering Lab (Navy) Port Hueneme CA, 1978.

[36] R. P. Aftring and B. F. Taylor, "Assessment of microbial fouling in an ocean thermal energy conversion experiment," *Applied and Environmental Microbiology,* vol. **38**, no. 4, pp. 734-739, 1979.

[37] C. Panchal and J. Knudsen, "Mitigation of water fouling: technology status and challenges," *Advances in Heat Transfer,* vol. **31**, pp. 431-474, 1998.

[38] J. Fetkovich, G. Grannemann, D. Meier, and C. Fette, "Novel method of measuring heat transfer coefficients with high precision," Carolina Power and Light Co., Raleigh, NC (USA); Pennsylvania State Univ ..., 1976.

[39] D. S. Sasscer, T. Morgan, T. R. Tosteson, and G. N. Grannemann, "In Situ Biofouling of Ocean Thermal Energy Conversion (OTEC) Evaporator Tubes," 1981.

[40] N. Farhat, M. Staal, A. Siddiqui, S. Borisov, S. S. Bucs, and J. S. Vrouwenvelder, "Early non-destructive biofouling detection and spatial distribution: application of oxygen sensing optodes," *Water research,* vol. **83**, pp. 10-20, 2015.

[41] E. Nebot *et al.*, "Marine biofouling in heat exchangers," *Biofouling: types, impact and anti-fouling. New York (NY): Nova Science,* pp. 65-104, 2010.

[42] P. S. Murthy, R. Venkatesan, K. Nair, and M. Ravindran, "Biofilm control for plate heat exchangers using surface seawater from the open ocean for the OTEC power plant," *International biodeterioration & biodegradation,* vol. **53**, no. 2, pp. 133-140, 2004.

[43] T. Nguyen, F. A. Roddick, and L. Fan, "Biofouling of water treatment membranes: a review of the underlying causes, monitoring techniques and control measures," *Membranes,* vol. **2**, no. 4, pp. 804-840, 2012.

[44] D. Nivens, R. Palmer Jr, and D. White, "Continuous nondestructive monitoring of microbial biofilms: a review of analytical techniques," *Journal of industrial microbiology and biotechnology,* vol. **15**, no. 4, pp. 263-276, 1995.

[45] J. T. Staley, "Growth Rates Of Algae Determined In-Situ Using An Immersed Microscope 1," *Journal of Phycology,* vol. **7**, no. 1, pp. 13-17, 1971.

[46] L. Fortunato, S. Bucs, R. V. Linares, C. Cali, J. S. Vrouwenvelder, and T. Leiknes, "Spatially-resolved in-situ quantification of biofouling using optical coherence tomography (OCT) and 3D image analysis in a spacer filled channel," *Journal of Membrane Science,* vol. **524**, pp. 673-681, 2017.
[47] G. Wolf, J. G. Crespo, and M. A. Reis, "Optical and spectroscopic methods for biofilm examination and monitoring," *Reviews in Environmental Science and Biotechnology,* vol. **1**, no. 3, pp. 227-251, 2002.
[48] H.-C. Flemming, A. Tamachkiarowa, J. Klahre, and J. Schmitt, "Monitoring of fouling and biofouling in technical systems," *Water science and technology,* vol. **38**, no. 8-9, pp. 291-298, 1998.
[49] M. Kögler *et al.*, "Real-time Raman based approach for identification of biofouling," *Sensors and Actuators B: Chemical,* vol. **230**, pp. 411-421, 2016.
[50] E. Fridjonsson, S. Vogt, J. S. Vrouwenvelder, and M. Johns, "Early non-destructive biofouling detection in spiral wound RO membranes using a mobile earth' s field NMR," *Journal of Membrane Science,* vol. **489**, pp. 227-236, 2015.
[51] J. Song *et al.*, "Enhanced Electrochemical Impedance Spectroscopy Analysis of Microbial Biofilms on an Electrochemically In Situ Generated Graphene Interface," *ACS sensors,* vol. **5**, no. 6, pp. 1795-1803, 2020.
[52] P. Janknecht and L. F. Melo, "Online biofilm monitoring," *Reviews in Environmental Science and Biotechnology,* vol. **2**, no. 2, pp. 269-283, 2003.
[53] D. I. Walker and C. W. Keevil, "Low-concentration diffusible molecules affect the formation of biofilms by mixed marine communities," *Cogent Biology,* vol. **1**, no. 1, p. 1103830, 2015.
[54] T. R. Bott, *Industrial biofouling*. Elsevier, 2011.
[55] T. Bott and L. Tianqing, "Ultrasound enhancement of biocide efficiency," *Ultrasonics Sonochemistry,* vol. **11**, no. 5, pp. 323-326, 2004.
[56] W. H. Avery and C. Wu, *Renewable energy from the ocean: a guide to OTEC*. Oxford University Press, 1994.
[57] J. S. Teodósio, M. Simoes, M. A. Alves, L. F. Melo, and F. J. Mergulhao, "Setup and validation of flow cell systems for biofouling simulation in industrial settings," *The Scientific World Journal,* vol. **2012**, 2012; J. Bryers and W. Characklis, "Early fouling biofilm formation in a turbulent flow system: overall kinetics," *Water research,* vol. **15**, no. 4, pp. 483-491, 1981.
[58] S. Rajagopal, H. A. Jenner, and V. P. Venugopalan, *Operational and environmental consequences of large industrial cooling water systems*. Springer Science & Business Media, 2012.

Acknowledgements

The present study was made possible through a funding from UMP-IIUM-UiTM Sustainable Research Collaboration 2020 grant (600-RMC/SRC/5/3 (028/2020)) and continuous support from School of Chemical Engineering and Biocatalysis and Biobased Material Technology Research Group, Universiti Teknologi MARA are gratefully acknowledged.

Wind, hydro and solar energy challenges for Technical Vocational and Training (TVET) electrical entrepreneur in Malaysia: A review

G Harun[1], S Sarip[2], A Y Abdul Fatah[2], H M Kaidi[2] and N Abd Rahim[2]

[1]Kolej Kemahiran Tinggi MARA, Majlis Amanah Rakyat, Malaysia
[2]Razak Faculty of Technology and Informatics, Universiti Teknologi Malaysia, Malaysia

*Corresponding author's email : shamsuls.kl@utm.my

Abstract. Renewable energy is one of the alternative energy resources in Malaysia to replace fossil fuel use, which is an important issue that needs to be established. Some of the possible renewable energy sources are wind, hydro and solar. Since 2019, various incentives announced by the Malaysian Ministry of Energy and Natural Resources (KeTSA) to enhance renewable energy development which is in line with Government Energy policy. However, today, in contrast to the participation of investors and international contractors, the presence of entrepreneurs and domestic workers in this sector is feeble. In this regard, the Technical Vocational and Training (TVET) institution is seen to have the potential to minimize this crisis by creating competent, skilled and competitive electrical entrepreneurs for the field of renewable energy. This paper explores the ability of TVET electrical entrepreneurs to participate in renewable energy businesses. Based on the literature on energy entrepreneur development, it was found that the TVET electrical entrepreneur faces four challenges, namely financial, technology costs, logistics and government support. The proposed future development of renewable energy is in Mini-hydro and solar photovoltaic (PV), while wind power does not seem viable to TVET electricity entrepreneurs.

1. Introduction

The world's current needs for affordable, renewable energy (RE) have been growing and taking the path of sustainable development, decarburization, and climate security [1]. Unlike in previous years, today's policymakers and regulators need a different market and economic context, including preparing for RE development. Wind, hydro and solar (WHS) REs has the potential to be developed. These resources are available in specific areas in Malaysia which must take into government consideration.

In 2018, Malaysia announced that it had set a target of 20% RE in its generation mix by 2025 [2]. The Southeast Asian nations, which thermal power plants mainly power, recorded a two per cent contribution from RE to its generation mix at the end of 2018. To achieve the target of 20% in 2025, it is estimated that Malaysia will need RM33 billion (US$8 billion) worth of investments in its RE sector[3]. The expected investments would come from the government and public-private partnerships and private financing [2] and [4].

There is an opportunity where young electrical entrepreneurs can take part RE development. However, knowledge and skills about RE be strengthened in their studies so that the challenges in entrepreneurship can be overcome well without directly affecting the investors and government [5]. The government should promote efforts to grow technical entrepreneurs in the field of RE through Technical

Vocational Education Training (TVET) institutions or any technical institution that will produce highly qualified technical entrepreneurs in preparation for 20% RE generation by 2025[6].

It was noticed that the graduates of TVET institutions still lack the capabilities to be an electrical entrepreneur in the region. To be exact, just 12% of the total number of TVET institutions in Malaysia is pursuing the fields of WHS [7]. This shows that the students in the electrical field are less ready to venture into electrical entrepreneurship, especially in renewable energy after graduation. Similar problems also have been observed in which the graduates have the entrepreneurial potential but lacked entrepreneurial knowledge, confidence, creativity and innovation in facing global challenges [8].

This study aims to identify the current issues on the development of RE in Malaysia and the challenges faced by TVET electrical entrepreneur to involve, participate, and get into the real business of WHS RE.

2. Overview of Sustainable Energy

2.1. Renewable Energy (RE) Development in Malaysia

As a developing country, Malaysia made a quick response to global warming with the initiation of RE development campaign in the National Energy Policy by setting a target of 5% for RE generation [2]. Commercial energy consumption in Malaysia has gradually risen and is increasing at the pace of national economic growth. Economic growth, industrialization and urbanization have been showing to be directly related to commercial energy use and to the consumption of energy products. Based on the Energy Malaysia Magazine in 2017, it showed that the total energy consumption in Malaysia had reached 21,773 MW and energy generated from various generation plants either from conventional thermal coal, gas or hydro were at 28,324 MW [2]. It is clearly found that the reserve margin of total energy is only 23.13%.

In Malaysia, several potentials RE resources have been developed, where the most considerable renewable energy came from solar, hydro and biomass. However, according to the 11th Malaysia Plan (2016-2020), the Malaysian government under the Ministry of Energy and Natural Resources (KeTSA) had launched the strategy and action plan that was aimed to achieve a target for RE of 7.8% or 2080 MW of the total electricity generation in 2020[2]. Hence, the incentives and encouragements through programmes such as Feed-in Tariff (FiT), Large Scale Solar (LSS) and Net Energy Metering (NEM) were introduced [9].

In December 2020, KeTSA again introduced NEM 3.0 programme to provide the opportunity for users to install the Solar Photovoltaic (PV) system into their premises [10]. This initiative is going to have an impact on the reduction of electricity bills, thus enhances the development of RE in this country. The quota of 500 Megawatt (MW) will be distributed to the consumer who is interested start from February 2021 until December 2023 to achieve the target of 20% RE electricity generation 2030[3],[10]

2.2. Wind, Hydro and Solar (WHS) in Malaysia

Another form of potential RE in Malaysia is wind power. Compared to other windy countries, Malaysia can be classified as a low wind speed country. In Malaysia, heavy winds are usually blowing from the Indian Ocean and the South China Sea. The mean monthly wind velocity is between 1.5 and 4.5 m/s. Areas at higher altitudes can harness wind power from 9 to 11 m/s. In Peninsular Malaysia, Mersing, Johor and Kuala Terengganu have been listed as high wind regions, while in the East Malaysia Kudat, Sabah recorded the highest. Generally, due to the monsoon season, the southern part of peninsular Malaysia has more wind capacity and more wind in January [4]. Recently, several wind powers developed in Malaysia but still in assessment to determine the wind energy capability for main electric generation. Table 1 shows the study on windmills that have been installed since 1995, where Pulau Terumbu, Sabah has the highest expected power generated with 150 kilowatts (kW) and Setiu, Terengganu has the smallest windmills with 3 kW based on the research by Universiti Malaysia Terengganu[11].

Table 1. The generation of wind energy in Malaysia [11].

Location	Power Generated (kW)	Owned by
Pulau Terumbu, Sabah	150	Tenaga Nasional
Pulau Perhentian, Terengganu	100	Tenaga Nasional
Kudat, Sabah	25	Tenaga Nasional
Setiu Terengganu	3	Universiti Malaysia Terengganu

Malaysia also has great potential from its natural hydropower resources. As Malaysia has 189 rivers with a total length of approximately 57,300 km, this is primarily due to its unique geographical advantage. In 2016, hydropower generation contributed to 13% of Malaysia's total power generation. Most of the large hydroelectric plants are owned by Tenaga Nasional Berhad (TNB), and two more plants under construction in Nenggiri (300 MW) and Tekai (156 MW) in the state of Kelantan. Small or mini-hydropower stations below 30 MW operate in a liberalized market, which means that they are open to any private developer and are mainly incentivized by the Sustainable Energy Development Authority (SEDA) in Malaysia Feed-in Tariff (FiT) scheme. Currently, these mini-hydropower stations contributed up to 200 MW of Malaysian energy generation [12].

At the initial stage of the deployment of wind and mini-hydropower in Malaysia, support and incentives from the government are needed in capital-intensive investment. Unlike solar, FiT incentive is still available for mini-hydro power and wind power plant [13]. However, the low wind speed condition is restricting the generation of wind energy, where suitable technologies for low wind conditions should be explored to harness wind power's untapped potential. Figure 1 shows the forecast mix of electricity generation until the year 2026. It was found that the coal energy generation will be decreased, and the presence of RE will take about 19% of the total electricity generated.

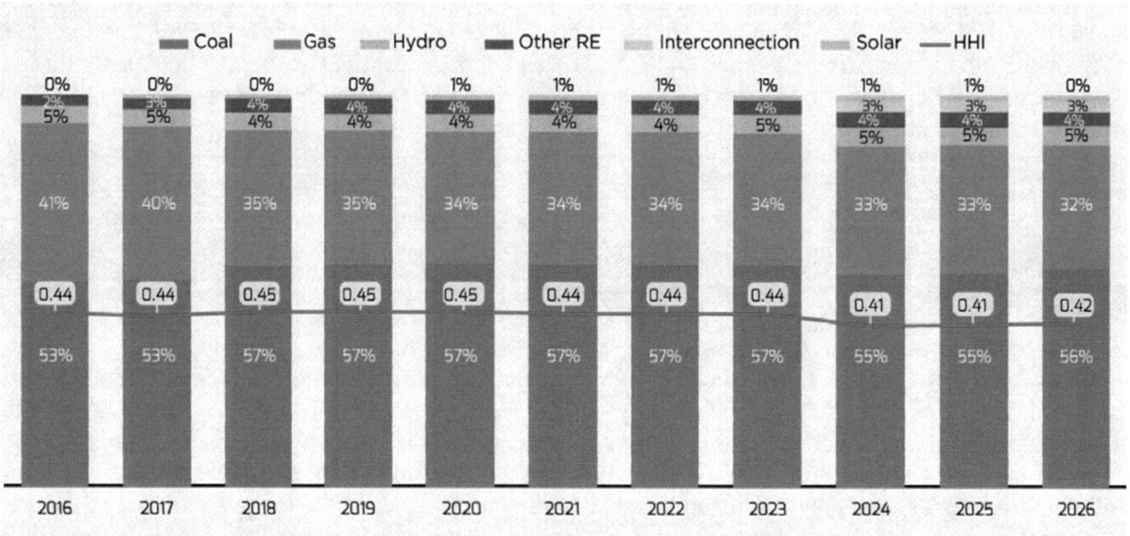

Figure 1. Generation mix in Peninsular Malaysia[14].

As a critical element in RE development, solar play a significant role in the RE programs initiated by the government, especially KeTSA. In order to encourage the generation of solar photovoltaic (PV) more widely, incentive schemes such as NEM, LSS and FiT were announced due to the high resource potential of solar in Malaysia [citation needed]. The low impact in initial costs and healthy market price competitions makes the development of solar PV energy generation ahead of other RE resources. Generally, the federal government is still struggling to achieve their target of 20% RE electricity generation in the year 2025 despite various incentives provided [3]. The involvement of local companies in the development of RE generation is one of the issues in the solar PV implementation. This is due to the fact that solar technology was owned by foreign companies. Thus the role of TVET electrical entrepreneur is required to drive the production of RE in accordance with the country's aspirations and desires.

2.3. Challenges for TVET Electrical Entrepreneur

The concept of "entrepreneur" or "entrepreneurship" was never consistent among economists (the word "entrepreneur" comes from the French verb entreprendre, meaning "to undertake") [15]. The classical and neoclassical economists left entrepreneurs out of their formal models, although the idea of an entrepreneur existed and was recognized for centuries: they believed that perfect knowledge would be known to entirely rational actors, leaving little space for risk-taking or experimentation. Economists did not actively integrate entrepreneurship into their models until the middle of the 20th century [16].

In Malaysia, the entrepreneurial sector is seen as the best mechanism for reducing the unemployment rate among young people, especially among Technical Vocational Education and Training (TVET) graduates [17]. Entrepreneurs can generate economic growth for the nation by serving as job producers. It is common knowledge that Malaysian young people, especially the TVET graduates, are currently facing a tough challenge to find a place in their career selection [18]. This is due to the increase of employer demands and career challenges, where graduates are evaluated in terms of academic credentials and skills and other value-added abilities. In developed countries, they are facing problems to absorb the rising number of graduates, where the employment system is seen as having constant pressure on the existing labour market.

The electrical field is one of the sectors that impacting the development of engineering entrepreneurship, where every single engineering fields nowadays must have the elements of electricity. As an electrical entrepreneur, the ability to overcome a new technology is the prior requirement. TVET graduates usually have the basis of high-skilled workers and problem-solving analysts, where strong technical values and productivities are loaded within their curriculum [19]. With the dawn of the IR 4.0 era, electrical entrepreneurs face new challenges that require them to get out of their normal practices to adapt to the advancement of new technologies such as the Internet of Things (IoT), Big Data analysis and the use of Virtual Reality (VR) for additional tools in the installation and maintenance purposes [20].

The development of RE from WHS indicates a gap to be fulfilled by the TVET electrical entrepreneurs. With the announced incentives by the federal government, TVET electrical entrepreneurs must participate and get involved in their fields. The elements and needs of WHS-based RE are needed to be exposed in their studies [16]. However, one of the problems faced by these entrepreneurs is the culture of the business that restricted them to explore more [21]. It is found that four challenges are faced by TVET electrical entrepreneurs that limit them in starting up a new business area, as shown in Table 2.

Table 2. Challenges faced by Technical Vocational Education and Training (TVET) electrical entrepreneur into RE business [21][22].

Challenges	Description
Financial	1. Difficulty obtaining credit with inadequate provision and access to official finance. 2. Secure the investor's interest.
Skills & knowledge	1. Electrical graduate's lack of skills related to RE business 2. Insufficient entrepreneurship education in the curriculum.
Prices of the technology	1. Failure to compete with fossil energy service or suppliers. 2. Lack of research and development transfer

Physical infrastructure and logistic	1. Lack of infrastructure and potential areas. 2. High cost on logistics due to no access in remote area
Governance support	1. Lack of policy, legislative and tax incentives. 2. Lack of enforcement on property rights, contracts

Since TVET electrical entrepreneurs have fundamental skills and proper operating methods, the government and related agencies should pay particular attention to their ability to adapt to solar technologies. However, the limited experiences and exposure in the RE business needed to be evaluated. Based on the challenges presented in Table 2, external variables must be considered to empower the involvement of TVET electrical entrepreneurs for RE generation in Malaysia [23].

3. Result and Discussion

Based on the literature review, it is found that the role of electrical entrepreneurs, especially the Technical Vocational and Training (TVET) graduates in RE generation, is very tough, whereas the potential of the three significant resources of WHS is different [7]. Wind energy generation in Malaysia faced the problem of low wind condition [11]. Several unsuccessful developments or the failure of wind turbines generator demonstration projects, such as the ones at Swallow Reef and Small Perhentian Island, indicated that it is not easy to build RE from wind power capacity in Malaysia, which is located in an equatorial area with low wind speed, seasonally variable monsoons and even inter-monsoon cycles with minimal wind and high humidity. From TVET institutions, there is a shortage of information and skills to be provided to the technical student. The number of RE plants is too limited for experiential learning on installation or maintenance.

It differs from mini-hydro generation, where the electricity generation technology is still slowly moving without rapid changes [24]. Therefore, TVET institutions are not faced with channelling the latest information on the generation, maintenance, associated facilities and even training facilities process. Furthermore, the incentives advertised by the government still available with long term agreements that attracted investors to involve with the development. Thus, TVET electrical entrepreneurs must cater to the vast installation and maintenance projects [25]. Figure 2 illustrates that the mini-hydro generation will always be available throughout the year as one of the component RE to achieved 20% RE generation goal.

The most potential renewable energy (RE) development from the literature review is from solar photovoltaic (PV), where the incentive from the government is desirable. With the solar PV equipment prices decreased drastically due to many resources throughout the world, especially from China, this leads to the rapid development of RE generation from solar PV [21] and [25]. Government energy policy helped to enhance this by providing special incentives starting from February 2021 until December 2023 with Net Energy Metering (NEM) 3.0 and Large Solar Scale (LSS) 4.0 with the quota of 1500 Megawatt (MW). Table 3 indicates that the demand for LSS increases year by year with the reduction of cost RM per kilowatt hour [9].

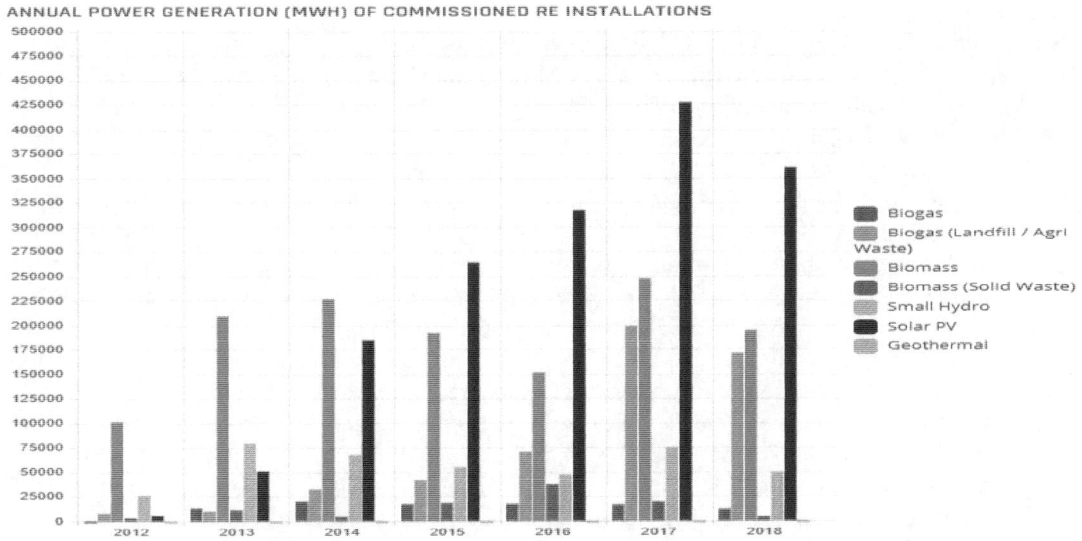

Figure 2. Renewable energy (RE) under FiT refer the Energy Commission updates [2].

Table 3. Total quota Large Solar Scale (LSS) awarded in Malaysian RE development [26].

Scheme	Year Announced	Quota (MW)	Award (MW)	*RM/kWh*	Project Commercial
LSS 1	2016	250	450	0.43	2018
LSS 2	2018	500	563	0.42	2020
LSS 3	2020	500	490	0.33	2022
LSS 4	2021	1000	-	0.17	2023

Referring to Table 3, TVET electrical entrepreneurs are seen to have a brilliant opportunity to penetrate the business market in solar PV with the highest quota offer announced by the government to achieve the target of 20% RE in 2025. TVET institutions in Malaysia need to be aware of the rise in solar PV growth, where skilled solar workers, particularly entrepreneurs, need to be equipped with sufficient knowledge to ensure that they are present as a passenger of technology and as a driver of technology [21].

4. Conclusion

In conclusion, the main challenges faced by TVET electrical entrepreneurs are financial support, government tax and policies, and the lack of skilled workers in the field of RE. With the ambitious goal of RE generation development announced and attractive incentives schemes, it will surely enhance the development of mini-hydro and solar PV and the development of skilled workers and entrepreneurs from the TVET institutions. Furthermore, a more comprehensive regulatory framework must be defined to establish a foundation upon which governance values and mechanisms can be imposed to support the systematic growth of the energy market and its associated consumers and industries.

Nowadays, TVET electrical entrepreneur is still unable to meet the government's policy requirements for the development of RE by 2024 because the challenges discussed above require support from TVET institutions and industry players and investors. An excellent incentive announced, the dream of TVET electrical entrepreneur development in the field of RE will undoubtedly be realised in the future.

5. References

[1] Suruhanjaya Tenaga, *Handbook Malaysia Energy Statistics*. 2019.

[2] S. S. Ahmad, C. Yap, M. Sueharti, Z. Mali, S. Mansor, and M. S. A. A. Rahim, "Energy Malaysia," *Suruhanjaya Tenaga*, vol. 12, pp. 10–13, 2017.

[3] M. Vaka, R. Walvekar, A. K. Rasheed, and M. Khalid, "A review on Malaysia's solar energy pathway towards carbon-neutral Malaysia beyond Covid'19 pandemic," *Journal of Cleaner Production*, vol. 273, p. 122834, 2020, doi: 10.1016/j.jclepro.2020.122834.

[4] M. A. Hannan, R. A. Begum, M. G. Abdolrasol, M. S. Hossain Lipu, A. Mohamed, and M. M. Rashid, "Review of baseline studies on energy policies and indicators in Malaysia for future sustainable energy development," *Renewable and Sustainable Energy Reviews*, vol. 94, no. August 2017, pp. 551–564, 2018, doi: 10.1016/j.rser.2018.06.041.

[5] A. Kasych, Z. Rowland, and O. Onyshchenko, "Evaluating Effectiveness of the Business Models of Electrical and Electronic Engineering Companies: Monitoring Methods, Experience and Characteristics," *Proceedings of the International Conference on Modern Electrical and Energy Systems, MEES 2019*, pp. 494–497, 2019, doi: 10.1109/MEES.2019.8896642.

[6] W. Syakirah, W. Abdullah, M. Osman, M. Zainal, and A. Ab, "The Potential and Status of Renewable Energy," *energies MDPI*, vol. 12, no. 2437, pp. 1–16, 2019.

[7] N. Asykin, M. S. Rasul, and N. Othman, "Teaching Strategies to Develop Technical Entrepreneurs," *International Journal of Innovation, Creativity and Change. www.ijicc.net*, vol. 7, no. 6, pp. 179–188, 2019, [Online]. Available: www.ijicc.net.

[8] H. Hasbolah, S. A. Mamat, Z. Abdullah, and S. Sidek, "A Review on Cyber Entrepreneurship In Malaysia: Past, Present And Future," *Journal Of Critical Reviews*, vol. 7, no. 14, pp. 2628–2636, 2020.

[9] Suruhanjaya Tenaga, *Guidelines On Large Scale Solar Photovoltaic Plant For Connection to Electricity Networks*, vol. 2016, no. 1–7. 2019.

[10] Bernama, "The Edge Market-KeTSA introduces three initiatives to encourage use of solar PV," *The Edge Markets*, pp. 1–5.

[11] A. Zaharim, S. Mat, K. Sopian, A. Jedi, N. Masseran, and M. A. M. Safari, "The impact of wind technology among rural community in Mersing Malaysia," *Energy and Sustainability in Small Developing Economies, ES2DE 2017 - Proceedings*, 2017, doi: 10.1109/ES2DE.2017.8015348.

[12] T. H. Oh, M. Hasanuzzaman, J. Selvaraj, S. C. Teo, and S. C. Chua, "Energy policy and alternative energy in Malaysia: Issues and challenges for sustainable growth – An update," *Renewable and Sustainable Energy Reviews*, vol. 81, no. May 2016, pp. 3021–3031, 2018, doi: 10.1016/j.rser.2017.06.112.

[13] M. Hossain *et al.*, "A state-of-the-art review of hydropower in Malaysia as renewable energy: Current status and future prospects," *Energy Strategy Reviews*, vol. 22, no. November, pp. 426–437, 2018, doi: 10.1016/j.esr.2018.11.001.

[14] K. H. Wong, "The future of the sun continues to shine in Malaysia: Outlook and prospects on the solar photovoltaic industry," *Asia Law Portal*, p. 8, 2018, [Online]. Available: https://asialawportal.com/2018/10/22/the-future-of-the-sun-continues-to-shine-in-malaysia-outlook-and-prospects-on-the-solar-photovoltaic-industry/.

[15] D. Dahalan, J. L. D'Silva, I. A. Ismail, and N. A. Mohamed, "Entrepreneurial mindset among students of Technical and Vocational Education and Training (TVET) institutions in Malaysia," *Journal of Social Sciences Research*, vol. 4, no. 11, pp. 303–311, 2018, doi: 10.32861/jssr.411.303.311.

[16] S. Saad, A. S. M. M. Hoque, and Z. Awang, "Technopreneurial Marketing (TM): A Construct for Integrating Emerging Technopreneurship and Marketing Perspectives," *Proceeding of International Seminar of Entrepreneurship and Business 2019*, no. November, 2019.

[17] A. Ismail, W. N. Adnan, A. Masek, R. Hassan, S. Hashim, and M. E. Ismail, "Effectiveness of entrepreneurship programmes in developing entrepreneurship skills towards quality TVET graduates," *Journal of Technical Education and Training*, vol. 11, no. 1, pp. 81–86, 2019, doi: 10.30880/jtet.2019.11.01.10.

[18] F. W. P. Vanduhe, V. v Apagu, and B. A. Wakili, "Technical Vocational Education and Training (TVET) and Entrepreneurship Training in Yobe State Technical Colleges for Sustainable Development," *ATBU Journal of Science ...*, vol. 8, no. 1, pp. 1–8, 2019, [Online]. Available: http://www.atbuftejoste.com/index.php/joste/article/view/901.
[19] N. H. Mohamad, A. Selamat, B. Ibrahim, and B. M. Salleh, "Exploration of spiritual elements in holistic-entrepreneur (Holistic-e) among TVET graduate students," *Journal of Technical Education and Training*, vol. 11, no. 3, pp. 73–83, 2019, doi: 10.30880/jtet.2019.11.03.010.
[20] V. Veleva, "The role of entrepreneurs in advancing sustainable lifestyles: Challenges, impacts, and future opportunities," *Journal of Cleaner Production*, vol. 283, p. 124658, 2021, doi: 10.1016/j.jclepro.2020.124658.
[21] C. A. Gabriel, "What is challenging renewable energy entrepreneurs in developing countries?," *Renewable and Sustainable Energy Reviews*, vol. 64, pp. 362–371, 2016, doi: 10.1016/j.rser.2016.06.025.
[22] C. A. Gabriel and J. Kirkwood, "Business models for model businesses: Lessons from renewable energy entrepreneurs in developing countries," *Energy Policy*, vol. 95, pp. 336–349, 2016, doi: 10.1016/j.enpol.2016.05.006.
[23] H. Saleh, "Employer satisfaction with engineering graduates employability: A study among manufacturing employers in Malaysia," *International Journal of Scientific and Technology Research*, vol. 8, no. 9, pp. 813–817, 2019.
[24] M. Pang, L. Zhang, A. B. S. Bahaj, K. Xu, Y. Hao, and C. Wang, "Small hydropower development in Tibet: Insight from a survey in Nagqu Prefecture," *Renewable and Sustainable Energy Reviews*, vol. 81, no. June 2017, pp. 3032–3040, 2018, doi: 10.1016/j.rser.2017.06.115.
[25] M. R. Shaharudin, A. I. Zainoddin, D. Abdullah, C. Hotrawaisaya, H. Soonthornpipit, and N. Norddin, "Factors that influence the green purchasing practices among suppliers of electrical components," in *AIP Conference Proceedings*, Oct. 2018, vol. 2020, doi: 10.1063/1.5062692.
[26] "Power Updates from Minister's session, MIDF Research," *MIDF Amanah Investment Bank Berhad*, no. March, 2019.

Acknowledgement
This work is supported by the Industry-International Incentive Grant (Vote: Q.K130000.3601.03M09) scheme Universiti Teknologi Malaysia and collaborates with UTM Ocean Thermal Energy Centre.

Computational Fluid Dynamics Model of Wells Turbine for Oscillating Water Column System: A Review

Abdul Settar, N.[1,2], Sarip, S.[1]* and Kaidi, H.M.[1]

[1] Razak Faculty of Technology and Informatics, Universiti Teknologi Malaysia, Malaysia
[2] Faculty of Engineering and Life Sciences, Universiti Selangor, Malaysia

***Corresponding author: shamsuls.kl@utm.my**

Abstract. Wells turbine is an important component in the oscillating water column (OWC) system. Thus, many researchers tend to improve the performance via experiment or computational fluid dynamics (CFD) simulation, which is cheaper. As the CFD method becomes more popular, the lack of evidence to support the parameters used during the CFD simulation becomes a big issue. This paper aims to review the CFD models applied to the Wells turbine for the OWC system. Journal papers from the past ten years were summarized in brief critique. As a summary, the FLUENT and CFX software are mostly used to simulate the Wells turbine flow problems while SST k-ω turbulence model is the widely used model. A grid independence test is essential when doing CFD simulation. In conclusion, this review paper can show the research gap for CFD simulation and can reduce the time in selecting suitable parameters when involving simulation in the Wells turbine.

1. Introduction

Ocean energy is among the primary sources of renewable energy other than Solar, Wind, Hydroelectric, Biomass, Geothermal [1]. Under ocean energy, wave energy is considered energy with a high potential as a renewable energy, but it is mainly untapped [2]. The main benefit of wave energy is having high energy density [3], high source availability [4], high load factor [3], source predictability compared to other renewable energy sources [3] and environmentally friendly [3]–[5]. A machine that absorbs wave energy and converts it into useful energy is called a wave energy converter (WEC).

The oscillating water column (OWC) is a wave energy converter (WEC) system with a simple working principle but has good efficiency. When the wave is at the crest, water will go into the water chamber and increase the water column. This water column will push and compress the air in the air chamber. The compressed air located in the air chamber then will push the blade at the air turbine, thus generating power. This is called exhalation process. Meanwhile, for the inhalation process, as the wave is at a trough, the water column will decrease, and the air will expand again. Even though the air is expanded, there will be no reverse rotation of the blade turbine, as it was designed to move only in one direction. This cycle is repeated continuously as wave crest and trough and will generate power. There are several types of turbines used in this OWC system.

The air turbine is the heart of the OWC system. Wells and Impulse turbines are the most common turbines used to compare to the cross-flow, Savonius [6], Hanna [7], or even Denniss-Auld [8] types. Although the operational flow range of the impulse turbine is wider than the Wells turbine, its peak efficiency hardly exceeds about 50 %, the main reason why the study focused more on the Wells turbine [8].

The early development of the Wells turbine created by Arthur Alan Wells in 1976 was based on the researchers' experiments, which was considered time and cost-consuming. Thus, in brief, this paper summarises that the time taken to select appropriate parameters in doing future CFD simulations can be reduced.

2. Wells Turbine

Wave energy converter can be divided into several types as shown in Figure 1. Includes fixed structure, floating, submerged, and floating structure with concentration i.e. wave dragon. All these types of WECs described have the unique advantage of producing the same function and purpose. Figure 2 shows the typical arrangement of the OWC system with the working principle of the system [6]. Inside the air chamber contains a very important device in a WEC which is a turbine. The Wells turbine as shown in Figure 3(a) was placed at the hub which is connected to a shaft. This shaft is responsible for rotating the generator to generate the power output. The Wells turbine was placed in an enclosure casing. The turbine can only rotate in one direction regardless of the air oscillating flow direction. This is due to the symmetrical blade profile used in the Wells turbine. The symmetrical blade profile usually belongs to the NACA 00XX series.

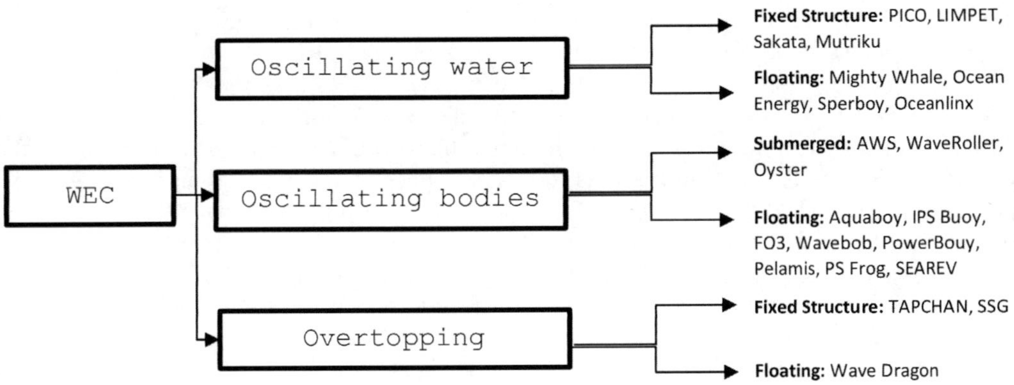

Figure 1. The various wave energy technologies [9]

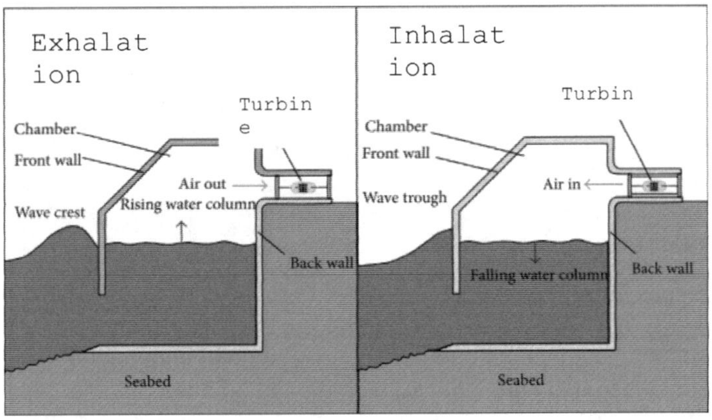

Figure 2. Typical schematic of an OWC wave energy converter [10]

Several types of Wells turbines were investigated by the researchers [8], [11]–[14]. The simplest is the single-plane Wells turbine as shown in Figure 3(a). This type of turbine just has one set of Wells turbines on one plane. The other type of Wells turbine is a biplane Wells turbine, which means that the turbine is arranged in two planes. Wells turbine with guide vanes is another type of Wells turbine as shown in Figure 3(b). The guide vanes can be placed for single plane and biplane Wells turbine. As for the biplane Wells turbine, the guide vanes can be placed in the outer or intermediate direction of the turbine. The biplane with both turbines moving in different directions is called a contra-rotating Wells turbine. The least concerned Wells turbine is the self-pitching Wells turbine, which means that the turbine does not have a fixed pitch angle, but the angle is set free to follow the direction of the oscillating flow. The fundamental theory for each of these types was mentioned briefly by Falcão and Gato [11].

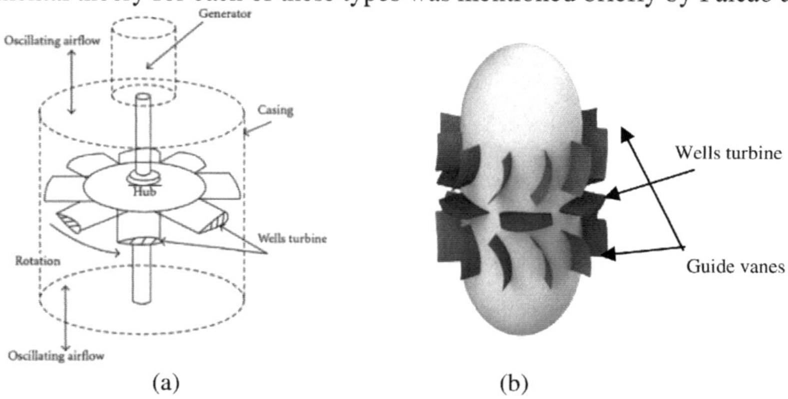

Figure 3. (a) Standard single plane Wells turbine [15] and (b) single plane Wells turbine with guide vanes [16]

Optimizing the Wells turbine is focused on the blade of the Wells turbine. There are eight important parameters in optimizing the Wells turbine which are the blade profile (thickness ratio), blade solidity, blade pitch angle, blade end plate, blade aspect ratio, blade sweep, blade tip clearance and blade tip ratio [17]. According to Das, Halder, and Samad, blade solidity and tip clearance are the most important parameters among the many parameters that influence the performance of the Wells turbine [6].

The main concern of optimizing the Wells turbine is to maximize the performance of the turbine. Power output, flow rate, operational range, efficiency and pressure drop are the main parameters used as an indicator for Wells turbine performance [17]. These factors relate to each other. Thus, the most important parameter should be focused on when designing and optimizing the Wells turbine. In most cases, having a wide operating range is preferable compared to the efficiency of the turbine [6].

3.Methodology

Journals on CFD simulation on the air and hydraulic Wells turbine in the past ten years have been selected. Theses and reports were not considered in the selection. From the selected journals, the turbulence model used with the corresponding Reynolds number together with the number of elements used during the simulation are tabulated in Table 1. The mesh and type of topology used during the simulation were placed at the last column of Table 1.

4. Discussion

Before running CFD simulation in the software, important settings should be incorporated into the software. These settings are important and essential to obtain accurate and correct results. Table 1 shows the settings used in the simulation by the researchers. The table consists of the year with reference, the turbulence model used in their CFD simulation, Reynolds number obtained for the simulation, number of elements, and mesh or topology used in the simulation. A brief review on each of the setting is mentioned in the next sub-section.

Table 1. Simulation internal setting used by the researchers.

Year [Ref]	Turbulence Model	Reynolds Number	No of Elements	Mesh / Topology
2010 [18]	R k–ε	0.74×10^5, 4.41×10^5	295,500	ST, HX, OG
2011 [19]	R k–ε	0.68×10^5, 4.41×10^5	295,500	ST, HX, OG
2011 [20]	R k–ε	2.4×10^5	2,120 60,000* 80,000* 119,000	NA
2011 [21]	SST k-ω	NA	4,000,000	ST, HX, MB

Year [Ref]	Turbulence Model	Reynolds Number	No of Elements	Mesh / Topology
2012 [22]	R k–ε	NA	735,540* 834,000 933,000 1,032,000 1,131,540*	ST, HX, OG, HG
2012 [23]	R k–ε	NA	1,236,576* 1,534,016	ST
2012 [24]	R k–ε*, RNG k–ε, SST k-ω	NA	424,686	ST, HX
2013 [25]	R k–ε	NA	1,194,320*	ST, HX
2014 [26]	SST k-ω	5×10^5	1,800,000	ST, OG
2014 [27]	R k–ε	2.4×10^5	2,120 60,000* 80,000* 119,000	NA
2014 [28]	SST k-ω	4.5×10^5	1,800,000	ST, OG
2014 [29]	R k–ε*, SST k-ω, S-A	6×10^4 - 2×10^5	312,951	ST
2014 [30]	SST k-ω	NA	1,425,044	ST, HX
2015 [31]	SST k-ω	NA	1,368,000	ST, HX, MB
2015 [32]	SST k-ω	NA	764,416 1,424,016* 1,589,250 1,784,000	ST, HX
2016 [33]	k–ε, R k–ε*, SST k-ω, R k–ε#	2.9×10^5	410,000 500,000* 530,000	HM
2016 [34]	SST k-ω	1.3×10^5 - 3.1×10^5	1,000,000	ST, CG, HG, MB
2016 [35]	k–ε, S-A*	2.2×10^5	3,000,000	PD
2016 [36]	SST k-ω	4×10^5, 8×10^5	1,300,000	ST, HX
2017 [37]	SST k-ω	NA	1,200,000 1,600,000* 2,200,000 5,400,000	UT
2017 [38]	k- ε, S-A, SST k-ω, γ-Re_θ*	0.8×10^5, 1.5×10^5	1,000,000	ST, CG, HG, MB
2017 [39]	k–ε, SST k-ω*	5.2×10^5	1,200,000 1,600,000* 2,100,000	UT
2017 [40]	SST k-ω	5.2×10^5	400,000 600,000* 800,000	UT
2017 [41]	R k-ε	NA	2,580,800	UT, TX
2017 [42]	S-A	NA	1,964,000	ST, CG, HG, MB
2017 [43]	SST k-ω	1.3×10^5, 3.1×10^5	1,000,000	ST, CG, HG, MB
2017 [44]	R k-ε	NA	266,756 990,204 1,571,704* 1,755,876	ST, OG, HG
2017 [45]	R k-ε	NA	799,594 941,732 1,114,358 1,307,852 1,550,154*	ST, OG, HG
2018 [46]	SST k-ω	1×10^5	24,500 35,000* 45,500 700,000 1,000,000*	CG, HG

			1,300,000	
2018 [47]	R k-ε	NA	1,400,000	UT, HX
2018 [48]	R k-ε	NA	1,440,182	UT, TX
			2,113,775	
			2,580,800*	
			3,470,241	
			4,057,831	
2018 [49]	SST k-ω	5.33 x 10^5	1,045,600	ST, MB
			1,460,250*	
			1,530,200	
			1,730,500	
			1,946,000	
			2,323,453	
2018 [50]	SST k-ω	NA	90,000*	ST, CG, MB
2018 [51]	S-A	NA	2,800,000	UT
			3,500,000	
2018 [52]	SST k-ω	NA	764,416	ST
			1,424,016	
			2,646,936	
2018 [53]	SST k-ω	2 x 10^5	1,000,000	ST, HG, CG, MB
2018 [54]	SST k-ω	NA	218,000	UT, TX
			307,000*	
			481,000	
2018 [55]	SST k-ω	NA	140,000	UT
			260,000	
			340,000*	
			480,000	
2019 [56]	SST k-ω	NA	1,200,000	UT
			1,600,000*	
			2,200,000	
			5,400,000	
2019 [57]	SST k-ω	NA	1,110,000	UT, TX
			2,540,000*	
			5,770,000	
2019 [58]	SST k-ω	6.5 x 10^5 - 6.6 x 10^5	2,413,000*	UT
			4,406,000	
			10,495,000	
2020 [59]	SST k-ω	5.25 x 10^5 - 5.43 x 10^5	1,600,000	UT
			3,700,000*	
			7,000,000	
			14,100,000	

Turbulence Model: Standard k–ε (k–ε), Realizable k–ε (R k–ε), Shear stress transport k-ω (SST k-ω), Re-Normalization Group k–ε (RNG k–ε), Spalart-Allmaras (S-A), Menter transitional model (γ-Re_θ).
Mesh: Unstructured (UT), Structured (ST), Hybrid-Mesh (HM), Hexahedral (HX), Tetrahedral (TX), Polyhedral (PD). Topology: O-Grid (OG), H-Grid (HG), C-Grid (CG), Multi-Block (MB). NA – Not Available, # - nonstandard wall function, * - used in the final simulation.

4.1 Turbulence model

Selecting the suitable turbulence model for simulation is important. An unsuitable turbulence model may lead to overprediction or underprediction of the results. Based on Table 1, 15 papers used R k–ε as the turbulence model corresponding to 36 %, while about 23 papers (55 %) used SST k-ω, three papers used the S-A model and one paper used γ-Re_θ.

Using previous data from other researchers, Mohamed et al. [6] used R k–ε. The previous researcher compared k–ε, RNG k–ε, R k–ε, SST k-ω and RSM with experimental data. Čarija et al. [24] suggested that the R k–ε model is the best fit for experimental work compared to RNG k–ε and SST k-ω models. While Shehata et al. [29] concluded that R k–ε is the best model compared to SST k-ω and S-A; however, the results between R k–ε and SST k-ω is similar, and R k–ε uses less computing time compared to SST k-ω. In this simulation, grid-independent test was not done. Cui and Hyun [33] also concluded that R k–ε is the best model compared to k–ε, SST k-ω and R k–ε with nonstandard wall function and SST k-ω giving the highest deviation from the experimental results. The author mentioned the grid-independent test, but the outcome of the simulation was not mentioned in detail.

In contrast, Ghisu et al. [34] mentioned that SST k-ω was chosen as the turbulence model based on the previous simulation) compared with several models. The researcher will use the same model for their future works [53]. Ghisu, Puddu, and Cambuli [38] suggested that γ-Re_θ is the best model compared to k- ε, S-A, and SST k-ω. They stated that this is because the Reynolds number for the simulation is under transition period and not in turbulence period.

Hu, Li, and Wei [35] compared between S-A and k–ε model. They concluded that the S-A model predicts better than the k–ε model. But it should be known that grid independence was not done in this study. Later they used this model for another research [51]. Mahboubidoust and Ramiar [42] also used S-A as the turbulence model by using the findings by Torresi et al. [60].

Most of the early simulation studies that used SST k-ω as the turbulence model stated that this model is widely used in turbomachine CFD simulations [26], [28], [30], [61], [62]. Gratton et al. [46] used the findings by Torresi et al. [63] to use SST k-ω as the turbulence model, while Nazeryan and Lakzian [49] found that simulation using SST k-ω as the turbulence model is in good agreement with the experimental data. This finding is the same with Halder, Samad, and Thévenin [39]. Das and Samad [58], [59] using SST k-ω based on research by Douvi [64] stated that this turbulence model is suitable for NACA aerofoil simulation.

In conclusion, SST k-ω is the best turbulence model for use in the Wells turbine CFD simulation, which was supported by Douvi's [51] research on NACA aerofoil and comparison data from simulation and experiment [39], [49]. S-A turbulence model is also in good agreement with simulation and experimental data. However, this model fails to predict stall point [65]. Another good model that can be used is R k–ε. This model also produces results in line with the experimental results. It should be considered that the flow of the model should exceed the critical Reynolds number so that the model can predict the result precisely [38].

4.2 Reynolds Number

Calculating Reynolds number is very crucial in CFD simulations. Knowing Reynolds number can differentiate the flow whether it is laminar, transition, or turbulent flow. Takao et al. stated that the critical Reynolds number is 4×10^5 [66]. Thus, a simulation with a Reynolds number bigger than 4×10^5 is suitable for the turbulence model. Table 1 shows that some of the researchers did not mention the value of the Reynolds number in their simulations.

Meanwhile, some researchers have flow in transition mode, but they used the turbulence model in their simulation, which may be why some of the turbulence model results differ between researchers. Ghisu et al. [38] also stated this in their research that had transition flow, and their transition model fit well with the results compared to the turbulence models. To differentiate the flow between transition and turbulence, the critical Reynolds number should be compared. But, up-to-date, only Takao et al. stated the value of critical Reynolds number [66] as 4×10^5. As Reynolds numbers is an important factor in CFD simulation, the critical Reynolds number in simulating Wells turbine is considered as one of the research gaps as no other research was done after Takao et al. [66] in 2006. This parameter is important as the newly suggested value of the critical Reynolds number can be lower or even higher than the value suggested by Takao et al. [66], which can lead to different output.

4.3 Grid Independence test

Another important test when dealing with CFD simulation is the grid independence test. This test is done to minimize the influence of grid size on the CFD simulation results [67]. As shown in Table 1, few researchers did not run the grid independence test in their research, giving some degree of ambiguity in their CFD simulation results. In the past few years, most of the researchers included the test in their simulation. There are also some researchers [50] that included this test in their research but did not document it properly.

Even though this test is important, there is still no a standardized test method or procedure to run the test [68]. Three common methods used in this test are the grid resolution method, general Richardson extrapolation method and grid convergence index [69]. In the reviewed paper, most of the tests used were based on the grid resolution method and only Das and Samad [58] used the grid convergence index method.

4.4 Meshing and topology

As for meshing, no researcher mentioned any advantages between the structured and unstructured or otherwise, as the meshing can be used according to suitability. It should be noted that despite the unstructured mesh being able to handle complex geometries, it requires higher computational time compared to structured mesh [70]. In general, structured mesh can be considered as sufficient to handle Wells turbine blade CFD simulation. Hybrid meshing can be used to compromise the advantages and disadvantages of structured and unstructured meshing.

It is common to use structured meshing with Hexahedral and unstructured meshing with Tetrahedral. Polyhedral meshing is the least concern meshing type. Only researcher that using Star-CCM+ used this meshing. No specific advantages are known in using this meshing. As for topology, in some complex geometry or to tackle critical parts of the simulation, the multi-block should be used together with O-Grid, H-Grid, or C-Grid. This can increase the result accuracy of the main concerning region.

5. Conclusion

In summary, 42 journal papers related to CFD simulation on Wells turbine were briefly summarised. FLUENT and CFX software were recommended to be used to simulate the Wells turbine flow problem with SST k-ω turbulence model. In order to avoid discrepancies in that result, Reynolds number of the flow should be higher than the critical Reynolds number, 4×10^5. Grid independence test is a must to reduce the error in the result. There is no specific recommendation for meshing and topology, as there is no significant improvement between structed and unstructured meshing. However, the capability of the computer should be considered when dealing with unstructured mesh.

References

[1] Turkenburg W C *et al.*, 2012, Chapter 11 - Renewable Energy, in *Global Energy Assessment - Toward a Sustainable Future*, Global Energy Assessment Writing Team, Ed. (Cambridge University Press, Cambridge, UK and New York, NY, USA and the International Institute for Applied Systems Analysis, Laxenburg, Austria: Cambridge University Press), p. 761–900.

[2] Gallucci M, 2019 At Last, Wave Energy Tech Plugs Into *IEEE Spectr.* **56** p. 8–9.

[3] Melikoglu M, 2018 Current status and future of ocean energy sources: A global review *Ocean Eng.* **148**, June 2017 p. 563–573.

[4] Tedeschi E Carraro M Molinas M and Mattavelli P, 2011 Effect of control strategies and power take-off efficiency on the power capture from sea waves *IEEE Trans. Energy Convers.* **26**, 4 p. 1088–1098.

[5] Rahman M Paul N Islam S and Rashed S, 2013 Power Generation from Sea Wave : An Approach **13**, 1.

[6] Das T K Halder P and Samad A, 2017 Optimal design of air turbines for oscillating water column wave energy systems: A review *Int. J. Ocean Clim. Syst.* **8**, 1 p. 37–49.

[7] Khare V Khare C Nema S and Baredar P, 2019, Introduction of Tidal Energy, in *Tidal Energy Systems*, p. 41–114.

[8] Falcão A F O and Henriques J C C, 2016 Oscillating-water-column wave energy converters and air turbines: A review *Renew. Energy* **85** p. 1391–1424.

[9] Falcão A F d. O, 2010 Wave energy utilization: A review of the technologies *Renew. Sustain. Energy Rev.* **14**, 3 p. 899–918.

[10] Cui Y and Liu Z, 2015 Effects of Solidity Ratio on Performance of OWC Impulse Turbine *Adv. Mech. Eng.* **7**, 1 p. 1–10.
[11] Falcão A F O and Gato L M C, 2012, Air turbines, in *Comprehensive Renewable Energy*, **8**, p. 111–149.
[12] Shehata A S Xiao Q Saqr K M and Alexander D, Jan. 2017 Wells turbine for wave energy conversion: a review *Int. J. Energy Res.* **41**, 1 p. 6–38.
[13] Takao M and Setoguchi T, 2012 Air turbines for wave energy conversion *Int. J. Rotating Mach.* **2012.**
[14] Karthikeyan T Samad A and Badhurshah R, 2014 Review of air turbines for wave energy conversion *Proc. - 2013 Int. Conf. Renew. Energy Sustain. Energy, ICRESE 2013* p. 183–191.
[15] Soltanmohamadi R and Lakzian E, Aug. 2016 Improved design of Wells turbine for wave energy conversion using entropy generation *Meccanica* **51**, 8 p. 1713–1722.
[16] Cui Y Hyun B S and Kim K, 2017 Numerical study on air turbines with enhanced techniques for OWC wave energy conversion *China Ocean Eng.* **31**, 5 p. 517–527.
[17] Gareev A, 2011, Analysis of Variable Pitch Air Turbines for Oscillating Water Column (OWC) Wave Energy Converters.
[18] Taha Z Sugiyono and Sawada T, 2010 A comparison of computational and experimental results of Wells turbine performance for wave energy conversion *Appl. Ocean Res.* **32**, 1 p. 83–90.
[19] Taha Z Sugiyono Tuan Ya T M Y S and Sawada T, 2011 Numerical investigation on the performance of Wells turbine with non-uniform tip clearance for wave energy conversion *Appl. Ocean Res.* **33**, 4 p. 321–331.
[20] Mohamed M H Janiga G Pap E and Thévenin D, 2011 Multi-objective optimization of the airfoil shape of Wells turbine used for wave energy conversion *Energy* **36**, 1 p. 438–446.
[21] Torresi M Pranzo D Camporeale S M and Pascazio G, 2011 Improved Design of High Solidity Wells Turbine *9th Eur. Wave Tidal Energy Conf.* January p. 5–9.
[22] Shaaban S and Abdel Hafiz A, 2012 Effect of duct geometry on Wells turbine performance *Energy Convers. Manag.* **61** p. 51–58.
[23] Shaaban S, 2012 Insight analysis of biplane wells turbine performance *Energy Convers. Manag.* **59** p. 50–57.
[24] Čarija Z Kranjčević L Banić V and Čavrak M, 2012 Numerical analysis of wells turbine for wave power conversion *Eng. Rev.* **32**, 3 p. 141–146.
[25] Mohamed M H and Shaaban S, 2013 Optimization of blade pitch angle of an axial turbine used for wave energy conversion *Energy* **56** p. 229–239.
[26] Starzmann R and Carolus T, Jan. 2014 Effect of Blade Skew Strategies on the Operating Range and Aeroacoustic Performance of the Wells Turbine *J. Turbomach.* **136**, 1.
[27] Mohamed M H and Shaaban S, Apr. 2014 Numerical optimization of axial turbine with self-pitch-controlled blades used for wave energy conversion *Int. J. Energy Res.* **38**, 5 p. 592–601.
[28] Starzmann R and Carolus T, Feb. 2014 Model-based selection of full-scale Wells turbines for ocean wave energy conversion and prediction of their aerodynamic and acoustic performances *Proc. Inst. Mech. Eng. Part A J. Power Energy* **228**, 1 p. 2–16.
[29] Shehata A S Saqr K M Shehadeh M Xiao Q and Day A H, 2014 Entropy generation due to viscous dissipation around a wells turbine blade: A preliminary numerical study *Energy Procedia* **50** p. 808–816.
[30] Halder P and Samad A, 2014 Effect of guide vane angle on wells turbine performance *ASME 2014 Gas Turbine India Conf. GTINDIA 2014* p. 1–7.
[31] Halder P and Samad A, 2015 Casing Treatment of a Wave Energy Extracting Turbine *Aquat. Procedia* **4**, Icwrcoe p. 516–521.
[32] Halder P Samad A Kim J H and Choi Y S, 2015 High performance ocean energy harvesting turbine design-A new casing treatment scheme *Energy* **86** p. 219–231.
[33] Cui Y and Hyun B S, 2016 Numerical study on Wells turbine with penetrating blade tip

treatments for wave energy conversion *Int. J. Nav. Archit. Ocean Eng.* **8**, 5 p. 456–465.
[34] Ghisu T Puddu P and Cambuli F, 2016 Physical Explanation of the Hysteresis in Wells Turbines: A Critical Reconsideration *J. Fluids Eng. Trans. ASME* **138**, 11 p. 1–9.
[35] Hu Q Li Y and Wei F, 2016 Preliminary results of numerical simulations of a bio-mimetic wells turbine *Proc. Int. Conf. Offshore Mech. Arct. Eng. - OMAE* **6**, August.
[36] Halder P and Samad A, 2016 Optimal Wells turbine speeds at different wave conditions *Int. J. Mar. Energy* **16** p. 133–149.
[37] Halder P and Samad A, Jun. 2017 Torque and efficiency maximization for a wave energy harvesting turbine: an approach to modify multiple design variables *Int. J. Energy Res.* **41**, 7 p. 1014–1028.
[38] Ghisu T Puddu P and Cambuli F, 2017 A detailed analysis of the unsteady flow within a Wells turbine *Proc. Inst. Mech. Eng. Part A J. Power Energy* **231**, 3 p. 197–214.
[39] Halder P Samad A and Thévenin D, 2017 Improved design of a Wells turbine for higher operating range *Renew. Energy* **106** p. 122–134.
[40] Halder P Rhee S H and Samad A, 2017 Numerical optimization of wells turbine for wave energy extraction *Int. J. Nav. Archit. Ocean Eng.* **9**, 1 p. 11–24.
[41] Hamed H Nawar M and Abd El-Maksoud R, 2017 A New Operating Concept to Enhance Wells Turbine Performance *J. Eng. Sci. Mil. Technol.* **17**, 17 p. 1–14.
[42] Mahboubidoust A and Ramiar A, 2017 Investigation of DBD plasma actuator effect on the aerodynamic and thermodynamic performance of high solidity Wells turbine *Renew. Energy* **112** p. 347–364.
[43] Ghisu T Puddu P Cambuli F and Virdis I, 2017 On the Hysteretic Behaviour of Wells Turbines *Energy Procedia* **126** p. 706–713.
[44] Shaaban S, Oct. 2017 Wells turbine blade profile optimization for better wave energy capture *Int. J. Energy Res.* **41**, 12 p. 1767–1780.
[45] Shaaban S, Mar. 2017 Wave energy harvesting using a novel turbine rotor geometry *Int. J. Energy Res.* **41**, 4 p. 540–552.
[46] Gratton T Ghisu T Parks G Cambuli F and Puddu P, 2018 Optimization of blade profiles for the Wells turbine *Ocean Eng.* **169**, July p. 202–214.
[47] González C M Fuertes G R Díaz M G García B P Castro F and Oro J M F, 2018 Wells turbine with variable blade profile for wave energy conversion *Proc. Int. Conf. Offshore Mech. Arct. Eng. - OMAE* **10** p. 1–7.
[48] Hashem I Abdel Hameed H S and Mohamed M H, 2018 An axial turbine in an innovative oscillating water column (OWC) device for sea-wave energy conversion *Ocean Eng.* **164**, June p. 536–562.
[49] Nazeryan M and Lakzian E, 2018 Detailed entropy generation analysis of a Wells turbine using the variation of the blade thickness *Energy* **143** p. 385–405.
[50] Licheri F Climan A Puddu P Cambuli F and Ghisu T, 2018 Numerical Study of a Wells Turbine with Variable Pitch Rotor Blades *Energy Procedia* **148** p. 511–518.
[51] Hu Q and Li Y, 2018 Unsteady RANS simulations of wells turbine under transient flow conditions *J. Offshore Mech. Arct. Eng.* **140**, 1 p. 1–11.
[52] Halder P Mohamed M H and Samad A, 2018 Wave energy conversion: Design and shape optimization *Ocean Eng.* **150**, September 2016 p. 337–351.
[53] Ghisu T Cambuli F Puddu P Mandas N Seshadri P and Parks G T, 2018 Numerical evaluation of entropy generation in isolated airfoils and Wells turbines *Meccanica* **53**, 14 p. 3437–3456.
[54] Kumar P M Halder P and Samad A, 2018 Performance analysis of wells turbine with radiused blade tip *Proc. Int. Conf. Offshore Mech. Arct. Eng. - OMAE* **10** p. 1–10.
[55] Madhan Kumar P Halder P Samad A and Rhee S H, 2018 Wave energy harvesting turbine: Effect of hub-to-tip profile modification *Int. J. Fluid Mach. Syst.* **11**, 1 p. 55–62.
[56] Halder P Das T K Samad A and Mohamed M H, 2019, Hysteresis Behavior for Wave Energy Conversion Device Under Alternative Axial Flow Conditions, in *Proceedings of the Fourth International Conference on Substorms*, **23**, (Springer Singapore), p. 717–723.
[57] Kumar P M and Samad A, 2019 Introducing Gurney flap to Wells turbine blade and

performance analysis with OpenFOAM *Ocean Eng.* **187**, May p. 106212.
[58] Das T K and Samad A, 2019 The effect of midplane guide vanes in a biplane wells turbine *J. Fluids Eng. Trans. ASME* **141**, 5 p. 1–13.
[59] Das T K and Samad A, 2020 Influence of stall fences on the performance of Wells turbine *Energy* **194** p. 116864.
[60] Torresi M Camporeale S M Strippoli P D and Pascazio G, 2008 Accurate numerical simulation of a high solidity Wells turbine *Renew. Energy* **33**, 4 p. 735–747.
[61] Ahmed N and Mueller M, 2013 Impact of varying clearances for the Wells turbine on heat transfer from electrical generators in oscillating water columns *2013 8th Int. Conf. Exhib. Ecol. Veh. Renew. Energies, EVER 2013* p. 3–8.
[62] Premkumar T M Ashish M A Banu Prakash T and Thulasiram D, 2014 Numerical analysis of wells turbine *Appl. Mech. Mater.* **592–594** p. 1125–1129.
[63] Torresi M Camporeale S M and Pascazio G, 2009 Detailed CFD analysis of the steady flow in a wells turbine under incipient and deep stall conditions *J. Fluids Eng. Trans. ASME* **131**, 7 p. 0711031–07110317.
[64] Douvi C. Eleni, Mar. 2012 Evaluation of the turbulence models for the simulation of the flow over a National Advisory Committee for Aeronautics (NACA) 0012 airfoil *J. Mech. Eng. Res.* **4**, 3.
[65] Cui Y Liu Z Zhang X and Xu C, 2019 Review of CFD studies on axial-flow self-rectifying turbines for OWC wave energy conversion *Ocean Eng.* **175**, December 2018 p. 80–102.
[66] Takao M Thakker A Abdulhadi R and Setoguchi T, 2006 Effect of blade profile on the performance of a large-scale Wells turbine for wave-energy conversion *Int. J. Sustain. Energy* **25**, 1 p. 53–61.
[67] Roychowdhury D G, 2020 *Computational Fluid Dynamics for Incompressible Flows* .
[68] Lee M Park G Park C and Kim C, 2020 Improvement of Grid Independence Test for Computational Fluid Dynamics Model of Building Based on Grid Resolution *Adv. Civ. Eng.* **2020**.
[69] Seeni A Rajendran P and Mamat H, 2019 A CFD mesh independent solution technique for low reynolds number propeller *CFD Lett.* **11**, 10 p. 15–30.
[70] Tu J Yeoh G-H and Liu C, 2013, Practical Guidelines for CFD Simulation and Analysis, in *Computational Fluid Dynamics*, (Elsevier), p. 219–273.

Acknowledgement
The authors would like to thank the Universiti Teknologi Malaysia under the Industry-International Incentive Grant scheme (Q.K130000.3601.03M09) for supporting this research.

The effect of corn stalk fiber loading on tensile properties and absorption ability on polybutadiene adipate terephalate composite.

M H H Mohamad Rosdi[1,*] and N A Ahad[1]

[1]Faculty of Chemical Engineering Technology, Kompleks Pusat Pengajian Jejawi 2, Taman Muhibbah, Universiti Malaysia Perlis (UniMAP), 02600 Arau, Perlis

*Corresponding author's email : hilmirosdi95@gmail.com

Abstract. The natural fiber have recently become attractive and been widely used as reinforcement material to replace synthetic fiber as a concern to environmental issue. However, there are some of natural fiber unable to perform well as reinforcement material due to their natural properties. In this research study, the corn stalk fiber was selected as natural fiber reinforced with polybutadiene adipate terephalate (PBAT) to form bio-composite materials. There are two type of PBAT used which are in pellets form and powder form. The objectives of using different type of PBAT are to identify their compatibility with filler and the dispersion of corn stalk fiber in both forms of matrix. Then, the both type of PBAT also compounded with difference loadings of corn stalk fiber. Lastly, their effect on tensile properties and absorption ability were identified. After the test is run, the composite of neat PBAT in powder form shows highest tensile strength and elongation at break but lowest in water absorption percentage compared to others. However, the composition of pellets PBAT with 30wt% of corn stalk fiber shows highest percentage of water absorption compared to others.

1. Introduction

Natural fiber also known as bio-fiber have been extensively used by researcher around the world in recent years. This is due to their collective valuable unique properties which includes low density, light weight, affordable cost, highly availability, recyclable, non-toxicity, eco-friendly, good thermal stability, lower abrasion of equipment during processing and bio-degradability [1]. Therefore, they have potential to use as reinforcement material to replace synthetic fiber in producing of green composite [2]. There are several example of natural fiber widely used as natural fiber are paddy straw, corn stalk fiber and sisal [3]–[5].

Usage of corn stalk fiber as reinforcement materials were increase day by days. They are highly available and categorized as hydrophilic materials because have large amount of porous cells making them excellent in absorption [6]. They were collected from agricultural-based industries which release to the environment without proper disposal procedure causing pollution and harmful effect on human and animal health [7]. Thus, the usage of corn stalk fiber were not only replacing conventional synthetic fiber, but also may help preventing environment pollution.

Polybutadiene adipate terephalate (PBAT) is one of bio-polymer. They are synthetic polymer based on fossil resources with high elongation at break and very flexible [8]. The PBAT commonly used in packaging fields and medicine application by blending with other polymer such as natural rubber (NR), polycaprolactone (PCL), poly(propylene carbonate) (PPC) and mostly blend with polylactic acid (PLA) [9]–[13]. However, there are less studies conducted by researcher before about the production of composite materials using PBAT with natural fiber. This is because the PBAT is relatively expensive [14]. Therefore one approach to reduce the cost of PBAT composite is to compound them with low cost natural materials such as natural fiber.

In this research study, there are two type of PBAT was used which are in pellet form and powder form to investigate their compatibility and dispersion with corn stalk fiber. The PBAT then compounded with varying corn stalk fiber weight which are 10wt%, 20wt% and 30wt% to evaluate their effect on different composition. The tensile properties and absorption ability in water of all specimens were investigated and compared among fiber loading and both types of PBAT form ad matrix.

2. Methodology

2.1. Materials

The natural fiber used in this study was corn stalk fiber taken from pith part (see figure 1). The corn stalk fiber used were randomly collected after harvesting by local farmer in Perlis, Malaysia. Furthermore, they were cleaned by using distilled water to remove dirt and dried by exposed to sunlight. Lastly, the corn stalk fiber was grinded by using a grinder and ready for compounded with bio polymer. The polymer used as a matrix to form bio-composite in this research is PBAT in pellet form and powder form as shown in figure 2.

Figure 1. Corn stalk fiber from pith part.

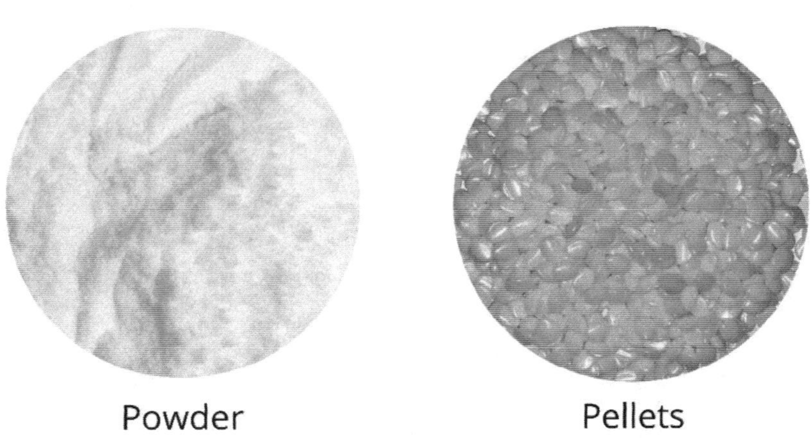

Figure 2. Two type of PBAT used.

2.2. Compounding

The two types of PBAT (pellets and powder) were supplied by Jiangsu (China) compounded with corn stalk fiber to form composite by using the heated roll mill and compression moulding followed composition in Table 1. The temperature used for heated roll mill used are 120°C with rotor speed is 6 rpm. However the temperature used for compression moulding also 120°C and the process within 10 minutes with pressure at 170 kg/cm^2.

Table 1. Composition of PBAT with corn talk fiber.

Type of PBAT	Composition	
	PBAT, wt%	Corn stalk fiber, wt%
Powder	100	0
	90	10
	80	20
	70	30
Pellet	100	0
	90	10
	80	20
	70	30

2.3. Characteristics

The tensile properties of the specimens was conducted by using a Universal Tensile Machine (UTM). The specimens were cut into dumbbell shape about 50 mm according to ASTM 638. The tensile test was performed with cross head speed of 50 mm/min at room temperature [15]. The absorption ability testing was conducted to prove the abilities of PBAT compounded with corn stalk fiber to absorb water and provide hydrophilic properties. The specimens with measurements of 20 mm x 20 mm x 1 mm were used as indicated by ASTM D570-98. Before testing, the specimens was dried by using vacuum oven to remove moisture content that may affected by environment. After drying process, the initial weight of the specimens then measured and recorded by using electronic weight balance. The specimens then immersed in water and weighed for everyday until equilibrium readings. After weighed, the specimens were calculated using the following equation (1) [7];

$$\text{Water Absorption (\%)} = \frac{W_2 - W_1}{W_2} X\ 100\% \quad (1)$$

Where, W_1 refer to the initial weight of corn stalk fiber and W_2 refer to the final weight of corn stalk fiber.

3. Results and Discussion

The tensile properties PBAT matrix containing different loading of corn stalk fiber are presented in figure 3 and figure 4. Based on both figures, the ultimate tensile strength and the elongation at break was drastically decrease with increasing the composition of fiber in composite for both powder and pellets form of PBAT. The composites with 30wt% of corn stalk fiber shows the lowest tensile strength for both PBAT in powder form and in pellets form with reading 4.622 MPa and 4.609 MPa, respectively. This result was explain that the strength and elasticity properties of the composite were worsen by increasing the number of corn stalk fiber. This result happen due to the compatibility between the polymer matrix which are PBAT and corn stalk fiber was weak then unable give better mechanical properties to the specimens. Similar behaviour also been reported by Moustofa and teams (2017) in their research study involving PBAT and coffee grounds fiber [16].

The figure 3 and figure 4 also show that the higher tensile strength and elongation at break on the PBAT in powder form compared to PBAT in pellets form. The neat PBAT in powder form shows elastic properties with highest tensile strength and elongation at break with reading 17.944 MPa and 948.9% respectively. This results was prove that the PBAT in powder form tends to well mixed with the fiber during compounding compared to PBAT in pellets form. Nassar and teams (2021) also conduct the testing by using different type of polymer matrix, which are using powder and pellets Polypropylene (PP) then have similar behaviour in with this research study [17].

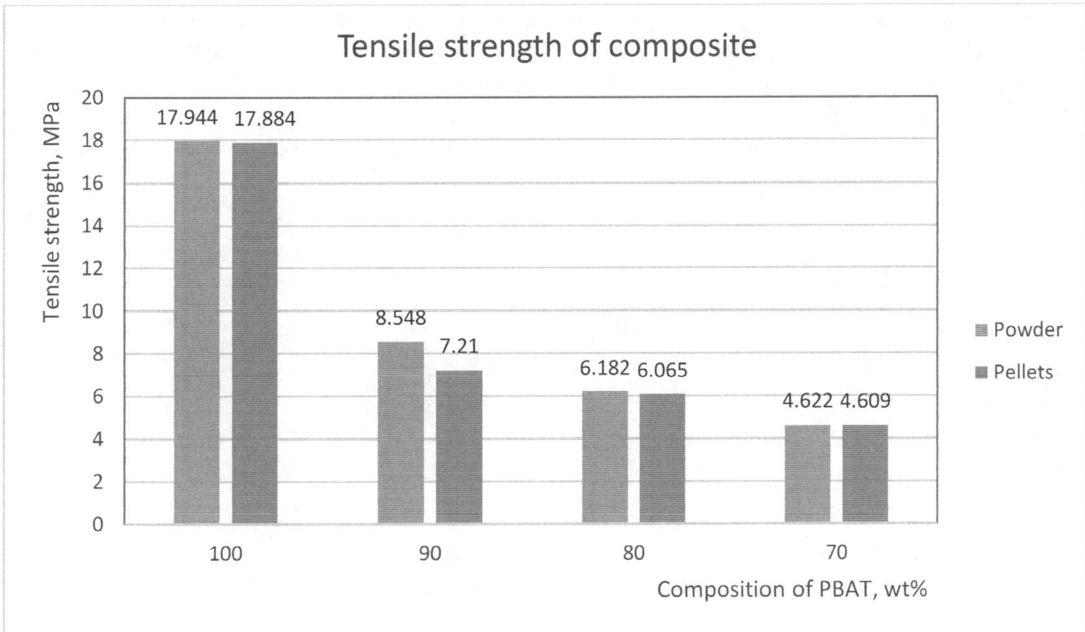

Figure 3. The comparison for tensile strength of PBAT compounded with corn stalk fiber.

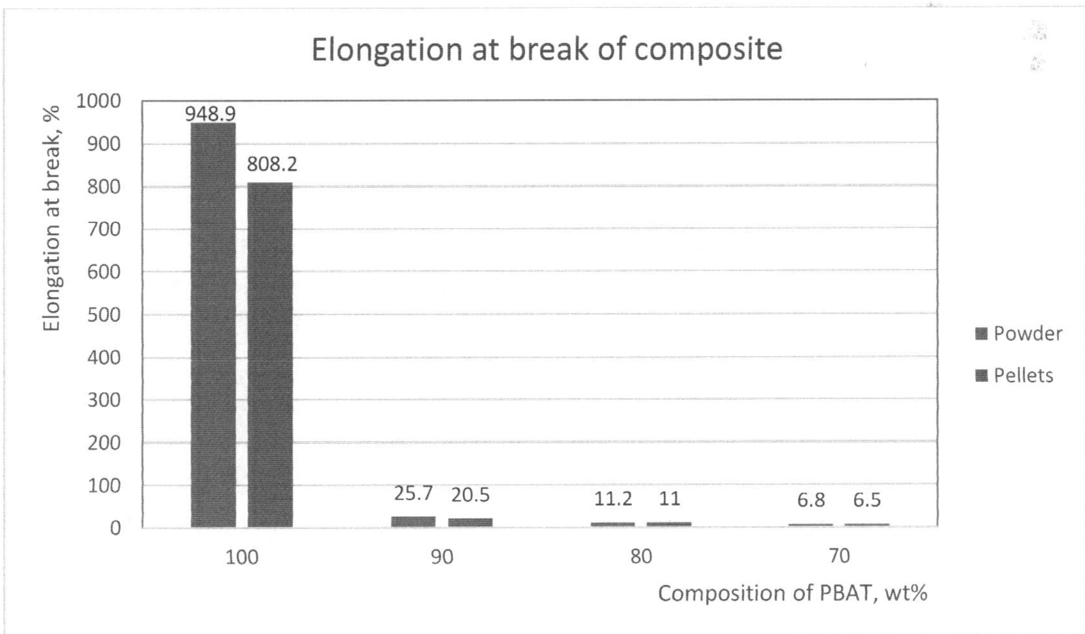

Figure 4. The comparison for elongation at break of PBAT compounded with corn stalk fiber.

The water absorption results for PBAT compounded with corn stalk fiber after 30 days were shown in figure 5. The neat PBAT for both powder and pellets form show the lowest percentage of water absorption compared to others with reading 0.02% and 0.05% respectively. That result were obtained due to the PBAT are hydrophobic bio-polymer [16], [18]. Therefore they would not absorb the water. Then, the percentage of water absorption of the specimens were increase by increasing the composition of corn stalk fiber. This is because bio-filler are polar materials with hydroxyl group which will absorb water. Liu and teams (2019) also mention in their research study that the corn stalk fiber have hydrophilic lignocellulosic molecules making them can absorb water [2].

Figure 5 also shows that the PBAT composites in pellets form have higher water absorption percentage compared to PBAT in powders form. This result manifest that the PBAT composites in powder form exhibit better interfacial bonding between polymer matrix with polymer matrix and polymer matrix with the filler. Similar behaviour also been reported by Nassar and teams (2021) which comparing water absorption of specimens by using powder and pellets polypropylene, PP [17].

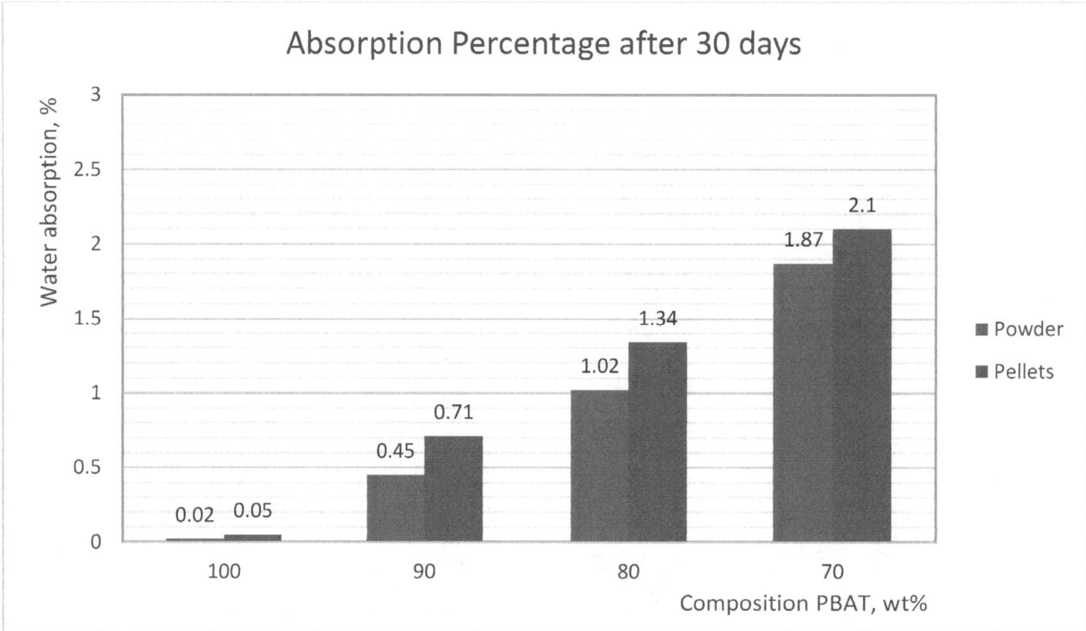

Figure 5. Absorption percentage of specimen after 30 days.

4. Conclusion

Based on investigation may conclude that the composition between corn stalk fiber with PBAT matrix may influence the tensile properties and water absorption abilities of the composites. Other than that, the type of PBAT used which are in powder and pellets form also can gives impact on the results in this research study. The neat PBAT in powder form show the highest value on tensile strength (17.944 MPa) and elongation at break (948.9%) due to the interaction only among the matrices particle, without any additives, filler and others. However the composites compounded between PBAT in pellets form and 30wt% of corn stalk fiber shows the lowest value on tensile strength (4.609 MPa) and elongation at break (6.5%) due to poor adhesion between polymer matrix and filler. This is because, higher filler loading will tends to decrease the interaction and adhesion between matric and filler. Therefore, there are further study should be conducted later to improve the compatibility of PBAT polymer matrix with corn stalk fiber such as chemical treatment or using coupling agent to achieve better tensile properties. For absorption ability testing, the neat PBAT in powder form shows lowest percentage of water absorption (0.02%) because of their hydrophobic properties and better interfacial bonding compared to neat PBAT in pellets form. The composites with PBAT in pellets form and 30 wt% of corn stalk fiber showed the highest percentage of water absorption causes by hydrophilic lignocellulosic molecules on corn stalk fiber and poor interfacial bonding between polymer matrixes with the fibers.

5. References

[1] P. Manimaran, P. Senthamaraikannan, M. R. Sanjay, M. K. Marichelvam, and M. Jawaid, "Study on characterization of Furcraea foetida new natural fiber as composite reinforcement for lightweight applications," *Carbohydr. Polym.*, vol. 181, no. November 2017, pp. 650–658, 2018.

[2] Y. Liu, J. Xie, N. Wu, Y. Ma, C. Menon, and J. Tong, "Characterization of natural cellulose fiber from corn stalk waste subjected to different surface treatments," *Cellulose*, vol. 26, no. 8, pp. 4707–4719, 2019.

[3] L. W. Lun, A. A. N. Gunny, F. H. Kasim, and D. Arbain, "Fourier transform infrared spectroscopy (FTIR) analysis of paddy straw pulp treated using deep eutectic solvent," *AIP Conf. Proc.*, vol. 1835, no. April, 2017.

[4] S. Liu, Y. Xu, Y. Zhao, L. Zou, and W. Lu, "Hydrothermal modification of lignocellulosic waste as microbial immobilization carriers for ethanol production," *Biochem. Eng. J.*, vol. 142, no. November 2018, pp. 27–33, 2019.

[5] J. S. S. Neto, R. A. A. Lima, D. K. K. Cavalcanti, J. P. B. Souza, R. A. A. Aguiar, and M. D. Banea, "Effect of chemical treatment on the thermal properties of hybrid natural fiber-reinforced composites," *J. Appl. Polym. Sci.*, vol. 136, no. 10, pp. 1–13, 2019.

[6] D. Kohli, "Thermal and Morphological Properties of Chemically Treated Barley Husk Fiber," *Int. J. Res. Mech. Eng. Technol. 153*, vol. 3, no. 2, pp. 153–156, 2013.

[7] M. H. H. Mohamad Rosdi and N. A. Ahad, "The Changes on Morphological and Absorption Ability of Treated Corn Stalk Fiber," *J. Phys. Conf. Ser.*, vol. 1529, p. 052009, 2020.

[8] F. V. Ferreira, L. S. Cividanes, R. F. Gouveia, and L. M. F. Lona, "An overview on properties and applications of poly(butylene adipate-co-terephthalate)–PBAT based composites," *Polym. Eng. Sci.*, vol. 59, no. s2, pp. E7–E15, 2019.

[9] E. Kalkornsurapranee *et al.*, "Influence of modified natural rubbers as compatibilizers on the properties of flexible food contact materials based on NR/PBAT blends," *Mater. Des.*, vol. 196, p. 109134, 2020.

[10] F. M. Sousa, A. Raffaela, M. C. Lízzia, F. B. Cavalcanti, L. H. Carvalho, and T. G. Almeida, "Rheological and thermal characterization of PCL / PBAT blends," *Polym. Bull.*, vol. 76, no. 3, pp. 1573–1593, 2019.

[11] F. Wang, S. Zhang, and H. Huang, "Structure and improved properties of PPC / PBAT blends via controlling phase morphology based on melt viscosity," vol. 48924, pp. 1–12, 2020.

[12] M. Zhang, H. Jia, Y. Weng, and C. Li, "International Biodeterioration & Biodegradation Biodegradable PLA / PBAT mulch on microbial community structure in di ff erent soils," *Int. Biodeterior. Biodegrad.*, vol. 145, no. September, p. 104817, 2019.

[13] S. Xiang, L. Feng, X. Bian, G. Li, and X. Chen, "Evaluation of PLA content in PLA / PBAT blends using TGA," *Polym. Test.*, vol. 81, no. October 2019, p. 106211, 2020.

[14] C. S. Wu, "Characterization of cellulose acetate-reinforced aliphatic-aromatic copolyester composites," *Carbohydr. Polym.*, vol. 87, no. 2, pp. 1249–1256, 2012.

[15] S. Husseinsyah, K. S. Chun, C. M. Yeng, and A. Ismail, "Tensile, thermal and water absorption properties of corn stalk filled low density polyethylene composites: Effect of coconut oil coupling agent," *Sains Malaysiana*, vol. 45, no. 11, pp. 1733–1739, 2016.

[16] H. Moustafa, C. Guizani, and A. Dufresne, "Sustainable biodegradable coffee grounds filler and its effect on the hydrophobicity, mechanical and thermal properties of biodegradable PBAT composites," *J. Appl. Polym. Sci.*, vol. 134, no. 8, pp. 1–11, 2017.

[17] M. M. A. Nassar *et al.*, "Polymer powder and pellets comparative performances as bio-based composites," *Iran. Polym. J. (English Ed.*, vol. 30, no. 3, pp. 269–283, 2021.

[18] L. B. Tavares, N. M. Ito, M. C. Salvadori, D. J. dos Santos, and D. S. Rosa, "PBAT/kraft lignin blend in flexible laminated food packaging: Peeling resistance and thermal degradability,"

Polym. Test., vol. 67, no. November 2017, pp. 169–176, 2018.

Acknowledgments
The authors would like to express their sincere thanks to the Ministry of Higher Education Malaysia (MOHE) for the research fund under the Fundamental research Grant Scheme (FRGS) of the Project Code: FRGS/1/2018/TK05/UNIMAP/02/3).

Towards the introduction of elements that contribute to the LRT operation sustainability through technology-based ancillary revenue

Fazarizul Hashidi Muhamad Pauzi[1*], Mohd Yusof Md Daud[2] and Ahmad Yusri Mohamed[3]

[1]Prasarana Integrated Development Sdn Bhd, Kuala Lumpur, Malaysia
[2]Razak Faculty of Technology and Informatics, Universiti Teknologi Malaysia, Malaysia
[3]Johor Corporation, Johor, Malaysia

*Conference author's email: fazarizul.hashidi@gmail.com

Abstract. Transportation and technology innovation is one of the Sustainable Development Goals (SDGs) elements, and it has been outlined in goal number nine. The high demand for LRT usage requires a large amount of electricity. It contributes to the increase in electricity costs that have to be borne by LRT operators, and it is one of the main contributors to the increase in LRT operating costs. Revenue from fares unable to offset the rising operating cost. Therefore, LRT operators need to explore ancillary revenue activities to cover LRT operating costs. This paper aims to identify the elements involved in generating ancillary income through technology-based businesses from fibre optic cable networks along LRT alignment for the LRT operation sustainability. This study adopted semi-structured interviews for data collection and thematic analysis to analyse the data collected to identify the elements involved in generating ancillary revenue. Eleven themes were identified from the thematic analysis using NVivo. The list of elements involved has been proposed in this study has the potential to be a driver in generating ancillary revenue for the sustainability of LRT operations and contribute to sustainable development in achieving one of the objectives outlined in the SDGs.

1. Introduction

Infrastructure systems such as energy, water, waste, transportation and telecommunications are essential needs of every society. All these infrastructures need to be sustainable to achieve the Sustainable Development Goals (SDGs) objectives. Infrastructure development of Public Transport is one of the critical and complicated issues for modern societies. Infrastructure investments are massive and expensive, and their impact can transform the landscape of cities and regions for decades, if not centuries [1]. To respond to this, transport planners and engineers have to implement the best strategy using knowledge, innovation, best practice and advanced technology to ensure all investment provides a good return. In recent years, the sustainability of a transportation system has become enormously important, mainly due to increasing environmental issues and climate change [2]. According to Hens, Melnyk [3], technology will reduce environmental risks while increasing energy and economic security and human capital. The LRT system in Malaysia uses full electrical power to contribute to the country's environmental sustainability. Powrie [4] states that Railway electrification offers benefits for both the environment (zero carbon dioxide (CO2) and particulate emissions at the point of use) and traction operation (reduced complexity and cost) but requires investment in fixed infrastructure. However, the high demand for the use of the LRT system requires a large amount of electricity. It contributes to the increase in electricity costs to the LRT operators. This cost is one of the main contributors to the increase in LRT operating costs. The energy consumption of urban rail transit systems increases rapidly with the expansion of transit networks and increased service frequency [5]. The development of sustainable infrastructure and promoting the economy of development can be realised in various ways. It is important since the operating costs of rail-based public transport are too high. The fare revenue is unable to offset the rising cost of operating the LRT. Therefore, it is important to explore the ancillary revenue in supporting the LRT operation sustainability. While new technology attributes to the growth of the global economy, the LRT operator faces the challenges of identifying the elements that can contribute to ancillary revenue increments in ensuring rail-based public transport becomes a sustainable infrastructure economically and financially. Therefore, this paper will focus on issues related to finance and economics in sustainable infrastructure through ancillary revenue. This paper aims to identify the elements involved in generating ancillary revenue through technology-based businesses from optical fibre cable networks deployed along with LRT Right of Way (ROW). This study adopted a business model canvas developed by Osterwalder and Pigneur [6] as the basis to identify other elements involve specifically to generate additional revenue through technology-based businesses for LRT operators.

2. Sustainable Development Infrastructure through Ancillary Revenue

One of the SDGs elements is transportation and technology innovation, and it has been outlined in goal number nine in the SDGs. There are three main components of sustainable development: environmental, economic, and social. Serebrisky, Watkins [7], Bhattacharya and Jeong [8] reported that the sustainable infrastructure framework encompasses four significant pillars: economic and financial, environmental, social, and institutional. Saghir [9] claims that the infrastructure development can be achieved by (i) improving access to key facilities; (ii) creating additional jobs and economic activities, (iii) expanding the overall production capacity; (iv) reducing production costs through improvements in transport and connectivity, and (v) connecting markets and other economic facilities. Therefore, sustainable infrastructure is economically sustainable if it generates a positive net economic return, considering all benefits and costs over the project lifecycle, including positive and negative externalities and spillovers. Considering that fare revenues are insufficient to cover operational and capital costs, some urban rail systems worldwide supplement their funding with non-fare commercial or ancillary revenues [10]. The fare revenue is income generated from fare ticket payments from users to board the LRT. In contrast, ancillary income is income generated from ancillary activities such as retail, parking, advertising, and technology-based business (telecommunications, mobile applications, smart transportation systems). The ancillary revenue is crucial to the financial viability of a mass transit railway without funding support from the Government – even one with high levels of patronage and excellent operating efficiency like MTR [11]. To date, various ancillary businesses have run along the rail-based public transport route. According to Chow [12], various streams can be envisaged to generate additional revenues to either compensate a deficit in the fare revenue or create additional investments resources: (i)Commercial & retail-related activities in stations, such as kiosks, shops and promotion space; (ii) Advertising in the station, on-board trains, on smartcards/token; (iii) Branding of infrastructure: a brand against a station; (iv)Infrastructure leasing: space, telecom assets; (v)Land value capture: TOD; and (vi)Others streams: Digital content, merchandising, Energy selling: solar, other ancillary revenue activities (e.g. in station events/filming, consultancy and parking).

3. Methodology

This study adopted semi-structured interviews to identify the LRT operator and technology provider's perspective on the elements involved. Respondents consisted of individuals involved in the railway and technology industries. The equation proposed by Galvin [13] will be used in this paper as a reference to determine the number of respondents involved will give the probability that the level of saturation will be achieved.

$$R = 1 - (1 - P)^{1/n} \tag{1}$$

In order to generate qualitative data through the use of open questions, semi-structured interviews were used in this study. The development phases of the semi-structured interview were prepared by adapting the procedure used by Kallio, Pietilä [14]. All interview questions developed are based on the nine main elements of the business model canvas by Osterwalder, A. and Y. Pigneur. The nine elements are customer segments, value propositions, channels, customer relationships, revenue streams, key resources, key activities, key partnerships, and cost structure. The semi-structured interview began with a free association task by asking respondents about their understanding of the business model canvas element and their opinion about the business model. It was followed by questions about the respondent perspective on each business model canvas element—the interview questions based on the study main research topic. In order to get a fixed set of responses, the closed question was used, and an open question was used to allow the respondent to express the answer in their own words. Thematic analysis was used to analyse the interview data. NVivo version 12 (NVivo) software is used to facilitate the process of coding and data analysis. Time for analysis and tedious manual and clerical tasks such as transcribing, searching for themes and producing the report can be simplified using NVivo. The software is also capable of handling immense amounts of data with validity and audit functions. The advanced feature of the software, such as identifying themes, gleaning insight and developing conclusions, are required in the study [15].

4. Results and Discussion

The respondent demographic questionnaire from the semi-structured interview required respondents to provide their background information to understand the background and knowledge of the respondents. All respondents involved are individuals with work experience in the rail-based public transport industry or technology providers. A total of ten respondents were interviewed, consisting of workers from the telecommunications and rail industries. The minimum number of populations required as per equation (1) to have a belief or attitude so that 95% ($P = 0.95$) are confident that ten interviews ($n = 10$) will be used, answering $R = 25.89\%$. Provided 25.89% of the population adheres to this belief, ten interviews will be enough to give 95% confidence to the researchers to discover it. According to Israel [16], in a normal distribution, approximately 95% of the sample values are within two standard deviations of the true population value (e.g., mean). 95% confidence level means 95 out of 100 samples will have actual population values within the specified accuracy range. According to Bernard and Bernard [17], 10 to 20 knowledgeable people are enough to uncover and understand the core categories in any well-defined cultural domain or study of lived experience. Therefore, Parse [18] recommends 2 to 10 participants to achieve redundancy or saturation. Table 1 shows the breakdown of the respondent profile obtained from the semi-structured interview.

Table 1. Summary of Respondent profile.

Respondent Code	Years of Experience	Position	Highest Education Level
R1	9	Associate	Bachelor's Degree in Technology Management
R2	13	Senior Associate	Bachelor's Degree in Business Computing
R3	2	Associate	Bachelor's Degree in Mathematics Technology
R4	13	Head of Planning	Master of Construction Contract
R5	15	Senior Associate	Master of Business Administration
R6	10	Project Manager	Bachelor's Degree in Controlling Instrumentation and Automation Engineering
R7	6	Network Engineer	Bachelor's Degree in Manufacturing Engineering
R8	16	Specialist	Master of Business Administration
R9	15	Senior Manager	Master of Business Administration
R10	18	Senior Engineer	Bachelor's Degree in Electrical Engineering

Eleven themes were identified as a result of the thematic analysis performed using NVivo. The findings obtained from the analysis conducted on all interview transcripts found that all nine elements created by

Osterwalder and Pignuer can contribute to the LRT operation sustainability with two additional elements identified, namely technology and risk. Table 2 shows the content analysis for all the themes involved that have been identified using NVivo. A code stating "yes" means that the respondent believes that the element is important for LRT operation sustainability, while "no" means that the respondent does not agree or has no opinion on the element not contributing to LRT operations' sustainability.

Table 2: The outcome from content analysis for all themes extracted from NVivo.

	R01	R02	R03	R04	R05	R06	R07	R08	R09	R10	Total
Activity	Yes	Yes	Yes	Yes	Yes	Yes	Yes	Yes	Yes	Yes	10
Resources	Yes	Yes	Yes	Yes	Yes	Yes	Yes	Yes	Yes	Yes	10
Cost	Yes	Yes	Yes	Yes	Yes	Yes	Yes	Yes	Yes	Yes	10
Customer	Yes	Yes	Yes	Yes	Yes	Yes	Yes	Yes	Yes	Yes	10
Value Proposition	Yes	Yes	Yes	Yes	Yes	Yes	Yes	Yes	Yes	Yes	10
Relationship	No	Yes	Yes	Yes	Yes	Yes	Yes	Yes	Yes	Yes	9
Revenue Streams	Yes	No	Yes	Yes	Yes	Yes	Yes	Yes	Yes	Yes	9
Technology	Yes	No	Yes	Yes	Yes	Yes	Yes	Yes	Yes	Yes	9
Key Partner	Yes	No	Yes	Yes	Yes	Yes	No	Yes	Yes	Yes	8
Channel	Yes	Yes	Yes	No	Yes	No	Yes	Yes	Yes	Yes	8
Risk	No	No	No	Yes	No	No	No	No	No	Yes	2

4.1. Key Activities

One significant finding is that marketing and sales are two main segments in key activities elements. These results match those observed in earlier studies. This finding was also reported by Silva and Maló [19], Štefan and Richard [20] and Dijkman, Sprenkels [21]. This study indicates that relevant data and sharing of views between marketing and sales teams might enhance performance and create a compelling experience for a prospective customer. Most respondents stated the use of technology to facilitate marketing and sales activities. The most widely coded for marketing components is online marketing. During R04 commented: "*With the internet, we can reach even more potential customers. By using marketing through the internet, we will be able to interact directly with users without time and space constraints. So, internet marketing is one of the effective ways nowadays to further expand the reach of the market.*" Similarly, R05 argues that: "*In my opinion, an effective marketing strategy is an online marketing strategy. We also have a social media account that can be used by technology partners in introducing products or services that will be provided.*" In ensuring the successful sales strategy, R09 argues that: "*Perform sales performance appraisals. Periodic performance appraisals can detect errors in existing sales methods.*" R10 highlighted the important strategy to ensure the success of the sales strategy during the interview. R10 mentioned that: "*And what is important is the cooperation between the marketing and sales team.*" In summary, the theme of the main activities is important to ensure that both parties have their respective responsibilities. As stated by R07: "*Through this main activity, all those involved in this collaboration can find out their respective roles in ensuring that the services provided to customers do not have any problems.*" R08 also mentioned that: "*By identifying the main activities involved, the direction of the collaboration carried out will be clear.*"

4.2. Resources

The results of this study agree with the statement made by Štefan and Richard [20], who points out that key resources consist of four major segments: physical resources, financial resources, intellectual resources, or human resources. There are several possible explanations for this result. Physical resources are critical to the organisation's success because it is challenging to operate a business without equipment, infrastructure, and inventory. R05 argues that: "*As one of the companies that operate rail-based public transport, our main physical resources are LRT tracks, LRT carriages, lands along the LRT line and the pillars that support the LRT line and station.*" R10 stated in the interview session: "*The physical resources we need are telecommunications-related assets. Among the assets involved in this collaboration are such as active equipment, passive equipment, and CME structure.*" Human resources play an important role in observing the whole process and avoiding mistakes through available skills and knowledge. According to R10: *"Among the experts we need are Engineers, Technicians and other support workers such as Customer Service Officer, Auxiliary Police and others."* R07 argues that: *"In my opinion, the human resources we need the most are skilled and knowledgeable workers in the field of technology. Our company needs skilled workers in the field of telecommunications, IT or electrical and electronics."* This study shows that intellectual resources are needed in making strategic planning of an initiative and accelerate the implementation of projects that have been planned. According to R03: *"The type of intellectual resources involved in ensuring the business runs through this collaboration is successful by involving ridership and patent data as well as copyrighted drawings."* Similarly, R07 argues that: *"the most important intellectual resources we need are technical details and LRT user data."* Another significant result is that the source of financial resources might come from various sources, namely: 1) From business operations such as revenue from products or services, 2) Capital funding: share and capital contributions issuance, and 3) External sources: vendor financing, bank loans and corporate bonds issuance. R10 argues that: *"The financial resources involved in this collaboration can be obtained either through borrowing or selling the company's equity."* R10 also mentioned that: *"If start-up company, I think, the best strategy to get start-up capital to run a business is through a government's financial grant."*

4.3. Cost Structure
The cost structure consists of two main segments: fixed cost and variable cost. During the interview session, R09 stated: *"In my opinion, the highest fixed cost is the rental cost to use the LRT's right of way. So rental costs, whether monthly or annual, will be a major contributor to our fixed costs."* R08 summarise that: *"expenses for the operation of the services provided such as overhead, site rental, bandwidth rental and telecommunication structure rental are also categorised as fixed costs that will be involved in this collaboration."* R05 mentioned that: *"For variable costs, I think the costs involved in the maintenance of technology-related equipment—utility costs such as electricity and water."* According to R05: *"Even general labour costs on a contract basis, I think, can be included as one of the variable costs."* R01 stated that: *"Variable costs involved, employee salaries, utility costs, possible asset rental can increase, and other unexpected costs such as vandalism, which we cannot control."* The result of the thematic analysis matched the theorised dimensions of Priyadi and Prasetio [22], Suwandiman and Mukti [23] and Sutrisno [24], in which are cost is divided into two main segments, namely fixed cost and variable cost. Another important finding was that an additional segment had been identified through the analysis conducted, namely start-up cost to complete the cost elements. A possible explanation for the start-up cost might be that setting a reasonable budget, in general, is crucial for every collaboration. It assists the LRT operator and Technology Provider in the collaboration staying on track and focusing on the end target. If the budget involved is never followed correctly in this collaboration, it can lead to serious financial issues.

4.4. Customer
The most prominent finding to emerge from the analysis is that the primary customers for this collaboration are the telecommunications operator, LRT user, enterprise, individual customer, and LRT operator. According to R04, *"the main customers who will use this product or service are LRT users."* The same goes for the opinions expressed by R07. He mentioned that: *"LRT users and companies along the LRT line are our main customers if we collaborate with the LRT."* While R10 views that the customer: *"can be telecommunication companies, companies that provide data centres, LRT users and others."* R06 commented: *"Companies close to the LRT line that need internet service have the potential to become our customers."* R05 commented that: *"In my opinion, if the product or service is based on technology, the main customers may consist of our passengers, retailers inside and outside our station, as well as corporate companies such as telecommunications companies, IT companies and others."* This finding corroborates the ideas of Štefan and Richard [20], who argue that customer segment is defined by five types of market: mass, segmented, niche, diversified and multi-sided. The present study raises the possibility that the potential customers to use the products resulting from this collaboration are not only focused on LRT users, but the customers can reach beyond the LRT colony under the coverage of

the deployed technology services. Therefore, it can be assumed that knowing the customers better is crucial because their information can help the business earns more revenue and sustains the business.

4.5. Value Proposition
This study found that the value proposition element provides many benefits to the consumer. One interesting finding is that this collaboration provides more savings in cost, time, and assurance of customer equipment safety. Telecommunication operators will further expand their network if they collaborate with LRT operators, as stated by R04: *"The most significant advantage of the service carried out through this collaboration is in terms of coverage of the fibre optic cable network."* R03 argues that: *"The LRT line makes a ring loop in the Klang Valley in most of the development-intensive areas, which enables our partners to expand their business network throughout the Klang Valley."* The safety of telecommunication equipment will also be safer if working with LRT operators as LRT lines are safe and guarded corridors that prevent unnecessary access to the public. R06 stated that during the interview session: *"The advantage of the service offered by us in collaboration with the LRT is in terms of safety and guarantee no service interruption to customers. Our fibre optic cable will be installed inside via the LRT line duct."* This study confirms that value capture is associated with added value to the customer, as mentioned by Silva and Maló [19], Bilgeri, Brandt [25] and Joha and Janssen [26]. Therefore, it can be assumed that the value proposition element is an element involved in ensuring the LRT operation sustainability.

4.6. Relationship
Silva and Maló [19], Štefan and Richard [20], Tseng and Wu [27], Jocovic, Melovic [28], Chuang, Lin [29], Chan [30] and Haaker, Bouwman [31] showed that the relationship element is revolving around the relationship with the customer. It differs from the findings presented here, where relationships in collaboration involve relationships with customers and collaboration partners, stakeholders, and shareholders. Stakeholders and shareholders involved in this collaboration include the ministry, local authorities, utility providers and the surrounding population. R10 argues: "*Creating a collaboration contract is the most effective strategy in maintaining collaboration. So, having a complete and detailed contract will make it easier for both parties to resolve any disputes that may occur during the collaboration period."* In maintaining relationships with stakeholders and shareholders, R07 and R10 argue that communication is the most effective strategy. R07 argues that: *"Communication is the most important method in maintaining relationships with stakeholders and shareholders."* Each of these identified relationship segments is likely to have a positive relationship with the performance of the collaborations undertaken. This means that if there is an improvement in each segment in the relationship element, it will also improve the performance of the collaboration carried out.

4.7. Revenue Streams
This study found that there are various sources of revenue that can be generated through this collaboration. Among the revenue sources are site rental, bandwidth leasing, fibre leasing, revenue share, Wi-Fi, end-to-end services, and commission. R07 argues: *"We generate the revenue from dark core fibre rental, bandwidth and site structure. But bandwidth sales and structural rentals are the biggest sources of revenue that can be generated, in my view."* R04 argues that revenue can be generated through revenue sharing, where R04 states that: *"The main source of revenue that is likely to be generated is through revenue sharing with partners."* R10 argues that: *"If we collaborate with other companies, the revenue sharing is also one of our revenue sources."* This finding is consistent with that of Štefan and Richard [20], who mentioned that revenue streams are usually multi-source, and the type of revenue is affected by the kind of business. These results may be limited to collaboration between telecommunication operators with LRT operators. Hence, all revenue streams are based on the primary sources of revenue generated from the telecommunications business.

4.8. Technology
An unexpected result was the listing of technology-related elements resulting from an analysis conducted using the NVivo12 software. According to R09: *"Technology can facilitate activity and may*

make the business run can be done better and more effectively. Incorporating technological elements in the business venture makes it possible for the business venture to reach a broader global market." While R04 states that: *"In this age of technology, it is necessary that we need to incorporate this technological element into the framework to be built."* R05 argues that: *"Digital transformation needs to be one of the main components of the technological element."* There are several possible explanations for this result. The first possibility is that this collaboration is related to technology-based businesses. It makes technology an essential element so that successful collaboration remains sustainable in every technological evolution and progression. Another possible explanation for this is to ensure that both collaborating parties are better prepared for future development. These findings related to the technology element must be interpreted with caution. If there is no proper and robust business model used to deploy the latest technology, successful collaboration is unlikely to succeed. The lifespan of established business models is declining as technology disrupts industry after industry [32]. Beqiri [33] suggests that it was more necessary than ever to change the business model toward a model based on competitive advantage in the last decade with the explosion of technology.

4.9. Key Partners
The current study found that selecting key partners consisting of technology-neutral players will increase flexibility in penetrating the market. Another important finding was that key partners with such infrastructure RoW could reduce the deployment cost. R03 argues that: *"Neutral companies, companies that can cooperate with other telco companies and any company that is seen as potential in generating additional revenue to us."* R03 commented on his preferred business partner and his expectation from the partnership: *"In my opinion, the criteria for a business partnership that need to be in ensuring that a collaboration carried out can be successful are companies that have an NSP (Network Service Provider) and NFP (Network Facilities Provider) license from SKMM."* R09 argues that: *"The main criteria I will choose our business partners who have infrastructures such as railways, highways, utility routes such as water, electricity and gas."* This finding is consistent with that of Nakano [34], who argues that infrastructure plays a fundamental role as a basis for economic aspects of both domestic and international activities. As such, the LRT has excellent potential to be a key partner for any technology provider company due to its infrastructure that meets the criteria mentioned. These findings suggest that selecting legitimate licensees such as NSPs and NFPs as key partners can provide an advantage in exploiting new market opportunities due to their experience in technology-related industries.

4.10. Channels
The results of this study indicate that element of the channel shall be part of the framework. This study found that the built framework's channel element is divided into network-based distribution and physical-based distribution. According to RO8: *"The product that can be delivered to customer Network-related distribution is one way of delivering products to customers. It can be done using virtual, telecommunication spectrum, fibre optic cable network, online and any network-connected medium."* While R10 argues that: *"Technically, the service provided by us is channelled through the optical fibre cable."* As mentioned by R03: *"We will implement the type of product distribution strategy to deliver the products or services we provide by using Customer Service Officer as a middleman to deliver the services provided and open as many kiosks or outlets that can be held at LRT stations, especially in penetrating as many markets as we can."* In contrast, R02 thinks that opening kiosks at LRT stations may facilitate the distribution of products. As a person who works in public transport industries, R02 states: *"We will usually offer to open as many kiosks as possible near the outlets at the LRT station to penetrate as many markets as possible."* This finding is consistent with Verhoef, Kannan [35] and Piotrowicz and Cuthbertson [36], who suggested that channel elements can be done either online or offline. A possible explanation for network-based distribution might be that as technology advances, more and more solutions are emerging to streamline the distribution process. In contrast, physical distribution involves various product distribution activities associated with the supply of finished products from technology operators to consumers. Therefore, combining the two segments found in this channel element must be present in this collaboration.

4.11. Risks

Another important finding from the analysis is related to the risk elements. According to R05: *"Fibre optic cables installed on the track have less risk of fibre cutting compared to fibre-optic cables installed on the road."* R02 explained why the optical fibre cable is less risky when installed inside LRT track's via duct compare to outside of the station: *"I see with this existing business that the fibre cable service using via duct will be safer because the issue of fibre cutting outside of the station area is much higher compared to the fibre cable inside the station."* For business risk, R07 argues that: *"Business partners are important in minimising the risk of running a business."* Similarly, R09 states: *"This business partnership is important in ensuring that business risks can be reduced."* The evidence from this study suggests that the element of risk is related to business risk and operational risk. A possible explanation for this might be that identifying risks can prevent the business from taking such risks if they have the high potential to occur and could result in substantial losses to both parties. Therefore, effective risk management is essential in ensuring that this collaboration remains sustainable in generating maximum profits. As technology changes and continues to evolve, it is also a risk, especially for technology-based businesses, to ensure their offerings are current and relevant to the customer base. All type of technology risk has the potential to trigger financial, reputational, legislative or strategic risks. Finch [37] claims that a business plan helps the organisation think well to mitigate any risk anticipated by the business in the future and optimise business growth and development according to priorities. Serebrisky, Watkins [7] claim that for infrastructure to be financially sustainable, it must generate an adequate risk-adjusted rate of return for project investors.

5. Conclusion

This paper aims to identify the elements involved in generating ancillary revenue through technology-based businesses from optical fibre cable for LRT operation sustainability. The list of elements involved has been proposed in this study to generate ancillary income for the sustainability of LRT operations and contribute to sustainable development and indirectly may be the driver to achieving one of the objectives outlined in the SDGs. The identified elements may be useful to initiate an ancillary revenue framework through collaboration with technology provider so that the public transport industry remains sustainable with effective cost and operation. The validation process of the identified elements needs to be done to ensure that it can be implemented in the industry. Therefore, subsequent studies should focus on the validation process of the identified elements and the development of the ancillary revenue framework. The expert elicitation method may be suitable for the validation process because this method will involve the views of industry experts on the effectiveness of the elements in the framework developed and the improvements that need to be made so that it is in line with industry requirements to ensure sustainability.

6. References

[1] Rossi-Hansberg, E. 2017 Briefing: Infrastructure projects and the flexibility of urban infrastructure. *Proceedings of the Institution of Civil Engineers-Smart Infrastructure and Construction.* **170(1):** p. 3-5.

[2] Haque, M.M., H.C. Chin, and A.K. Debnath 2013 Sustainable, safe, smart—three key elements of Singapore's evolving transport policies. Transport Policy. 27: p. 20-31.

[3] Hens, L., et al. 2019 Transport Economics and Sustainable Development in Ukraine. *Marketing and Management of Innovations* **Issue 3**: p. 272-284.

[4] Powrie, W., D.J. Richards, and V.K. Mootoosamy 2020 The design of railway overhead line equipment mast foundations. *Proceedings of the Institution of Civil Engineers-Geotechnical Engineering.* 173(5): p. 428-447.

[5] Gao, Z. and L. Yang 2019 Energy-saving operation approaches for urban rail transit systems. *Front. Eng. Manag.* **6(2)**: p. 162-163.

[6] Osterwalder, A. and Y. Pigneur 2013 Designing business models and similar strategic objects:the contribution of IS. *Journal of the Association for information systems.* **14(5)**: p. 237.

[7] Serebrisky, T., et al. 2018 *IDBG Framework for Planning, Preparing, and Financing Sustainable Infrastructure Projects: IDB Sustainable Infrastructure Platform. Inter-American Development Bank.*
[8] Bhattacharya, A. and M. Jeong, 2018 *Driving The Sustainable Infrastructure Agenda In Emerging Markets.*
[9] Saghir, J., 2017 *Sustainable Infrastructure Development in Sub Saharan Africa: A View from the Ground.* Institute for the Study of International Development (ISID): Quebec, QC, Canada.
[10] Pulido, D., et al., 2018 *The Urban Rail Development Handbook.* Washington: The World Bank.
[11] Tang, S. and H.K. Lo, 2016 *Property Models for Financing Railway Development, in Sustainable Railway Futures: Issues and Challenges*, B.P.Y. Loo and C. Comtois, Editors. Routledge. p. 201-218.
[12] Chow, G. 2017 Non-fare revenue for Metro Networks. *10th UMI & CODATU XVII Conference.* Hyderabad, India: CODATU.
[13] Galvin, R., 2015 How many interviews are enough? Do qualitative interviews in building energy consumption research produce reliable knowledge? *Journal of Building Engineering*, **1**: p. 2-12.
[14] Kallio, H., et al., 2016 Systematic methodological review: developing a framework for a qualitative semi-structured interview guide. *Journal of advanced nursing.* **72(12)**: p. 2954-2965.
[15] Sotiriadou, P., J. Brouwers, and T.-A. Le, 2014 Choosing a qualitative data analysis tool: A comparison of NVivo and Leximancer. *Annals of Leisure Research.* **17(2)**: p. 218-234.
[16] Israel, G., 2013 Determining sample size. *IFAS Extension.* University of Florida.
[17] Bernard, H.R. and H.R. Bernard, 2013 *Social research methods: Qualitative and quantitative approaches.* Sage.
[18] Parse, R.R., 1990 Parse's research methodology with an illustration of the lived experience of hope.*Nursing Science Quarterly.* **3(1)**: p. 9-17.
[19] Silva, E.M. and P. Maló, 2014 IoT testbed business model. *Advances in Internet of Things.* **4(04)**: p. 37-45.
[20] Štefan, S. and B. Richard, 2014 Analysis of Business Models, *Journal of Competitiveness* **6(4)**: p. 19-40.
[21] Dijkman, R.M., et al. 2015 Business models for the Internet of Things. *International Journal of Information Management*, **35(6)**: p. 672-678.
[22] Priyadi, Y. and A. Prasetio. 2018 Implementation of supply chain business application through business model canvas and waterfall framework collaborations for fish farmers SMEs in ulekan market bandung. *Journal of Physics: Conference Series.* IOP Publishing.
[23] Suwandiman, K.Z. and G.W. Mukti, 2018 Business Model Analysis of Organic Rice Farmers Group (Study Case in Sarinah Organic Farmers Group, Bumiwangi Village, Ciparay, Bandung Regency, West Java Province. *Sustainable Collaboration. Business, Technology, Information and Innovation (SCBTII).* **1(1)**: p. 1-6.
[24] Sutrisno, W. 2019 Analysis of provider training business model development based on canvas business model approach. *IOP Conference Series: Materials Science and Engineering.* IOP Publishing.
[25] Bilgeri, D., et al. 2015 The IoT business model builder *A White Paper of the Bosch IoT Lab in collaboration with Bosch Software Innovations GmbH.* University of St. Gallen: St. Gallen, Switzerland.
[26] Joha, A. and M. Janssen 2014 Factors influencing the shaping of shared services business models: Balancing customization and standardization. *Strategic Outsourcing: An International Journal,* **7(1)**: p. 47-65.
[27] Tseng, S.-M. and P.-H. Wu 2014 The impact of customer knowledge and customer relationship management on service quality. *International journal of quality and service sciences.* **6(1)**: p. 77-96.
[28] Jocovic, M., et al. 2014 Modern business strategy Customer Relationship Management in the area of civil engineering. *Applied Mechanics and Materials.* **678(2014)**: p. 644-647.

[29] Chuang, H.-M., et al. 2015 Elucidating the Merits of Customer Relationship Management in Cloud Computing. *Applied Mathematics & Information Sciences.* **9(4)**: p. 2001 - 2013.
[30] Chan, H.C. 2015 Internet of things business models. Journal of Service Science and Management **8(04)**: p. 552 - 568.
[31] Haaker, T., et al. 2017 Business model stress testing: A practical approach to test the robustness of a business model. *Futures* **89**: p. 14-25.
[32] Euchner, J. 2016 Business Model Innovation. *Research-Technology Management.* **59(3)**: p. 10-11.
[33] Beqiri, G. 2014 Innovative Business Models and Crisis Management. *Procedia Economics and Finance.* **9**: p. 361-368.
[34] Nakano, H. 2017 A Study on the Features of the Evolution Processes and Business Models of Global Enterprises in the Transport Sector. *Transportation Research Procedia* **25(Supplement C)**: p. 3769-3788.
[35] Verhoef, P.C., P. Kannan, and J.J. Inman 2015 From multi-channel retailing to omni-channel retailing: introduction to the special issue on multi-channel retailing. *Journal of retailing.* **91(2)**: p. 174-181.
[36] Piotrowicz, W. and R. Cuthbertson 2014 Introduction to the special issue information technology in retail: Toward omnichannel retailing. *International Journal of Electronic Commerce.* **18(4)**: p. 5-16.
[37] Finch, B. 2016 *How to write a business plan.* Fifth Edition ed. London, United Kingdom: Kogan Page Publishers. 192.

Acknowledgement
The authors would like to express their sincere thanks to the Ministry of Higher Education Malaysia (MOHE) for the research fund under the UTM Encouragment Research (ER) of the Project Cost Centre Q.K130000.2656.18J24.

Journal of Physics: Conference Series 2053 (2021) 012016 doi:10.1088/1742-6596/2053/1/012016

Effect of corrugated minichannel variable width on entropy generation for convective heat transfer of alpha-Alumina-water nanofluid

N M Muhammad [1,3*], N A C Sidik [1,2], A Saat [1], Y Asako [2], W M A A Japar [2] , G H Musa [3] and S N A Yuof [2]

[1] Faculty of Engineering, Universiti Teknologi Malaysia, 81310 Skudai, Johor, Malaysia.
[2] Malaysia-Japan International Institute of Technology. Universiti Teknologi Malaysia, Jalan Sultan Yahya Petra, 54100 Kuala Lumpur, Malaysia.
[3] Faculty of Engineering, Kano University of Science and Technology, Wudil. PMB 3244. Nigeria.

*Corresponding author's email : nuramuaz@gmail.com

Abstract. Energy management and sustainability in thermal systems require maximum utilization of resources with minimal losses. However, it is rarely unattainable due to the ever-increasing need for a high-performance system combined with device size reduction. The numerical study examined convective heat transfer of an alpha-Alumina-water nanofluid in variable-width corrugated minichannel heat sinks. The objective is to study the impact of nanoparticle volume fractions and flow area variation on the entropy generation rate. The determining variables are 0.005 – 0.02 volume fractions, the fluid velocity 3 – 5.5 m/s and heat flux of 85 W/cm^2. The numerical results show an acceptable correlation with the experiment results. The results indicate the thermal entropy production drop with an increase in nanoparticles volume fraction. Contrastingly, the frictional resistance entropy suggests the opposite trend due to the turbulence effect on the fluid viscosity. The induction of Alumina-Water nanofluid with enhanced thermal conductivity declined the entropy generation rate compared to water alone. The increase in width ratio by 16% between the cases translates to at least a 9% increase in thermal entropy production. The outcome of this study can provide designers and operators of thermal systems more insight into entropy management in corrugated heatsinks.

1. Introduction

The recent integration of compact, high-performance electronic devices posed a severe challenge to cooling efficiency due to an overwhelming high heat flux. The dissipation of this heat flux is necessary to reduce the mean time between failure ad increase the lifespan of the devices. The improvement of heat transfer in thermal systems has received much attention recently. The researchers recommended using high thermal conductive liquid-like nanofluids and changes to the flow path [1]. Nevertheless, entropy generation has not been given such preference despite its significance in thermal management and sustainability of systems. Few studies on entropy based on the second law of thermodynamics indicates the varying outcome of thermal and frictional resistance entropy generations [2, 3]. Bejan [4], in his classical work, studied the thermodynamic efficiency of a system. He found entropy as a combined effect of thermal irreversibility and frictional losses. Sohel defined system entropy production as the entropy measure formed by irreversibilities like a chemical reaction, mixing and heat transfer through a finite temperature gradient [5]. Mir et al. [6] observed an increment in volume fraction of Ag nanoparticles in water enhanced the temperature of the central flow path. For the natural convection, the entropy generation rate was significant at low Gr because of the high-temperature gradient near the elliptical curved minichannel wall. Shahsavar et al. [7] also reported that Fe_3O_4 and CNT concentrations enhance the total entropy generation and convective heat transfer coefficient of the inner and outer walls of the concentric annulus. Mahian and his co-researchers [8] explored numerical analysis of various nanofluids including Cu/water, Al_2O_3/water, SiO_2/water and TiO_2/water on the thermal performance of minichannel flat plate solar collector. They observed minimization of entropy production due to different volume fraction of nanofluids and significant mass flow rate. Ahammed et al. [9] corroborated the findings of Mahian and his colleagues. They observed Re rise from 200 to 1000 at the maximum heat flux of 25 kWm^{-2} leads to the decline of total entropy production by 96% from 0.0361 W/K. Due to recent advantages of computing in machine learning, Bahiraei and Abdi [10] proposed an ANN model to examine the entropy production of TiO_2-water nanofluid in a circular minichannel. The main parameters include Re 500-2000, average concentrations 1% - 4% and particles diameter of 20-80 nm. They found that at a constant heat flux of 10 kW/m^2, the influence of particle migration is substantial on entropy production, especially at high concentration and largest particles diameter. However, the effect is negligible at a low volume fraction of 0.01 and the smallest diameter of 20 nm. They observed that the Bejan number is higher than 0.8 at all concentrations, which implies that heat transfer is responsible for more than 80% of the generated entropy. Other researchers that explored entropy production include [11, 12]. Recently, Huminic and Huminic [13] presented a comprehensive review of works on entropy production using nanofluids in various geometries. They affirmed the advantage of nanofluids in micro/minichannels over the conventional thermal system in achieving effective cooling and reduced entropy generation. Divergent-convergent geometry can recirculate flow and significantly augment heat transfer with moderate pressure loss [14, 15]. Based on the authors' knowledge, the literature has no account of the study on entropy generation in a divergent-convergent minichannel heatsink. The objective is to study the impact of nanoparticle volume fractions and flow area variation on the entropy generation rate using water-based nanofluid in a corrugated minichannel. The outcome of this study can provide designers and operators of thermal systems more insight into entropy management in corrugated heatsinks.

2. Methodology

2.1. Physical models

The study involved divergent-convergent minichannel geometry made of Aluminium substrate. Thus, only a complete and symmetrical unit of the heat sink of the seven parallel minichannels formed the computational domain to minimize the computational cost. For the analysis, two cases C1 and C2, with dimensionless width (w/W) of 0.4 and 0.5, respectively, are considered. Here, w and W represent the minimum and maximum width of the minichannel, as shown in figure 1(a). Thus, the computational domain has 30mm, 3mm, and 2.25 mm as the length, width, and height. Table 1 expressed the geometrical specifications of the cases.

Table 1. Dimensions of the minichannels cases.

Case	w (mm)	W (mm)	Channel height (mm)	Dh (mm)
Case 1	1.0	2.3	1.25	1.42
Case 2	1.0	2.0	1.25	1.36

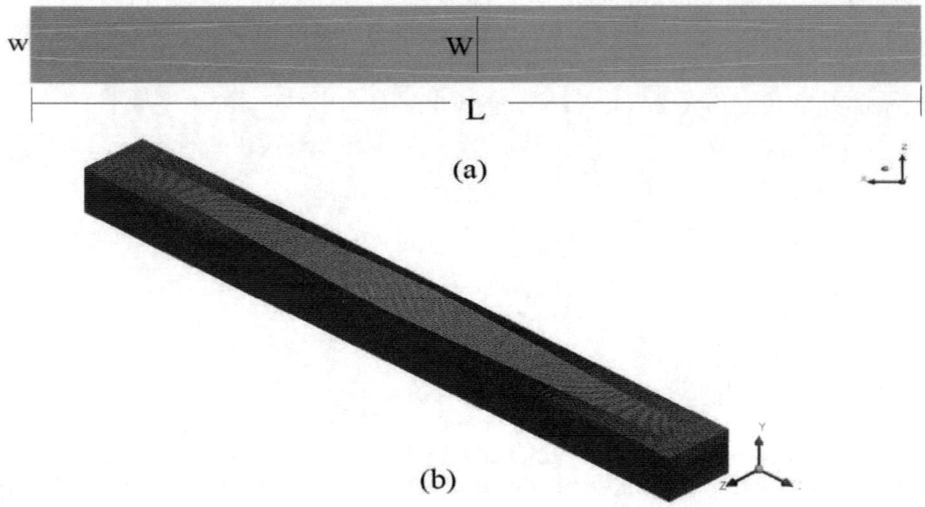

Figure 1. (a) Mesh of the computational domain (b) The top view of the minichannel showing the dimensions.

2.2. Mathematical models

The mathematical modelling made some assumptions to establish the governing equations for the numerical analysis. First, the nanofluid behave as Newtonian fluid flowing in a turbulent continuum through the minichannels. Second, the nanofluid, due to the small volume fraction, dissolve entirely in the water to form a homogenous single-phase mixture. Third, the fluids and solid particles move with the same velocity and exist in thermal equilibrium [16]. Viscous dissipation and radiation are neglected. The thermophysical properties of fluids depend on temperature. The analysis adopted polynomial values of the thermophysical properties used by Bahiraei et al. [17]. Considering the assumptions, the study expressed the Navier-Stokes conservation and energy equations as follows:

Continuity equation:

$$\nabla \cdot (\rho \cdot \vec{V}) = 0 \quad (1)$$

$$\nabla \cdot (\rho \cdot \overrightarrow{V \cdot} \vec{V}) = -\nabla P + \nabla(\mu \cdot \nabla\vec{V}) \quad (2)$$

$$\nabla \cdot (\rho \cdot \overrightarrow{V \cdot} Cp \cdot T) = \nabla \cdot (k\nabla T) \quad (3)$$

Additional equations formulate the turbulent model. The realizable k-ε turbulent model proposed by Shih et al. [18] can effectively simulate the analysis. It has two different equations: the turbulent kinetic energy k and dissipation of this energy ε expressed as follows:

$$\frac{\partial}{\partial t}(\rho k) + \frac{\partial}{\partial x_j}(\rho k v_j) = \frac{\partial}{\partial x_j}\left[\left(\mu + \frac{\mu_t}{\sigma_k}\right)\frac{\partial k}{\partial x_j}\right] + G_k + G_b - \rho\varepsilon - Y_M + S_k \quad (4)$$

$$\frac{\partial}{\partial t}(\rho\varepsilon) + \frac{\partial}{\partial x_j}(\rho\varepsilon v_j) = \frac{\partial}{\partial x_j}\left[\left(\mu + \frac{\mu_t}{\sigma_\varepsilon}\right)\frac{\partial \varepsilon}{\partial x_j}\right] + \rho C_1 S\varepsilon - \rho C_2\frac{\varepsilon^2}{k + \sqrt{v\varepsilon}} + C_{1\varepsilon}\frac{\varepsilon}{k}C_{3\varepsilon}G_b + S_\varepsilon \quad (5)$$

G_k and G_b represent turbulence kinetic energy generation arising from average velocity gradient and buoyancy, respectively. Y_m is the contribution of the fluctuating dilatation incompressible turbulence to the overall dissipation rate. From equation (5), the values of the model constants are: $C_{1\varepsilon}$, C_2, σ_k and σ_ε are 1.9,1.0 and 1.2, respectively.

The following boundary conditions apply to the current study:

(i) At the inlet of the heatsink minichannel, labelled as "velocity-inlet": $u = u_{in} = 3$ - 5.5 ms^{-1} equivalent to Reynolds number 5000 – 10000. $T = T_{in}$ =303K.

(ii) At the outlet of the minichannel referred to "Pressure outlet": P= 0 Pa.

(iii) A constant heat flux imposed on the bottom wall: $-k_s \frac{\partial T_s}{\partial n} = q = 85\ W/cm^2$

(iv) All other contact surfaces of fluid and substrate experienced no-slip condition and coupled walls: $u_x = u_y = u_z = 0;\ T_f = T_s;\ -k_s \frac{\partial T_s}{\partial n} = k_f \frac{\partial T_f}{\partial n}$

(v) The two parallel sides of the computational domain are considered as symmetry: $-k_s \frac{\partial T_s}{\partial y} = 0$

(vi) All remaining solid and fluid surfaces are insulated (adiabatic); $\frac{\partial T_s}{\partial x} = 0;\ \frac{\partial T_f}{\partial x} = 0$

2.3. Numerical code and grid sensitivity check

The ANSYS CFD code v17, according to the finite volume method, solved the governing equations in a three-dimensional domain. The COUPLE scheme discretized the equations of pressure and velocity for faster and effective convergence. Convergence set when all the residual diminished below 10^{-6}. The hexahedral mesh mapped the computational domain with a finer grid in the minichannel periphery. The work tested the grid sensitivity to numerical result using three structured mesh (240000, 480000 and 960000) against the finest grid G4 of 1920000 elements. The error of G3 relative to G4 shows 0.48% in terms of the average Nusselt number for water flowing at Re 10000. Hence, the study adopted the third mesh due to its optimal computational time and accuracy, shown in figure 1(b).

2.4. Thermophysical properties of the fluids

In this work, the nanofluid composed of alpha-Alumina nanoparticles dispersed in a distilled water. The volume fractions include 0.005 to 0.02. Table 2 expressed the nanofluids and the base fluid thermophysical properties.

Table 2. Thermophysical properties of base fluid and nanoparticles referenced T = 303K.

Materials	Density (kg/m^3)	Specific heat (J/kg K)	Thermal conductivity (W/mK)	Dynamic Viscosity (kg/ms)
Water (H_2O)	995.1	4182.8	0.603	799×10^{-6}
Alumina (α-Al_2O_3)	3970	765	40	-

The nanofluid modelling estimates the effective thermophysical properties appropriately. They include density (ρ), specific heat capacity (C_p), thermal conductivity (*k*), and viscosity (μ), corresponding to equations (6) – (9) as follows:

$$\rho = (1-\phi)\rho_{bf} + \phi\rho_p \tag{6}$$

$$Cp_{nf} = [\phi(\rho Cp)_p + (1-\phi)(\rho Cp)_{bf}]\rho_{nf}^{-1} \tag{7}$$

$$k_{nf} = \left[1 + 4.4\phi^{0.66}\left(\frac{T}{T_{freez}}\right)^{10}\left(\frac{k_p}{k_{bf}}\right)^{0.03} Pr^{0.66} Re_p^{0.4}\right] k_{bf} \tag{8}$$

where Rep is the particle Reynolds number, its expressed as : $Re_p = \frac{\rho_{bf} k_B T}{3\pi\mu_{bf}^2 l_{bf}}$

The equation is valid for 10 ≤ dp ≤150 nm, 0.2 ≤ ϕ ≤9 % and 294 ≤ T ≤324 K.

$$\mu_{eff} = \left[1 - 34.87\left(\frac{d_P}{d_{bf}}\right)^{-0.3}\phi^{1.03}\right]^{-1} \tag{9}$$

where $d_{bf} = 0.1\left(\frac{6M}{N\pi\rho_{fo}}\right)^{1/3}$ where ρ, Cp, k, and ϕ are the density, heat capacity, thermal conductivity, viscosity, and nanoparticle concentration. Whereas the base fluid, the nanoparticle and nanofluid were denoted by subscripts bf, p, and nf, respectively.

2.5. Validation of numerical results

The numerical analysis validates the result with the experimental data for the fully developed turbulent condition of Phillips et al. [19] and Blasius relation [20] for local Nusselt number and average frictional factor, respectively. The analysis used distilled water at Reynolds number 10000. In figure 2(a), the local Nusselt number deviates by about 3%. However, in figure 2(b), the experimental friction factor values vary by around 20% and 4% lower at Re 5000 and 10000. The deviation was primarily attributed to the assumptions proposed during the simulation.

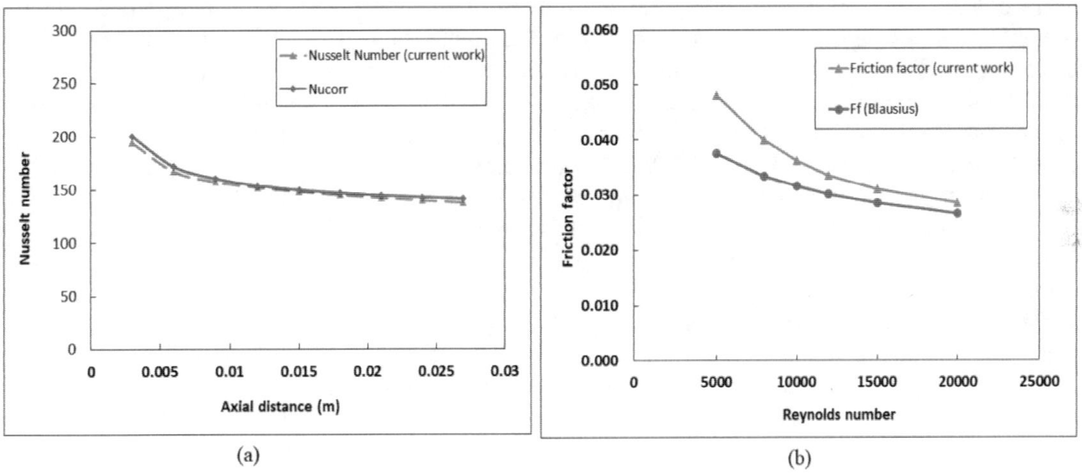

Figure 2. Validation of (a) Local Nusselt number and (b) friction factor.

2.6. Data Processing

In this section, the parameters used to evaluate the thermal performance of the minichannel are presented with appropriate equations as follows:

Equations set (10) evaluates the Reynolds number (Re), average heat transfer coefficient (h), Nusselt number and frictional resistance (f), respectively:

$$Re = \frac{\rho u D_h}{\mu} \; ; \; h_{av} = \frac{q_w}{\left(T_w - T_f\right)}; \; Nu_{av} = \frac{hD_h}{k}; \; f = \frac{2D_h \Delta P}{2L\rho u^2} \tag{10}$$

where ΔP is the pressure differential obtained as the P_{in} - P_{out}. D_h, and L represents the hydraulic diameter, channel length.

The formulation of thermal ($\delta_{gen,th}$) and friction entropies ($\delta_{gen,f}$) and dimensionless Bejan no (Be) are as follows [5]:

$$\delta_{gen,th} = \frac{Q\Pi D_h L}{Nu\, k\, T_b} \text{ and } \delta_{gen,f} = \frac{\dot{m}^3 f L}{\rho T_b D_h A_{con}^2} \tag{11}$$

$$Be = \frac{\delta_{gen,th}}{\delta_{gen,th} + \delta_{gen,f}} \tag{12}$$

Q is the sensible heat absorbed by the fluid. A_{con}, ṁ and Π denote the convection area, fluid mass flow rate and the channel perimeter.

3. Results and Discussion

The present paper examined the production of entropy of nanofluids due to thermal and frictional effects in two similar corrugated minichannel heat sinks with a variable maximum width. It involves using a single-phase model for volume concentrations of 0.5 to 2 % with Reynolds number range 5000 – 10000 and constant heat flux of 850 kW/m^2.

3.1. Effect of variable width on thermal entropy production

Figure 3(a) shows that the entropy generation due to thermal irreversibilities decreases with an increase in fluid velocity. The velocity increment dislocates the thermal boundary layer due to the thickening of the hydraulic boundary layer and results in heat transfer enhancement. Therefore, the thermal entropy production is higher in case 1 than in case 2. The reduction of surface area and lower hydraulic diameter in case 2 results in a higher heat transfer coefficient (HTC). It is also clear that an increase in volume fraction of Al_2O_3 in water provides better thermal augmentation due to thermal conductivity enhancement and a rise in mass density. Therefore, the higher the nanofluid volume concentration, the higher the thermal boost and the lower the thermal entropy production. For instance, at 2% Al_2O_3-H_2O nanofluid, the thermal irreversibility reduces by 2.11% and 0.3% compared to water for Case 1 at 3 ms^{-1} and 5.5 ms^{-1}, respectively. Similarly, at the same condition, the reduction is 11.81% and 9.77% for case 2. Comparing the cases, at 2 vol%, the thermal irreversibility is 9.4% and 9.6% higher in case 1 than in case 2 for the corresponding fluid velocities of 3 ms^{-1} and 5.5 ms^{-1}. Hence, the increase in width ratio of 16% translate to at least a 9% increase in thermal entropy production. Therefore, the reduction of surface area can have a significant effect on thermal entropy generation.

3.2. Effect of variable width on Frictional entropy production

Higher fluid velocity depreciates the flow friction resistance and increases hydraulic boundary layer thickness. Therefore, frictional entropy production rises with the increase in the fluid rate for all cases. Also, an increase in Al_2O_3-H_2O volume concentration increases the viscosity significantly. Hence, frictional irreversibility increases with an increase in nanofluid volume fraction. Water indicated the lowest frictional entropy production due to its lower viscosity. However, the viscosity effect in turbulent flow is minor due to insufficient frictional resistance. Hence, it results in the lower contribution of frictional entropy losses. For case 1, the frictional entropy increment between water and 2 vol.% Al_2O_3-H_2O was 18.5% and 18.01% for 3 ms^{-1} and 5.5 ms^{-1}, respectively. For case 2, the frictional entropy production increases by 32.8% and 33.1%, as shown in figure 3 (b). The high values account for reducing hydraulic diameter in case 2, which leads to increased pressure loss and, consequently, a rise in frictional resistance.

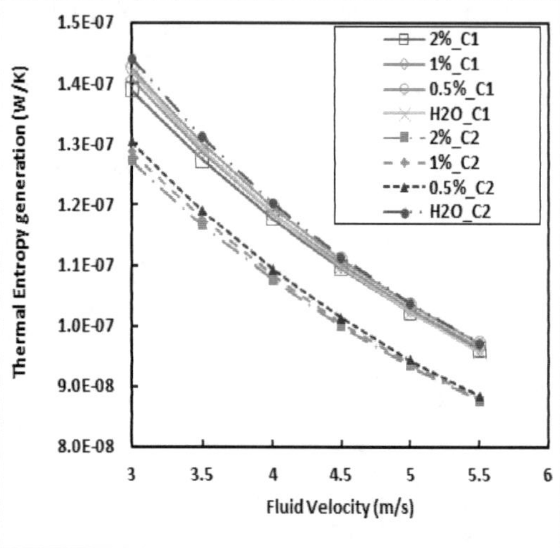

(a)

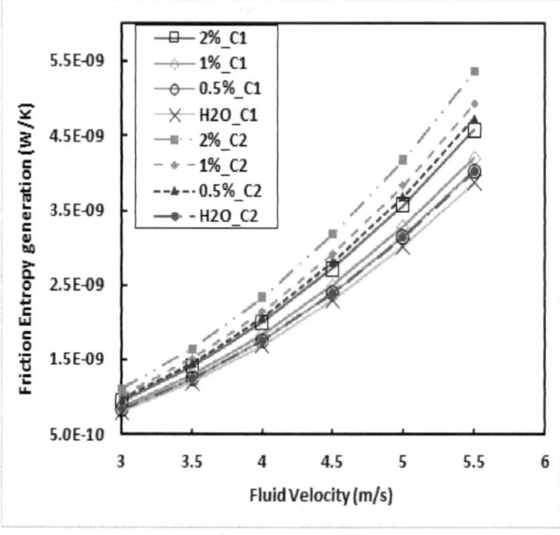

(b)

Figure 3. Entropy generations against fluid velocity for the different cases: (a) Thermal (b) Frictional.

3.3. The comparative effect between the thermal and frictional entropies
The study used Bejan no, expressed as a ratio of thermal entropy to total entropy production, to determine the most influential parameter between the two entropies. The Bejan number decline with the increase in volume concentration and fluid velocity, as shown in figure 4. For all cases, it fell below unity, though it is lower in case 2 for all the volume fractions. The Be values of 0.93 to 0.99 signify that thermal augmentation was responsible for over 90% of the total entropy production. Hence, thermal entropy production was found to be accountable for overall entropy production in the two heatsinks. Therefore, thermal irreversibility should be given much consideration during thermal systems design.

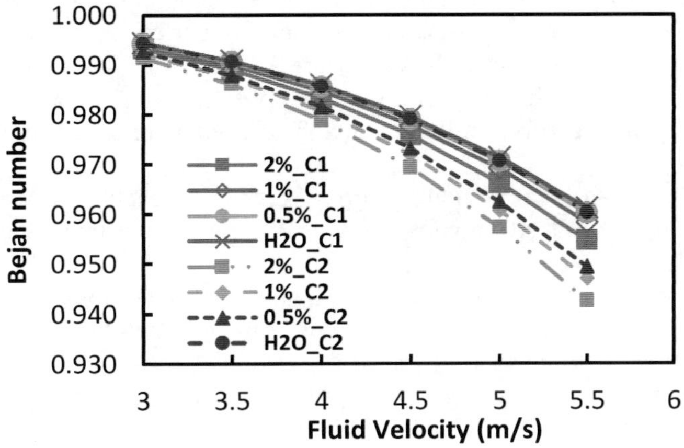

Figure 4. Dimensionless Bejan no as a function of fluid velocity for the cases.

4. Conclusion
The study presented the analysis of irreversibilities in a thermal system due to the application of water and nanofluid in two variable-width minichannel heat sinks. It was observed that entropy generation and dimensionless Bejan number are lower in case 2 due to the lower surface area of the flow domain. The increase in volume fraction of nanofluid results in heat transfer enhancement, especially at lower flow surface area. At 2 vol%, the thermal irreversibility is 9.4% and 9.6%, comparatively higher in case 1 than in case 2 for the corresponding fluid velocities of 3 ms^{-1} and 5.5 ms^{-1}. Since turbulence affects viscosity significantly, the frictional entropy losses are lower with 18.5% and 32.8 % at 5.5 ms^{-1} for case 1 and 2, respectively.

5. References

[1] Liang G and Mudawar I, Review of single-phase and two-phase nanofluid heat transfer in macro-channels and micro-channels, International Journal of Heat and Mass Transfer, 136, 2019. 324-354
[2] Faridi Khouzestani R and Ghafouri A, Numerical study on heat transfer and nanofluid flow in pipes fitted with different dimpled spiral center plate, SN Applied Sciences, 2, 2, 2020. 298
[3] Jiang Y, Zhou X and Wang Y, Effect of nanoparticle shapes on nanofluid mixed forced and thermocapillary convection in minichannel, Int Commun Heat Mass, 118, 2020. 104884
[4] Bejan A and Kestin J, Entropy generation through heat and fluid flow, 1983.

[5] Sohel M R, Saidur R, Khaleduzzaman S S and Ibrahim T A, Cooling performance investigation of electronics cooling system using Al2O3-H2O nanofluid, Int Commun Heat Mass, 65, 2015. 89-93
[6] Mir S, Akbari O A, Toghraie D, Sheikhzadeh G, Marzban A, Mir S and Talebizadehsardari P, A comprehensive study of two-phase flow and heat transfer of water/Ag nanofluid in an elliptical curved minichannel, Chinese Journal of Chemical Engineering, 28, 2, 2020. 383-402
[7] Shahsavar A, Moradi M and Bahiraei M, Heat transfer and entropy generation optimization for flow of a non-Newtonian hybrid nanofluid containing coated CNT/Fe 3 O 4 nanoparticles in a concentric annulus, J Taiwan Inst Chem E, 2018.
[8] Mahian O, Kianifar A, Sahin A Z and Wongwises S, Performance analysis of a minichannel-based solar collector using different nanofluids, Energ Convers Manage, 88, 2014. 129-138
[9] Ahammed N, Asirvatham L G and Wongwises S, Entropy generation analysis of graphene–alumina hybrid nanofluid in multiport minichannel heat exchanger coupled with thermoelectric cooler, International Journal of Heat and Mass Transfer, 103, 2016. 1084-1097
[10] Bahiraei M and Abdi F, Development of a model for entropy generation of water-TiO2 nanofluid flow considering nanoparticle migration within a minichannel, Chemometr Intell Lab, 157, 2016. 16-28
[11] Qiu S, Xie Z, Chen L, Yang A and Zhou J, Entropy generation analysis for convective heat transfer of nanofluids in tree-shaped network flowing channels, Thermal Science and Engineering Progress, 5, 2018. 546-554
[12] Ebrahimi-Moghadam A, Mohseni-Gharyehsafa B and Farzaneh-Gord M, Using artificial neural network and quadratic algorithm for minimizing entropy generation of Al2O3-EG/W nanofluid flow inside parabolic trough solar collector, Renewable Energy, 129, 2018. 473-485
[13] Huminic G and Huminic A, Entropy generation of nanofluid and hybrid nanofluid flow in thermal systems: A review, Journal of Molecular Liquids, 302, 2020. 112533
[14] Muhammad N and Sidik N 2019 Numerical analysis on thermal and hydraulic performance of diverging-converging minichannel heat sink using Al2O3-H2O nanofluid. In: IOP Conference Series: Materials Science and Engineering: IOP Publishing) p 012046
[15] Muhammad N M a, Sidik N A C, Saat A and Abdullahi B, Effect of nanofluids on heat transfer and pressure drop characteristics of diverging-converging minichannel heat sink, CFD Letters, 11, 4, 2019. 105-120
[16] Ying Z, He B, He D, Kuang Y, Ren J and Song B, Comparisons of single-phase and two-phase models for numerical predictions of Al2O3/water nanofluids convective heat transfer, Adv Powder Technol, 31, 7, 2020. 3050-3061
[17] Bahiraei M and Ahmadi A A, Thermohydraulic performance analysis of a spiral heat exchanger operated with water–alumina nanofluid: Effects of geometry and adding nanoparticles, Energ Convers Manage, 170, 2018. 62-72
[18] Shih T-H, Liou W W, Shabbir A, Yang Z and Zhu J, A new k-ϵ eddy viscosity model for high reynolds number turbulent flows, Computers & Fluids, 24, 3, 1995. 227-238
[19] Phillips R J, Microchannel heat sinks, Lincoln Laboratory Journal, 1, 1988. 31-48
[20] Abdolbaqi M K, Mamat R, Sidik N A C, Azmi W H and Selvakumar P, Experimental investigation and development of new correlations for heat transfer enhancement and friction factor of BioGlycol/water based TiO2 nanofluids in flat tubes, International Journal of Heat and Mass Transfer, 2017.

Acknowledgement
The authors thankfully acknowledge the supports funder by the Ministry of Higher Education under FRGS, Registration Proposal No: FRGS/1/2019/STG07/UTM/02/18 & UTM R.K130000.7843.5F273.

Journal of Physics: Conference Series **2053** (2021) 012017 doi:10.1088/1742-6596/2053/1/012017

Field measurement of indoor environment of a room with single-sided sliding glass opening

Z Wellun[1], W F Mohammad Yusoff[*1], M F Mohamed[1], K A Mat Sulaiman[1], and M R Mohammad Rasani[2]

[1]Department of Architecture and Built Environment, Faculty of Engineering and Built Environment, Universiti Kebangsaan Malaysia, 43600 UKM Bangi, Selangor, Malaysia
[2]Department of Mechanical and Manufacturing Engineering, Faculty of Engineering and Built Environment, Universiti Kebangsaan Malaysia, 43600 UKM Bangi, Selangor, Malaysia

*Corresponding author's email: wardahyusoff@ukm.edu.my

Abstract. The indoor environment of a room is also affected by the opening type and configuration. The application of an opening in a building is mainly influenced by its aesthetic look rather than its environmental performance. Understanding of an opening's impact towards the indoor environment is essential. The impact of a single-sided sliding glass opening to the indoor environment of a room discussed using field measurement. A room located in a two-storey detached house was selected. Total of ten days from 7 a.m. to 7 p.m., the sliding glass was slid to one side to allow natural ventilation into the room, hence, a cavity was created in between the two glasses. The results of the study indicated that high heat accumulation was detected at the glass surface area which faced the cavity and the air velocity in the room was found to be lower than the outdoor wind speed. This study provides important findings on the performance of a single-sided sliding glass opening and how it impacts the indoor environment of a room then help the building owner, designer, and people in the built environment industry to further understand the performance of sliding glass openings.

1. Introduction

Understanding the opening behaviour and its impact towards the indoor environment is challenging. The information is found to be still lacking until today [1]. One of the most valued factors to be addressed regarding the opening application is choosing the right design in relation to the climate condition [2], besides the importance of energy efficiency due to high energy consumption by the occupants and the building itself [3]. Nowadays, many building owners mainly focus on the aesthetic value rather than the exact functions of the opening as well as its benefits towards the people and environment [4]. The application of an opening with improper design might cause the increase of heat gain in hot and humid climates, though its purpose is expected to be vice versa [5]. Losing heat in a cold climate is also not an option when the purpose of having a suitable opening or window design is to gain heat [6], therefore an opening replacement into the latest, upgraded design is always the best choice to be done [7].

In this study, the outdoor wind speed in relation to the air flow entering the indoor space was examined along with the heat accumulated within the single-sided sliding glass opening. Previous studies highlighted that the prediction of natural ventilation rates of a building is challenging due to many opening varieties and flow factors [8] without the investigation on heat accumulation between the opening gap created during the field measurement conducted. Researchers mainly investigated the indoor environmental quality as an important measure in determining the quality of a life [9] and internal air quality which contributes to the health of the occupants [10]. Indoor thermal comfort of the building occupants are also essential, which consists of the radiant temperature, air velocity, air temperature, and relative humidity [11]. The thermal comfort of the indoor environment of a space or room is determined by many factors such as the building orientation, the building materials, and the building components. Note that this study mainly focuses on the building component, namely the sliding glass opening which is located at one side of the building facade. Hence, the ventilation type that is created by this configuration is a single-sided ventilation. Even though in many cases where single-sided ventilation is less favourable compared to cross ventilation, some modifications on certain criteria can help to improve its performance [12]. Moreover, the single-sided ventilation is also widely used, especially for buildings with a deep layout plan.

To date, many studies have investigated the performance of square or rectangular openings, and less studies have focused on specific openings, especially the window types that are mainly used in the market. In conjunction with this, due to many factors affecting the indoor environment, such as opening size and location, the temperature decreasing process is still considered sophisticated even today [13]. Hence, this study investigates the performance of one type of opening that is available in the market and actively used in buildings, which is the sliding glass window. The understanding of the opening performance and its impact towards the indoor environment is pivotal in order to help the right selection of opening in relation to the climatic condition.

2. Materials and Methods

Different types of research methodologies are available and can be adopted to achieve the objective, which includes experiment, wind tunnel investigation, and numerical simulation. All these methods possess their own advantages and drawbacks [14] such as the inadequacy of verification with computational analysis [15]. However, the inadequacy issue can be dealt with by conducting efficient experimental procedures beforehand, then comparing the results gained with data analyses using computational simulation [16].

In this study, the field measurement method was adopted, where it was conducted in the room of a two-storey detached house. The room, which had the dimension of 3.8-metre length times 2.9-metre width (3.8m x 2.9m), and 2.5-meter (2.5m) height, was located on the upper floor of the house. The upper floor was chosen instead of the ground floor due to its location which allowed minimal distraction of wind flow by the natural and man-made elements around the house. The sliding glass opening was located at one side of the façade, which faced the North direction. The effects of an opening located in other directions such as East, West, and South will be investigated in the future research. The investigated opening had the industrial standard dimension of 1.2 metre width times 1.2 metre height (1.2m x 1.2m), with two aluminium framed clear glass panels. One panel was static, and the other panel could be slid to allow natural ventilation into the indoor space. It is important to note that when one of the glass panels was slid to one side, an air cavity was created in the area sandwiched between the two glass panels.

There were three types of instruments involved: the portable weather station, namely the HOBO U30 Station; the multifunction indoor environmental instrument, namely the Delta Ohm thermocouple infrared thermometer and Testo 400 Universal IAQ; and the k-type thermocouple, namely the Onset HOBO data logger thermocouple. The portable weather station was located at a distance of 6.67 metres from the house. The indoor environmental instruments, namely the Delta Ohm and Testo, were positioned near the opening and at the centre of the room, respectively. Meanwhile, the data of heat accumulation at the opening were recorded using the k-type thermocouples, where they were placed at the outside and inside surfaces of the inactive (static) glass as shown in figures 1 and 2. Hence, when the movable glass was slid to the side of the static glass, thermocouple A faced the air cavity that was created in between the glasses. The multifunction indoor environmental instrument and the k-type thermocouples were placed at a height of 1.5 metres from the floor level. All the instruments had been calibrated prior to the measurement. The measurement was executed for ten days, from 7 a.m. to 7 p.m. For this initial study, priority was given to the daytime duration, as normally in the night-time, the opening in a hot and humid climate is closed due to some factors such as preventing mosquitoes and ensuring the occupant's safety. However, in the future research, the measurement duration can also be prolonged to the night-time.

3. Results and Discussion

The results are presented and discussed regarding two variables, namely the air velocity and the glass surface temperature. Comparisons between the air velocity values recorded at the weather station, the sliding glass opening, and the centre of the room are made. Meanwhile, for the glass surface temperature, comparisons are made for the glass surface that faces the outdoors and the glass surface that faces the air cavity. This air cavity is created when one of the glass is slid to allow air flow into the room.

3.1 Air velocity data

The air velocity investigations were carried out at three different points, namely the weather station, the sliding glass opening, and the indoor area of the room, which represented the outdoor wind speed, opening air velocity, and indoor air velocity, respectively. As mentioned before, the data collected represented the average wind speed and air velocity for every hour, starting from 7 a.m. to 7 p.m. in a total of ten consecutive days with rainy days on day fifth, sixth, tenth and others were clear weather. During the measurement, the air flow occurred naturally via pressure differences, and no mechanical medium, such as a fan, was operating.

Figure 1. Position of the measuring instruments.

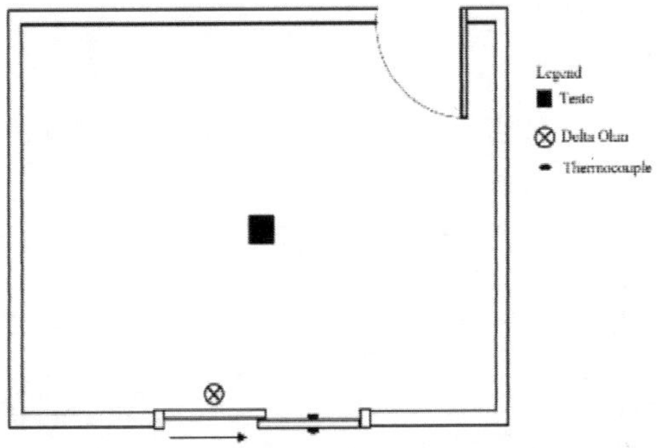

Figure 2. Plan view of the measuring instruments.

Figure 3 shows that the outdoor wind speed had higher value compared to the air velocity at the sliding glass opening and the indoor area. The blue line indicates the wind velocity value recorded at the weather station, while the black line and the red line are the air velocity values measured at the opening and the centre of the room, respectively. The comparative analysis of the three locations indicates that the air velocity value decreased further when it reached the indoor area. There were also high reductions in the velocity values between the data recorded at the weather station and the opening. The air velocity value measured at the opening barely achieved 0.1m/s although the wind speed recorded was more than 0.4m/s most of the time.

In the first two hours of the field measurement period, the outdoor wind speed had a slight increment from 0.10m/s to 0.12m/s, while the air velocity at the opening and the indoor area recorded a decrement from 0.04m/s to 0.03m/s for both areas. Huge increase in outdoor wind speed was recorded at 9, 10, and 11 a.m. with a value of 0.45m/s, 0.72m/s, and 0.89m/s, respectively. Despite the increase of outdoor wind speed from 9 to 11 a.m., the air velocity measured at the opening still indicated low values, which were 0.04m/s, 0.07m/s, and 0.08m/s. The air velocity continued to decrease further when it reached the indoor area, which was 0.04m/s at 9 and 10 a.m., and 0.06m/s at 11 a.m.

Before reaching the highest value of outdoor wind speed at 3 p.m., the weather station's wind speed at 12, 1, and 2 p.m. were 0.95m/s, 1.07m/s, and 1.11m/s, respectively. Similarly, during these hours, the air velocity decreased when reaching the opening and the indoor area. The data shows that at 12, 1, and 2 p.m., the opening's air velocity was 0.09m/s, 0.08m/s, and 0.11m/s, respectively.

At 3 p.m., when the outdoor wind speed recorded the highest value of 1.14m/s, the air velocity at the opening was found to be only 0.08m/s, which indicated a 92.9% decrement. The air velocity decreased further when it reached the indoor area, in which the measured value was only 0.05m/s. Hence, the results indicated that the total decrement of velocity from the weather station to the indoor area was around 95.6%.

Finally, the wind speed started to decrease during the last four hours, which were from 4 p.m. onwards. The air velocity at the opening and the indoor area remained low during these hours. The recorded air velocities at the opening were 0.09m/s, 0.07m/s, 0.07m/s, and 0.04m/s at 4, 5, 6, and 7 p.m., respectively. Meanwhile, at the indoor area, the measured air velocities were 0.06m/s at 4 p.m. and remained at 0.05m/s from 5 p.m. until the end of field the measurement period.

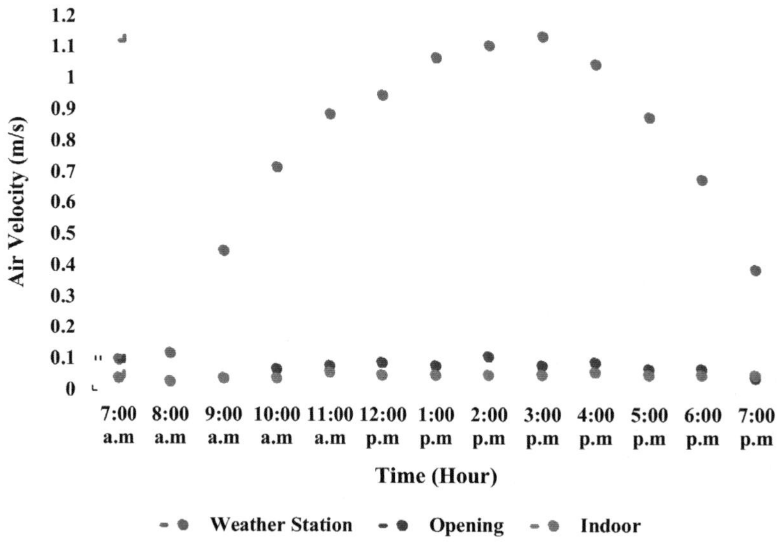

Figure 3. Velocity-Time graph of weather station, sliding glass opening and indoor area.

Overall, the results indicate a common scenario where the air velocity is normally lower at the indoor space compared to the inlet and outlet areas [17]. By referring to the air velocity values at the indoor area, none of the data shows a pleasant result in achieving indoor thermal comfort for a hot and humid climate, where the air velocity values are supposed to be within the range of 0.9m/s to 1.8m/s [18]. In the measurement, the indoor air velocity ranged from 0.03m/s to 0.06m/s. The highest indoor air velocity recorded was 0.06m/s, which was too far from the recommended air velocity for thermal comfort in hot and humid climates.

3.2 Surface temperature of sliding glass opening

From the experiment, it is found that there was high heat accumulation recorded on the glass surface which faced the air cavity. Data shows that high heat accumulation occurred around 10 a.m. until 5 p.m. as depicted in figure 4.

Data shows that at 9 a.m., the surface temperature recorded by thermocouple A was 32.19°C compared to thermocouple F which was 29.85°C. This resulted in the difference of approximately 2.31°C. At 10 a.m., higher surface temperature difference was recorded by thermocouple A and F compared to the surface temperature data at 9 a.m. Figure 4 shows that at 10 a.m., the surface temperature measured by thermocouple A was 34.94°C, while for thermocouple F, it was 30.55°C. Hence, the surface temperature difference between both thermocouples at 10 a.m. was 4.39°C, which was higher than at 9 a.m.

Table 1. Indoor Air Temperature.

Time (Hour)	7:00 am	8:00 am	9:00 am	10:00 am	11:00 am	12:00 pm	1:00 pm	2:00 pm	3:00 pm	4:00 pm	5:00 pm	6:00 pm	7:00 pm
Temp (°C)	29.2	29.9	31.2	31.7	31.8	32.3	32.8	33.3	33.6	33.4	33.2	32.6	32.1

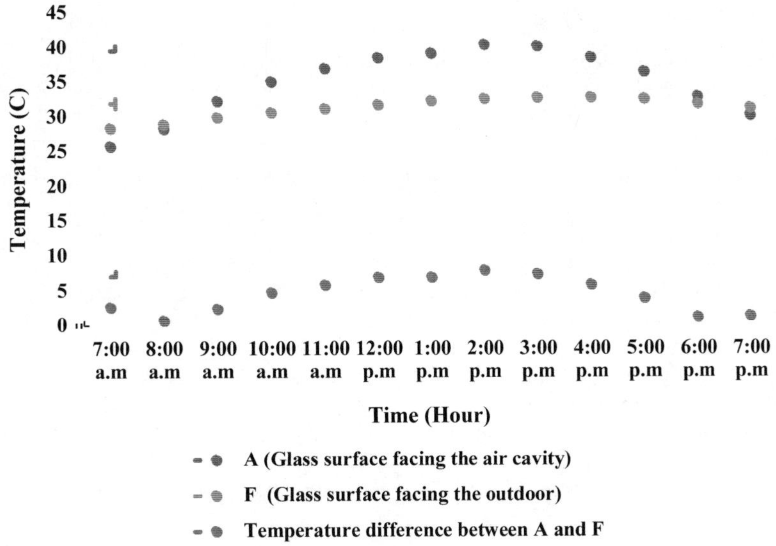

Figure 4. Surface temperature of the sliding glass opening.

In figure 4, the line in blue colour represents the surface temperature of the glass that faces the air cavity, while the green line represents the surface temperature of the glass that faces the outdoors. In the same graph, the red line represents the temperature differences between both data. At 5 p.m., thermocouple A and F recorded surface temperatures of around 36.24°C and 32.38°C, respectively, which resulted in a surface temperature difference of 3.86°C. The highest recorded surface temperature of thermocouple A throughout the measurement was 40.24°C which was at 2 p.m., while for thermocouple F, the highest measured surface temperature was 32.66°C which was at 4 p.m. In figure 4, the bottom line marked by the red colour represents the difference in surface temperature between A and F, starting from the beginning until the end of the field measurement. The data resulted in high heat accumulation between 10 a.m. and 5 p.m.

This finding is significant as less (considering the time this article was written) previous studies have mentioned the heat accumulation of a sliding glass opening when one of the glass is slid to allow air flow. Three of the latest studies that investigated the sliding opening mainly focused on air pollution [19], airborne particles [20], and window orientation with its position related to natural ventilation [21]. Although the double-pane glass manages to decrease the hot air temperature during summer from entering the indoor space (in the condition of closed windows) [22], the heat accumulation at the cavity causes an unpleasant indoor environment during the hot season. The sliding glass opening condition during the measurement, which also creates air cavity, most likely adopts the same criteria as the double pane glass. Hence, the question arises whether the accumulated heat at the sliding glass opening provides a significant effect to the indoor environment of a room or space. Does it also significantly contribute to the increase of indoor air temperature?

Table 1 shows the indoor air temperature that was recorded throughout the measurement. The indoor air temperature increased from morning to afternoon and remained high during late afternoon. It only decreased when approaching evening. Although the indoor air temperature is actually influenced by many factors, such as the building materials and orientation, this study's focus is based on the variables of air velocity and glass surface temperature. This is to achieve the objective of examining how the glass sliding opening affects the indoor environment of a space. However, it is important to note that there is a limitation in this initial study where factors, such as glass materials and properties, are not investigated. Hence, further investigation on this finding is necessary in the future to provide more detailed information regarding the heat transfer that occurs at the sliding glass opening. Further detailed investigation helps to provide deeper understanding on the effects of sliding glass opening to the environment of indoor space.

4. Conclusion

The analysis of the collected data shows that around 10 a.m. to 5 p.m., the value of k-type thermocouple (thermocouple A), that was located at the glass surface facing the air cavity, was higher compared to the glass surface that faced the outdoor environment. Meanwhile, the air velocity was found to be lower at the opening and the centre of the room compared to the outdoor wind speed. Throughout the measurement, it was also found that the indoor air temperature of the room was around 29°C to 33°C from 7 a.m. to 7 p.m., on average. Based on these results, it can be summarised that although the sliding glass opening allows natural ventilation into the room, the air velocity is found to decrease when approaching the centre of the room. Meanwhile, the character of the opening which creates air cavity in between the glass panes when one of the glass is slid, has caused heat accumulation. This cavity traps heat, which results in higher glass surface temperature of the glass surface that faces the cavity. The results and findings of this study are able to embark further investigations such as the impacts of cross ventilation and different glass properties to the indoor environment of a space or room.

5. References

[1] Wang J, Zhang T, Wang S and Battaglia F 2018 Numerical investigation of single-sided natural ventilation driven by buoyancy and wind through variable window configurations *Energy Build.* **168** 147–64

[2] Wang J, Wang S, Zhang T and Battaglia F 2017 Assessment of single-sided natural ventilation driven by buoyancy forces through variable window configurations *Energy Build.* **139** 762–79

[3] Ramli N A, Fairuz M, Hamid A, Gimang H and Richard A 2019 Energy Efficiency Measures on Two Different Commercial Buildings in Malaysia *J. Build. Perform.* **10** 17–29

[4] Lick E, Bargenda A and Trabelsi D 2020 Windows to the sold: verbo-visual multimodality in storefront windows *Int. J. Retail Distrib. Manag.* **48** 501–15

[5] Geetha Ramesh K and Ramachandraiah A 2020 A fractal approach for optimization of daylighting by exploring the window geometry *Int. J. Adv. Res. Eng. Technol.* **11** 36–47

[6] Chen Z, Hammad A W A, Kamardeen I and Haddad A 2020 Optimising window design on residential building facades by considering heat transfer and natural lighting in nontropical regions of Australia *Buildings* **10** 1–27

[7] Carlos J S 2017 The impact of refurbished windows on Portuguese old school buildings *Archit. Eng. Des. Manag.* **13** 185–201

[8] Gough H L, Barlow J F, Luo Z, King M F, Halios C H and Grimmond C S B 2020 Evaluating single-sided natural ventilation models against full-scale idealised measurements: Impact of wind direction and turbulence *Build. Environ.* **170** 106556

[9] Langer S, Ramalho O, Derbez M, Ribéron J, Kirchner S and Mandin C 2016 Indoor environmental quality in French dwellings and building characteristics *Atmos. Environ.* **128** 82–91

[10] Cone J E 1998 Indoor environmental quality. *West. J. Med.* **169** 373–4

[11] Abdul Mujeebu M 2019 Introductory Chapter: Indoor Environmental Quality *Indoor Environ. Qual.* 1–13

[12] Mohamed M F 2017 Computational analysis on the effects of façade modifications on wind-driven natural ventilation performance of a single-cell room *ARPN J. Eng. Appl. Sci.* **12** 3093–100

[13] Bazdidi-Tehrani F, Masoumi-Verki S and Gholamalipour P 2020 Impact of opening shape on airflow and pollutant dispersion in a wind-driven cross-ventilated model building: Large eddy simulation *Sustain. Cities Soc.* **61** 102196

[14] Yusoff W F M 2020 The effects of various opening sizes and configurations to air flow dispersion and velocity in cross-ventilated building *J. Teknol.* **82** 17–28

[15] Harun Z, Azman F N, Lotfy E R and Abbas A A 2018 Flow Structure in Modern Cities: Wind Tunnel Investigation *J. Kejuruter.* **30** 39–46

[16] Prakash D and Ravikumar P 2012 Simulation of indoor air flow for a room with windows at their adjacent walls under various wind flow direction using CFD *ARPN J. Eng. Appl. Sci.* **7** 1387–94
[17] Muhsin F, Wardah Fatimah Mohammad Yusoff, and Mohd Khairul Azhar Mat Sulaiman 2019 Preliminary Evaluation Of Air Flow In Atrium Of Building In Hot And Humid Climate *Engineering Green Build. Alam Cipta* **12** 43–52
[18] Mihara K, Sekhar C, Takemasa Y, Lasternas B and Tham K W 2019 Thermal comfort and energy performance of a dedicated outdoor air system with ceiling fans in hot and humid climate *Energy Build.* **203** 109448
[19] Wang W, Kato N, Kimoto S, Matsui Y and Yoneda M 2020 Simulation and evaluation of sheltering efficiency of houses equipped with ventilation systems *Build. Environ.* **168** 106491
[20] Sadrizadeh S, Pantelic J, Sherman M, Clark J and Abouali O 2018 Airborne particle dispersion to an operating room environment during sliding and hinged door opening *J. Infect. Public Health* **11** 631–5
[21] Liu T and Lee W L 2019 Using response surface regression method to evaluate the influence of window types on ventilation performance of Hong Kong residential buildings *Build. Environ.* **154** 167–81
[22] González-Julián E, Xamán J, Moraga N O, Chávez Y, Zavala-Guillén I and Simá E 2018 Annual thermal evaluation of a double pane window using glazing available in the Mexican market *Appl. Therm. Eng.* **143** 100–11

Acknowledgements
The authors would like to thank the Ministry of Education Malaysia (MOE) through the Fundamental Research Grant Scheme (FRGS/1/2019/TK10/UKM/02/4 -Air flow characteristics and behaviour around openings with various types of shading elements), and the National University of Malaysia (UKM) through the research grant GUP-2019-017.

Solar and biomass potential of renewable energy in selected ASEAN countries and Japan

Ku Nur Afrina Ku Azman Shah[1], Mohd Zamri Mohd Yusop[1], Jafri Mohd Rohani[1*], Nor Akmal Fadil[1], Mohd Faizal Hasan[1], Mohd Fairus Mohd Yasin[1], Natrah Kamaruzaman[1], Engku Mohd Nazim Engku Abu Bakar[1], Syahrullail Samion[1], Mohd Yazid Yahya[1], Norhuda Abdul Manaf[2], Budi Hartono[3], Nguyen Duc Tuyen[4], Tanemura Masaki[5], Abdul Samad Ahmad[6] and Ashaari Ramli[6]

[1] School of Mechanical Engineering, Faculty of Engineering, Universiti Teknologi Malaysia, Johor, Malaysia
[2] Department of Chemical Process Engineering, Malaysia-Japan International Institute of Technology, Universiti Teknologi Malaysia, Kuala Lumpur, Malaysia
[3] Mechanical and Industrial Engineering Department, Gadjah Mada University, Yogyakarta, Indonesia
[4] School of Electrical Engineering, Hanoi University of Science and Technology, Hanoi, Vietnam
[5] Department of Physical Science and Engineering, Nagoya Institute of Technology, Showa-ku, Aichi, Japan
[6] MyBamboo Sdn. Bhd., Subang Jaya, Selangor, Malaysia

*Corresponding author's email : jafrimr@utm.my

Abstract. ASEAN region is believed to have a great potential for biomass and solar energy generation due to its abundance of sunlight, rain, and fertile land. However, these resources are still underutilized as the current rate of renewable energy is relatively low, whereby most countries are still highly emphasizing fossil fuels. The use of fuel without replenishing the fuel source will ultimately deplete the resource in very much the same as fossil fuels currently utilized. While the concept of fuel replenishment seemed simple, the implementation posed a challenge to meet. This is where free energy in the form of solar, wind, wave, underground thermal, and biomass is being harnessed. Therefore, these forms of energy also have certain limitations depending on the geographical and the availability of the sources throughout the region. This study aims to analyze the current potential renewable energy scenario and future prospect of renewable energy resource for biomass and solar energy in selected ASEAN countries and Japan.

1. Introduction

Energy resources are essential for livings, where the energy demand is expected to rise annually to fulfil the rapid growth of the economy and industrialization. However, the dependence on non-renewable energy sources, including fossil fuels, will be depleted over time. The transition from non-renewable energy to renewable energy resources could significantly impact the common environmental effect and attenuate the energy shortages. ASEAN countries benefited from a geographical advantage in terms of natural resource ability for renewable energy production. Current potential renewable energy in this region varies from geothermal, hydropower, biomass, solar, and wind energy, depending on the geographic condition of the region. However, some countries are not associated with geothermal and wind energy where there is no substantial amount of that resource.

Most of the areas in ASEAN countries are covered with 15% of the world's tropical forests [1]. The abundance of biomass resources from forestry, agricultural residues, industrial, animal, and municipal solid waste could potentially contribute to biomass energy in ASEAN countries. ASEAN is known as the most productive agricultural in the world. Moreover, the ASEAN region is covered with an average of 12 hours of sunlight due to its location that lies astride the equator, which is substantial for solar energy generation [2]. The state of the art of this region's geographic condition brings solar and biomass renewable energy as the most promising resource for electricity generation, which was selected in this study to be the potential solution as a long-term energy supply in the ASEAN region.

ASEAN energy ministers have agreed to set a target of the total primary supply of renewable energy share in the region by 23%, and ASEAN installed power capacity of 35% by 2025 [3]. On the other hand, Japan targeted on renewable energy is 22 to 24% of its energy mix by 2030 as gazetted on its 5th Strategic Energy Plan. With the tremendous amount of renewable energy potential, there are a relatively large amount of resources still underutilized by these countries. Towards achieving the goal, Malaysia, Indonesia, Vietnam, and Japan have taken the initiative to adopt the Feed-in Tariff (FiT) for the past few years to enhance the growth of renewable energy generation percentage. In this policy mechanism, the renewable energy electricity generated is purchased by electric power companies at a fixed rate per kWh for a certain period of time. The country's renewable resource generation will govern the variation of FiT rate and duration. Therefore, the objective of this study is to compare the percentages of renewable energy and compare the energy mix of solar and biomass energy installed capacity for respected countries.

2. Methodology

This study is based on descriptive statistics, and graphical analysis comparing Japan and selected ASEAN countries of Malaysia, Indonesia, and Vietnam energy generation. These data are generated based on secondary data from each country's government reports with yearly trends from 2003 to 2020. This variation is due to the insufficient accessibility of the latest reports released by the countries. The data represent the total energy installed capacity by mixed fuels and the total installed capacity of renewable energy. The trend is then analyzed and compared in terms of availability, utilization, and potential of solar and biomass renewable energy generation.

3. Result & Discussion

Malaysia, Indonesia, Vietnam, and Japan are highly dependent on fossil fuels. Japan had the highest total energy installed capacity of 305.5 GW, followed by Indonesia with 64.5 GW, Vietnam with 48.6 GW, and Malaysia with 34.0 GW. The report from the government of the ASEAN countries and Japan for power generation installed capacity by fuel type as for 2018 is illustrated in figure 1. Therefore, large-scale hydropower is excluded from renewable energy percentage in this report.

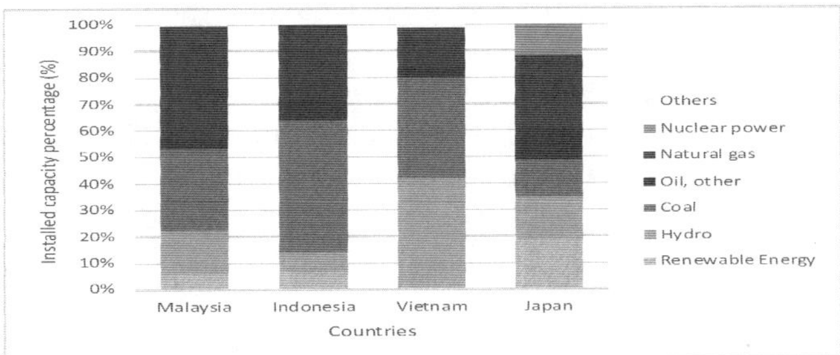

Figure 1. Comparison of renewable energy percentages of total installed capacity generation 2018 [4] [5] [6] [7].

In recent years, the government is interested in initiating and promoting renewable energy development to cater to the growth of energy demand without depending on fossil fuels. According to Malaysia Energy Commissions [4], Malaysia's total renewable energy capacity is 6% (2.1 GW), 2.2% biomass, and solar with 2.1%. Whereby, as referred to Secretariat General of National Energy Council [5], Indonesia resulted in 7% (4.5 GW) of installed renewable energy with the majority of 3% is contributed by geothermal followed with biomass of 2.7% Therefore, based on the data from National Load Dispact Center [6], Vietnam has recorded a minimum value of total renewable energy with less than 1% (0.4 GW). The installed renewable energy generated by Japan constitutes 19% (58.0 GW) of its total power generation capacity, according to Japan Electric Power Information Center, INC (JEPIC) [7]. This consisted a large number of solar by 16% and followed by wind 1%. Therefore, the utilization of renewable energy resources in the ASEAN region up until 2018 is relatively low compared to Japan which recorded the highest amount of renewable energy capacity generation. Thus, Japan has high potential in achieving its renewable energy target by 2030 of its energy mix.

3.1. Malaysia potential of solar and biomass energy

Malaysia currently has limited renewable energy resource types compared to other countries. There is no substantial amount of wind and geothermal energy in this region that could be extensively harvested. Therefore, Malaysia could fully utilize its abundance potential of solar and biomass energy. Figure 2 shows the pattern of the renewable energy installed capacity generation from the year 2012 to 2018.

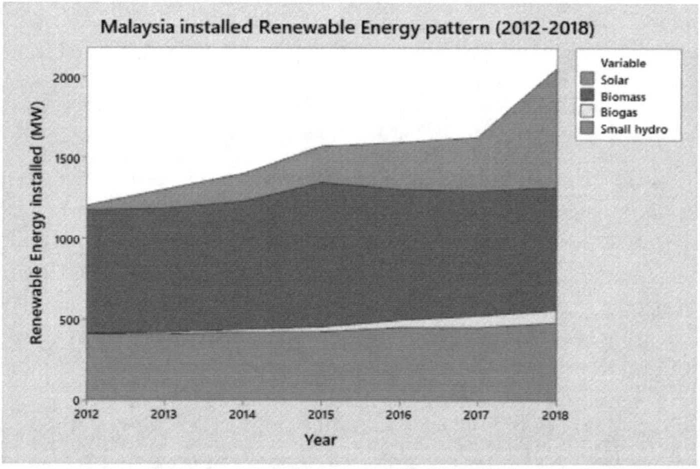

Figure 2. Malaysia renewable energy installed capacity generation changes from year 2012-2018 [4] [8] [9] [10] [11] [12] [13].

Feed-in Tariff (FiT) scheme for Malaysia obliges Distribution Licensees (DLs). The contracts for solar PV are set for 21 years with FiT rate between 0.087 to 0.140 USD per kWh. The implementation of FiT resulted to the increase of renewable energy capacity for solar energy over the years. Recently, the Malaysia government, however, is pushing for rooftop solar using a Net Energy Metering (NEM) aimed at self-consumption together with the implementation of Large Scale Solar (LSS) [14]. In 2016, Malaysia's Energy Commission started the bidding process for LSS. The total capacity allocation for LSS is 1000 MW and a levelized tariff for the auction ranging from 0.096 to 0.110 USD per kWh [15]. According to year 2017 to 2018, solar energy recorded a drastic changed by 125% increased on installed capacity of solar resource. The implementation of LSS has contributed in 42% with approximately 310 MW of the total solar capacity on 2018. The estimated potential for solar generation can reach up to 6500 MW [16].

The forest area in Malaysia has covered around 62% of the total land area, and its agriculture covers approximately 4,890,000 hectares of the region [17]. The capacity of biomass energy in Malaysia is reducing from year 2016 to 2018 by 15%. Therefore, biomass has an advantage over other types of renewable energy in terms of availability, cheaper cost, and sustainability. Malaysia's available biomass resources are from the oil palm, empty fruit bunches, palm kernel shells, sugarcane, and rice husk. In addition, Malaysia is the second-largest palm oil producer globally, which accounted a large amount of oil palm waste to be processed into biomass energy production. With the abundance of biomass resources in Malaysia, agricultural and forestry resource utilization is relatively low. The FiT rate for biomass energy is between 0.064 to 0.074 USD per kWh which covers a period of 16 years [15]. According to the National strategy, from a supply-side perspective, by 2020, Malaysia's palm oil industry is expected to generate about 100 million dry tonnes of solid biomass [18].

3.2. Indonesia potential of solar and biomass energy

Indonesia is the largest country by geographical area compared to Malaysia, Vietnam, and Japan. This country has the most renewable energy resource due to its location as the country spans the longest equator line. solar, biomass, small or medium hydropower, wind, geothermal, and biogas are the available renewable energy resource discovered in Indonesia. Figure 3 shows the Indonesia renewable energy installed capacity generation changes from the year 2009 to 2019.

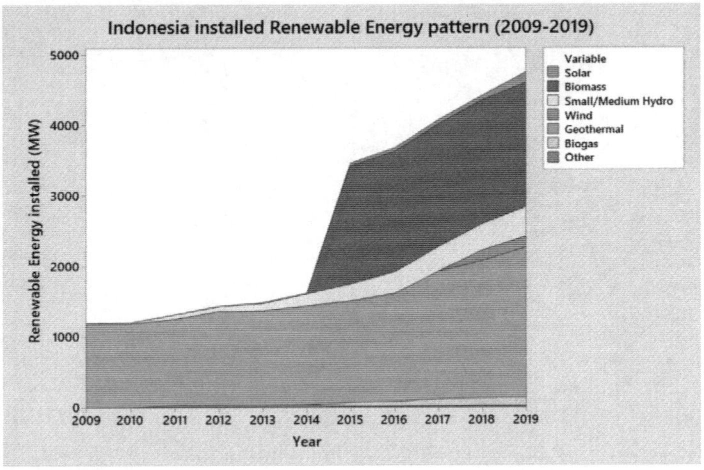

Figure 3. Indonesia renewable energy installed capacity generation changes from year 2009-2019 [5].

The trend for solar energy in Indonesia is relatively low compared to other resources. Indonesia's initiatives for solar are mainly focused on residential usage. Therefore, the estimated potential of solar energy is reported around 4.8 kWh/m2 in the entire region [19]. According to the National Energy Plan, the potential capacity for solar power plants is 207,898 MW [20]. Meanwhile, the share of the solar resource in renewable energy installed capacity generation for Indonesia constituted only 145.8 MW in 2019. The FiT in Indonesia for solar PV initially introduced in 2012, ranging from 0.25 to 0.30 USD per kWh. The FiT rate started to reduce in 2016 due to the rapid development of solar power technology [21]. To extend, the Indonesian government had initiated the Rooftop Photovoltaic Solar Systems (RPVSS) policy in 2018 to promote the solar energy system. However, this policy is constrained by both consumer and institutional barriers.

Biomass had recently produced a high capacity of installed renewable energy, approximately around 1.8 GW, the second-largest renewable energy contributed in Indonesia next to geothermal energy. Indonesia is known for its abundance of agriculture, and the forest area covered 94,432,000 hectares in the region [22]. Indonesia's significant biomass sources available are from the agricultural, which is the oil palm residues, bagasse, rice husk, rice straw, and corn cob. Furthermore, the country had explored using bamboo as biomass energy to generate electricity in rural areas. Indonesia's potential biomass as renewable energy is estimated to reach 146.7 million tons of biomass per year, approximately 50 GW [23].

3.3. Vietnam potential of solar and biomass energy

Apart from other countries, Vietnam biomass energy is currently treated as a non-commercial energy resource which used locally. Figure 4 shows the pattern of installed renewable energy in Vietnam from the year 2018 to 2020. Vietnam recorded a large installed capacity for large hydropower and coal. Therefore, renewable energy is relatively minimal in the last decade.

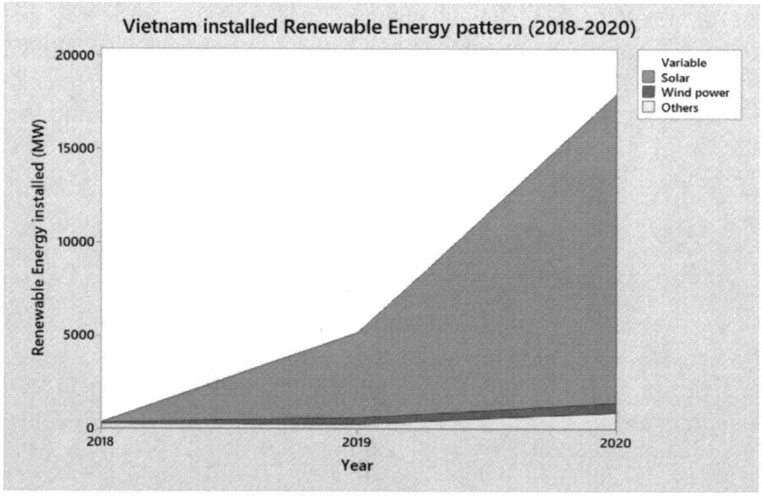

Figure 4. Vietnam renewable energy installed capacity generation changes from year 2018-2020 [6].

Renewable energy in Vietnam is increased dramatically from less than 1% in 2018 with a total installed renewable energy of 0.4 GW to more than 25% in 2020 with 18.0 GW. Solar PV is the last resource that is introduced to FiT in 2017 with the enacted tariff of 0.0935 USD per kWh. Prior to the expiration of the solar promotion in the middle of the year 2019, the pattern of solar energy capacity in Vietnam has drastically changed recorded a 12% of renewable energy increment. The progress continues to the year 2020, where the implementation of solar PV rooftop installation recorded growth by 12%, resulting in more than 25% renewable energy achieved by the country on 2020. This achievement has made Vietnam the world's third-largest solar market in 2020 and eventually lead the ASEAN to reach its renewable energy target.

Biomass energy in Vietnam recorded a slow growth. The country has abundance of biomass energy resources from agriculture and forestry, such as bagasse, rice husk, coffee husk, bamboo, and wood chip The total installed capacity for biomass is 560 MW up to 2020. Therefore, the FiT for biomass has increased from 0.058 USD per kWh to 0.0703 USD per kWh for combine heat production (CHP), while Non-CHP with 0.0728 USD per kWh. Electricity of Vietnam (EVN) purchases all renewable energy generated from powerplants with a validation of purchase agreement for 20 years. The policies also include the incentives on investment capital and taxes, as well as land incentives. Forecast for installed capacity for 2030 is 3384 MW, and 2040 is 4509 MW. Vietnam has forecasted the 2030 installed capacity for biomass to be around 3384 MW, and 2040 with 4509 MW [24]. These numbers are large compared to current capacity recorded, thus Vietnam is predicted to have a great future in renewable energy and has the possibilities to make renewable energy as the country main energy with greater generation compared to fossil fuels.

3.4. Japan potential of solar and biomass energy

Japan is one of the most advanced countries in the world with remarkable technologies in energy production. Japan is ahead of ASEAN countries by introducing its first nuclear power reactor. Therefore, Japan's government had the pledge to increase the renewable energy resource, especially solar. Figure 5 illustrated the changes in Japan's renewable energy installed capacity from the year 2003 to 2019. Japan's renewable energy resource availability is solar, wind, geothermal, biomass, and small or medium hydropower.

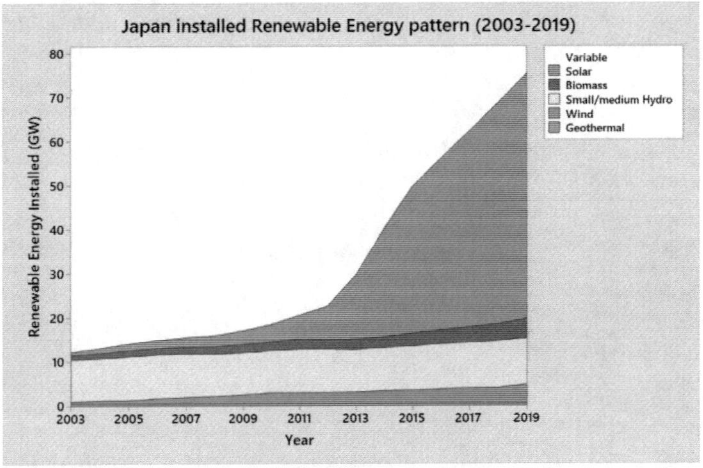

Figure 5. Japan renewable energy installed capacity generation changes from year 2003-2019 [7].

Regarding Japan's catastrophic nuclear energy power plant, the government has become more environmentally conscious. The installed renewable energy capacity is proliferating with the introduction of FiT in the year 2012. Japan has recorded a rapid inclined growth on the average of 18% annually of renewable energy installed capacity generation starting from 2012 to 2019. The average yearly irradiance for Japan region ranges from 125 to 162 W/m2 [25], which is sufficient for solar generation. Solar energy recorded the largest share among other renewable energy resources. Japan planned to invest more than 100 billion USD from 2020 to 2030 in solar and wind generation which expected to drive the renewables share of generation mix to 27% by 2030, exceeding the country's target [26]. Thus, solar energy is considered the most promising potential for renewable energy generation.

Japan's land is covered by almost 70% of forests. Biomass renewable energy is increasing over the years but relatively low in the renewable energy percentage compared to other resources. The agricultural residue from the domestic resource can only contribute to a small amount of energy. Japan has set a target for biomass energy of 4% in the energy mix by 2030. There is a significant potential for biomass energy production from waste wood. However, the crop potential in this region is limited due to the climate in northern and southern Japan and it has to be imported [27].

4. Conclusion

The potential of renewable energy in ASEAN countries is enormous, whereby the utilized renewable energy is relatively low. Solar energy and biomass have rich potential across all ASEAN countries and Japan. These replenished resources could significantly impact the economic growth, environmental, and socioeconomic of the regional communities. Despite the fact that the generated large scale hydropower leads the renewable energy resources in ASEAN countries, this renewable energy source is dealing with issues related to environmental and social threats. Due to the concern, the countries began to exploit the underutilized potential in other renewable energy resources, especially solar and biomass energy, for future electricity generation. This paper has discussed the current potential renewable energy scenario and future prospect of renewable energy resource for biomass and solar energy in selected ASEAN countries and Japan. Based on the findings, the potential renewable energy for these four countries has been determined. Malaysia, Indonesia and Vietnam has abundance potential on solar and biomass energy, whereby Japan has large potential on solar energy, but low potential on biomass. The scheme and policies as a driving force in promoting renewable energy has successfully implemented on these regions with a remarkable turn of renewable energy percentage. Vietnam and Japan could become the role model for other ASEAN countries to follow as these countries recorded a drastic development in renewable energy and reached country's target. The future of biomass and solar energy is expected to increase with solar PV and biomass potential acceleration and cost reduction.

5. References

[1] H J Stibig, F Achard, S Carboni, R Rasi and J Miettinen 2014 Change in tropical forest cover of Southeast Asia from 1990 to 2010 *Biogeosciences* ***11*** pp. 247-258

[2] R Nepal, H Phoumin and A Khatri 2021 Green Technological Development and Deployment in the Association of Southeast Asian Economies (ASEAN)—At Crossroads or Roundabout? *Sustainability 2021* ***13*** *758*

[3] AMEM 2020 *Energy Transition Towards Sustainable Development* Joint Ministerial Statement of The 38th ASEAN Ministers on Energy Meeting Viet Nam

[4] Sustainable Energy Development Authority (SEDA) 2020 *Malaysia: Net Energy Metering* Sustainable Energy Development Authority (SEDA), Malaysia

[5] Secretariat General of the National Energy Council 2019 *Indonesia Energy Outlook 2019* Secretariat General of the National Energy Council

[6] Vietnam Initiatives for Energy Transition 2020 *Share of renewable energy in total installed capacity increased rapidly in recent years* Vietnam Initiatives for Energy Transition

[7] Japan Electric Power Information Center INC 2020 *The Electric Power Industry in Japan 2020* Japan Electric Power Information Center, INC, Tokyo, Japan

[8] Suruhanjaya Tenaga (Energy Commission) 2020 *Malaysia Energy Statistics Handbook 2019* Suruhanjaya Tenaga (Energy Commission) Malaysia

[9] Suruhanjaya Tenaga (Energy Commission) 2019 *Malaysia Energy Statistics Handbook 2018* Suruhanjaya Tenaga (Energy Commission) Malaysia

[10] Suruhanjaya Tenaga (Energy Commission) 2018 *Malaysia Energy Statistics Handbook 2017* Suruhanjaya Tenaga (Energy Commission) Malaysia

[11] Suruhanjaya Tenaga (Energy Commission) 2017 *Malaysia Energy Statistics Handbook 2016* Suruhanjaya Tenaga (Energy Commission) Malaysia

[12] Suruhanjaya Tenaga (Energy Commission) 2015 *Malaysia Energy Statistics Handbook 2015* Suruhanjaya Tenaga (Energy Commission) Malaysia

[13] Suruhanjaya Tenaga (Energy Commission) 2014 *Malaysia Energy Statistics Handbook 2014* Suruhanjaya Tenaga (Energy Commission) Malaysia

[14] Sustainable Energy Development Authority (SEDA) *National Renewable Energy Policy* [Online]. Available: http://www.seda.gov.my/policies/national-renewable-energy-policy-and-action-plan-2009/. Assessed 25 May 2021

[15] Timo 2020 From 2% to 20% – Accelerating Renewable Energy Development in Malaysia part 3 *The Energy Bit*

[16] S Ahmad, M Kadir and S Shafie 2011 Current perspective of the renewable energy development in Malaysia *Renew. Sustain. Energy Rev* vol **15** pp 897-904

[17] M M Tun, D Juchelkova, M M Win, A M Thu and T Puchor 2019 Biomass Energy: An Overview of Biomass Sources, Energy Potential, and Management in Southeast Asian Countries *resources 2019* vol **8** no 81

[18] Kementerian Tenaga Teknologi Hijau dan Air (KeTTHA) 2019 *National Renewable Energy Policy & Action Plan* Malaysia

[19] I Rahardjo and I Fitriana 2005 Analisa Potensi Pembangkit Listrik Tenaga Surya di Indonesia *Pusat Pengkajian dan Penerapan Teknologi Konversi dan Konservasi Energi*, BPPT Jakarta

[20] D Setyawati 2020 Analysis of perceptions towards the rooftop photovoltaic solar system policy in Indonesia *Energy Policy*

[21] B Suryadi 2018 *ASEAN Feed-in-Tariff (FIT) Mechanism Report* ASEAN Centre for Energy

[22] A B Office Forest Resources in ASEAN Countries [Online]. Available: https://www.asiabiomass.jp/english/topics/1111_04.html. Assessed 26 May 2021

[23] S Dani and A Wibawa 2018 Challenges And Policy For Biomass Energy In Indonesia *International Journal of Business, Economics and Law* vol **15** no 5

[24] E S P GIZ 2020 Overview of biomass development in Vietnam *GIZ, Energy Support Programme*, Vietnam.

[25] B L Stoll, T A Smith and M R Deinert 2013 Potential for rooftop photovoltaics in Tokyo to replace nuclear capacity *Environmental Research Letters*

[26] L Woellwarth 2020 Japan invests in wind and solar power plants [Online]. Available: https://www.energyglobal.com/special-reports/19082020/japan-invests-in-wind-and-solar-power-plants/. Accessed 10 June 2021

[27] N Pambudi, K Itaoka, A Chapman and D H Nguyen 2017 Biomass energy in Japan: Current status and future potential *International Journal of Smart Grid and Clean Energy*

Acknowledgments
The work reported on this paper is part of the AUN/SEED-Net Collaborative Education Program Research Grant (UTM CEP 2103). The authors wish to acknowledge the support given by Universiti Teknologi Malaysia, Malaysia Japan International Institute of Technology, Gadjah Mada University, Hanoi University of Science and Technology, Nagoya Institute of Technology and MyBamboo Sdn. Bhd. This grant registered in Universiti Teknologi Malaysia under the project (R.J130000.7351.4B673).

Organic Rankine Cycle (ORC) as waste heat recovery system for micro-cogeneration: A preliminary study

S Saadon[1,2]

[1]Department of Aerospace Engineering, Faculty of Engineering, Universiti Putra Malaysia, 43400 Serdang, Selangor, Malaysia
[2]Aerospace Malaysia Research Centre, Faculty of Engineering, Universiti Putra Malaysia, 43400 Serdang, Selangor, Malaysia

Corresponding author's email: mimisaadon@upm.edu.my

Abstract. Although the energy consumption pattern has been optimized, the main problem facing by energy revolution is the low conversion efficiency of renewable energy and waste heat. Among the utilization technologies of these waste heat, organic Rankine cycle (ORC), which has characteristics of simple structure and high efficiency is chosen as the potential candidate to overcome this situation. However, ORC is facing several limitations including having low turbine performance leading to low power production. Additionally, the design of ORC and the heat recovery system should be improved to reduce the investment cost. One of the ways to increase the performance of this ORC is to design ORC waste heat recovery system for cogeneration. Thus, this study will be focused on to identify the effect of different heat exchanger positions to the ORC thermal performance and to investigate the effect of integrating ORC for cogeneration. In order to achieve this, identification of the suitable heat exchanger for ORC will be performed by modelling and simulation to predict the effect of the new design to the ORC thermal performance. Then, the improved novel ORC design will be integrated to a cogeneration system to simulate the power produced by the combined system.

1. Introduction

Many sectors have been destroyed by the COVID-19 pandemic, and solar energy is no exception. Under the 7th Sustainable Development Goals for affordable and clean energy, the world seeks to ensure access to effective, secure, safe and modern energy for everyone. Energy access is a very significant pillar for people's well-being as well as for economic growth and poverty alleviation. This goal has five goals to be accomplished, one of which is to double the increase in energy efficiency. Being a developing country, Malaysia is part of the plan, in particular for infrastructure support. Furthermore, in order to provide a cleaner environment for future generations, Malaysia plans to reduce carbon emissions by 40% in the coming years, 45% by 2030, and be carbon neutral by 2050 **Error! Reference source not found.**.

The need for safe and affordable electricity in health centres has been illustrated by the COVID-19 pandemic. This pandemic has also triggered a decrease in transport energy demand due to the global distance response that makes individuals rely more on electricity for telecommunications and virtual meetings at their home. Based on exergy-based analysis done by S. Saadon et al [1], the use of a solar collector integrated to a building is very promising in terms of economy and environment. However, while switching to use more energy in a residential building, rather than at the exterior, a large share of energy is also wasted because the demand is much greater than the request of the consumer [3] and Malaysia being in the equatorial zone goes through this same situation as well. Thus Organic Rankine

Cycle (ORC) has emerged as a relevant waste heat recovery system to save this energy. However, this system still going through some challenges.

ORC is a well-known thermodynamic cycle that is practically used to convert heat to work [4]. It employs organic fluid as the main working fluid, which makes it a slightly different from a conventional steam Rankine cycle. This cycle could be presented in a micro-size that can be used in a distributed generation system by taking advantage of low-temperature heat source [5]. Industrial waste heat, biomass energy, geothermal energy or solar energy are examples of renewable or waste heat sources for the device [6], [7].

Solar energy is one of Malaysia's most promising clean energy sources. Due to the abundance of solar energy, Malaysia has the capacity to use it to produce heat and power through solar thermal systems. The combined heat and power system can be used in both urban and rural areas of Malaysia when moderate solar radiation is taken into account. A cogeneration system is a combined heat and power system. A single heat source in the device produces both usable heat and energy at the same time. Since heat and energy are generated at the same time, the combined heat and power system's overall plant efficiency can be increased. In recent years, the integrated heat and power system dependent on renewable energy sources has piqued people's interest.

Despite the very interesting potential of micro-cogeneration systems in households to minimise CO_2 emissions, to date, their adoption, due to economic viability, has been minimal. Ultimately, it must be mentioned that the financial feasibility of ORCs for the recovery of waste heat is promising. A payback period of 2.7 years has been found, which is a low and reasonable value. In fact, the investment cost relates only to the ORC and the method of heat recovery. However, for the financial feasibility of the system, a proper construction of residential building micro-cogeneration systems, the annual operating hours and the electricity price are important criteria for their achievement in economics.

One of the ways to increase the performance of this ORC is to design ORC waste heat recovery system for cogeneration. Thus, this study will be focused on to identify the effect of different heat exchanger positions to the ORC thermal performance and to investigate the effect of integrating ORC for cogeneration with optimum power. In order to achieve this, identification of the suitable heat exchanger for ORC will be performed by modelling and simulation to predict the effect of the new design to the ORC thermal performance by using MATLAB. Then, the improved novel ORC design will be further study and integrated to a cogeneration system to simulate the power produced by the combined system through a thermal building simulation software TRNSYS, in order to ensure an output of a novel micro-cogeneration ORC system.

2. Methodology

2.1. Design description

Solar photovoltaic (PV) collector and power conversion systems are the two primary subsystems of a solar cogeneration system. Aside from power generation, the solar PV collector thermal energy can be used for heating and a combined heat and power system. The solar collector captured thermal energy from the Sun. The heat from this energy is then pumped to exchange heat with organic fluid in the ORC through evaporator. The low-pressure working fluid produced by the turbine is then used for heat demand applications. Figure 1 below shows the schematic diagram of micro-cogeneration system based on the ORC. The details process occur in the ORC is explained in the next sub-section.

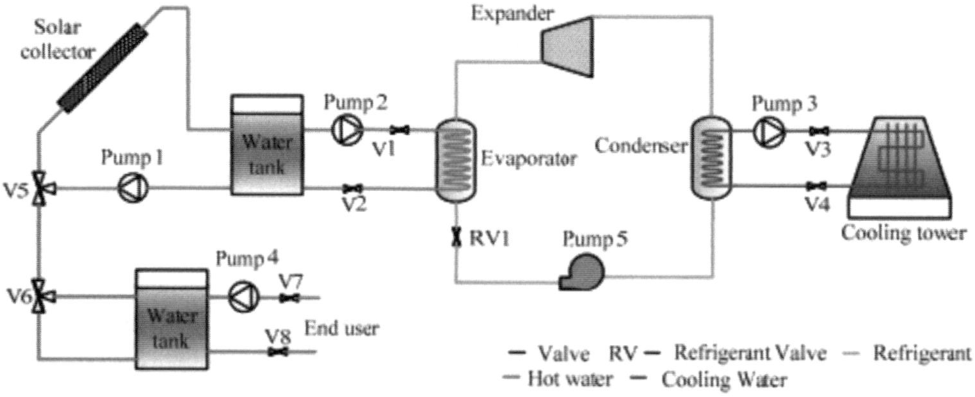

Figure 1. A small-scale cogeneration system with ORC [8].

As Malaysia is a developing country, applying a micro size of cogeneration system would be more suitable for the majority of residential buildings. The micro-cogeneration system is also more feasible in nations near the equator due to the availability of modest solar energy [9]. Figure 2 below shows the annual ambient temperature and solar radiation received in Kuala Lumpur (3.133° N, 101.683° E).

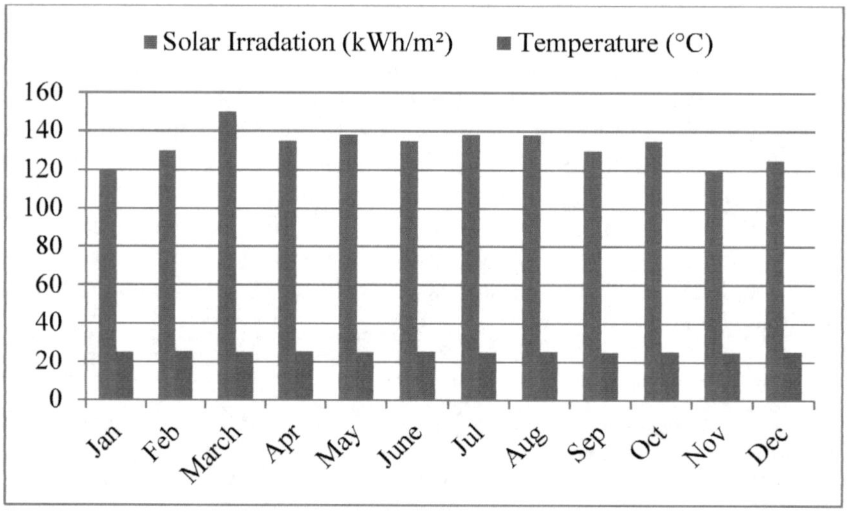

Figure 2. Annual ambient temperature and solar radiation in Kuala Lumpur.

2.2. Organic Rankine Cycle power model

A schematic diagram of a basic ORC system with preheater and superheater are shown below in figure 3 and 4. A condenser, an expander, an evaporator, a preheater or superheater and a working fluid pump are all parts of this device. Organic fluid is chosen due to its low liquid-vapor phase change, which suits well with the characteristics of waste heat (low pressure and low temperature). As for the cycle with preheater, at first, the pre-heated organic fluid is transferred to the evaporator, where it exchanges heat with the heat source which is produced by solar collector. The fluid is eventually converted into vapor. The organic vapor is then expanded to generate power, which is then transformed into electricity by the generator. Whereas for the ORC with superheater, the organic fluid after being pumped, is transferred to the evaporator and eventually the rest of the process is similar to previous cycle. At the end, the exhaust organic gas from the turbine is condensed using cooling water in the condenser.

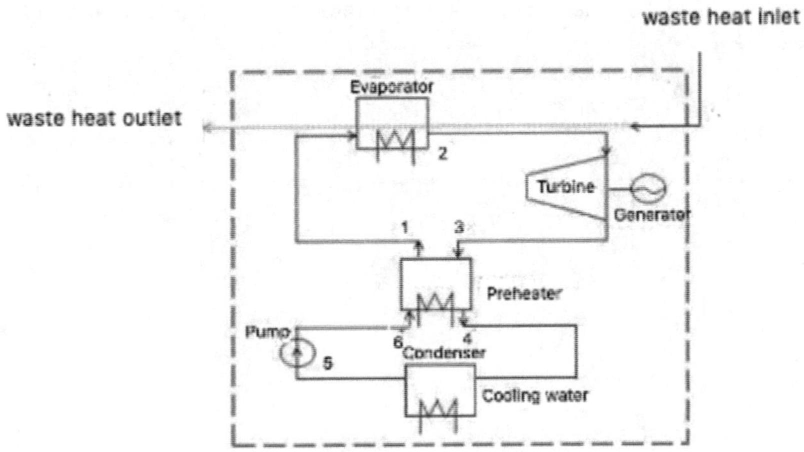

Figure 3. Schematic diagram of a basic ORC system with preheater.

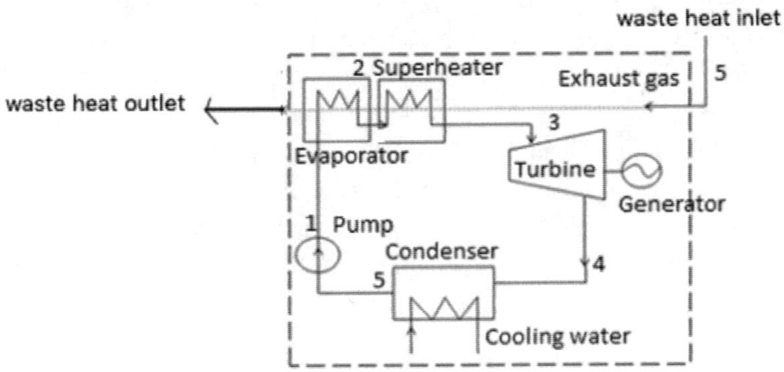

Figure 4. Schematic diagram of a basic ORC system with superheater.

First, in the pump part, W_{pump} is the pump power consumption provided by

$$W_{pump} = \frac{\dot{m}_{wf} c_{p_{wf}} T_{cond,wf\_out} (\Pi_{pump}^{\frac{\gamma-1}{\gamma}} - 1)}{\eta_{pump}} \tag{1}$$

where $\Pi_{pump} = \frac{P_{pump,out}}{P_{pump,in}}$.

$$Q_{evap} = \dot{m}_{wf} C_{p_{wf}} (T_{evap,wf\_out} - T_{evap,wf\_in}) \tag{2}$$

The output power of the turbine W_{exp} is expressed as follows in order to evaluate the process in the turbine

$$W_{exp} = \dot{m}_{wf} c_{p_{wf}} \eta_{exp} T_{exp,in} (1 - \Pi_{exp}^{\frac{1-\gamma}{\gamma}}) \tag{3}$$

where $\Pi_{exp} = \frac{P_{exp,in}}{P_{exp,out}}$.

Meanwhile, inside the condenser, the total heat transfer rate between the working fluid and the cooling water is expressed as follows

$$Q_{cond} = \dot{m}_{wf} c_{p_{wf}} [T_{cond,wf\_in} - T_{cond,wf\_out}] \tag{4}$$

As for the cycle with superheater, the total heat transfer rate within the superheater is calculated using the equation (4) below

$$Q_{sup} = \dot{m}_{wf} C_{p_{wf}} (T_{exp,in} - T_{sup,out}) \tag{5}$$

The ORC system's net power production is,

$$W_{net} = W_{exp} - W_{pump} \tag{6}$$

3. Results and Discussion

3.1. Modeling of ORC with preheater and superheater

The modelling of the basic ORC system has been validated previously [10]. It is then applied to a suitable waste heat source. The results of the first objective of identifying the effect of different heat exchanger positions to the ORC thermal performance is being illustrated in figure 5 below. The heat source is the waste heat that is being recovered and transferred to the evaporator of the ORC. From these results, it can be said that the power output will generally increase as the temperature of the heat source increases. The ORC with preheater exhibits a higher net power output compared to the one with superheater since the demand of pump power is lower. This is basically thanks to the preheater which regenerates the heat coming out from the turbine and uses it to preheat the fluid exiting the pump, thus reducing the amount of work used by the pump.

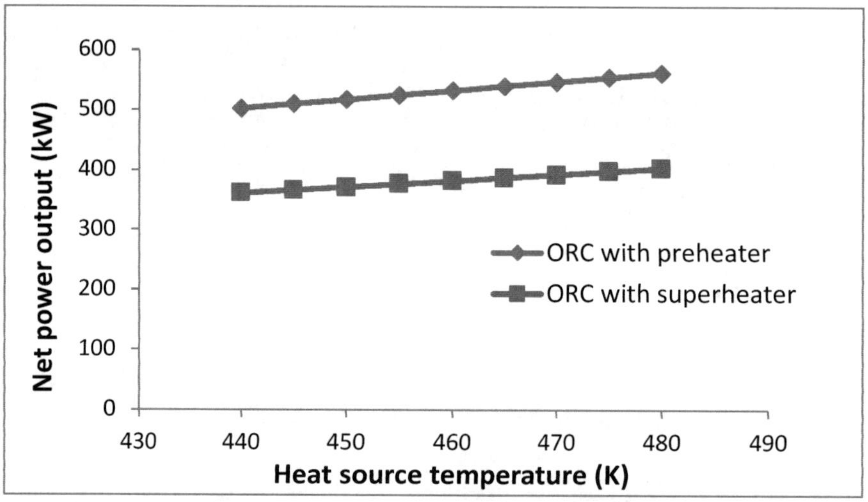

Figure 5. Net power output produced by ORC with preheater and superheater.

Another point worth to be noted, is that the preheater could be combined with the superheater at the same time as both of these integrations will add more beneficial impact towards the potential power produced. However, one should also note that by having extra components to the basic design, will eventually increase the installation's cost. Thus an economic and payback period analysis would be very important.

3.2. Building integrated solar PV collector

In this section, a residential building which is a single-family house is studied first without ORC system in order to understand the electric consumption normally needed by this type of house. The building is integrated with a solar PV collector which is located on top of the roof south surface wall of the building. The building model is simulated by using TRNSYS software. The house is modelled as four thermal zones consist of a living room, a kitchen, a bedroom and a bathroom. The results of simulation of electric consumption due to cooling needs of the residential house situated in Kuala Lumpur are shown in table 1 and figure 6.

Table 1. Solar PV production and electric consumption in a residential house at Kuala Lumpur

	PV production (kWh/m2)	Electric consumption (kWh/m2)
with solar PV	244	513
without solar PV	0	556

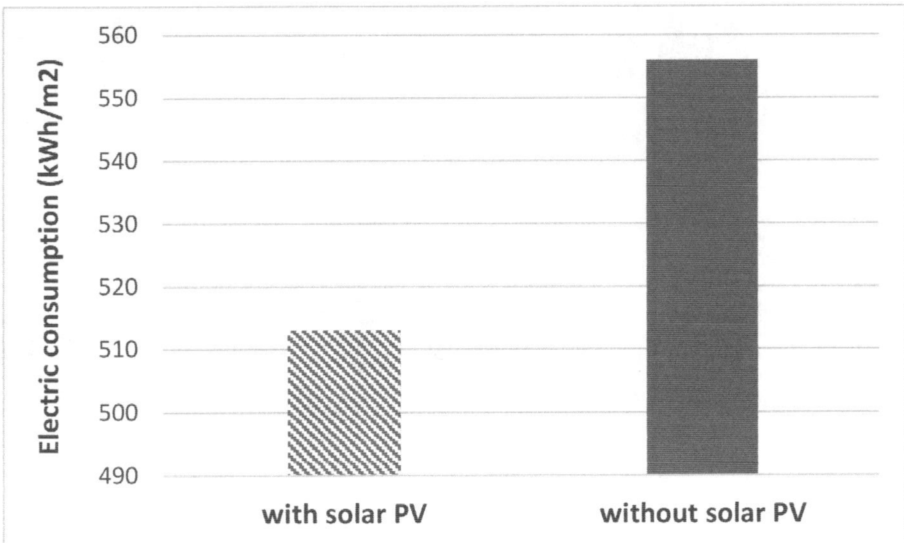

Figure 6. Electric consumption due to cooling for Kuala Lumpur throughout the year.

For a residential building, the solar PV collector does contribute to reducing the electric consumption compared to the one without solar PV collector. Thus, this study will be further continued to look into perspective where the waste heat of the solar collector is recovered by using micro-ORC cogeneration system to reduce more the electric consumption.

4. Conclusion

There are several advantages and drawbacks of various ORC designs as a heat recovery system. When working under various heat transfer and thermal conditions, they differ in size and building complexity. The study's findings are likely to serve as a starting point for Malaysians considering a cogeneration system that uses ORC for waste heat recovery. It is believed that ORC technology can effectively extract a significant amount of energy from waste heat to produce electricity, increasing productivity while also reducing primary fuel demand and consumption, lowering CO_2 emissions and the greenhouse effect.

5. References

[1] Vaka, M., Walvekar, R., Rasheed, A. K., & Khalid, M. (2020). A review on Malaysia's solar energy pathway towards carbon-neutral Malaysia beyond Covid'19 pandemic. Journal of Cleaner Production, 273, 122834.

[2] Saadon, S., Gaillard, L., Menezo, C., & Giroux-Julien, S. (2020). Exergy, exergoeconomic and enviroeconomic analysis of a building integrated semi-transparent photovoltaic/thermal (BISTPV/T) by natural ventilation. Renewable Energy, 150, 981–989.

[3] Arteconi, A., Del Zotto, L., Tascioni, R., & Cioccolanti, L. (2019). Modelling system integration of a micro solar Organic Rankine Cycle plant into a residential building. Applied Energy, 251, 113408.

[4] Martinez, S., Michaux, G., Salagnac, P., & Bouvier, J.-L. (2017). Micro-combined heat and power systems (micro-CHP) based on renewable energy sources. Energy Conversion and Management, 154, 262–285.

[5] Delgado-Torres, A. M., & García-Rodríguez, L. (2010). Analysis and optimization of the low-temperature solar organic Rankine cycle (ORC). Energy Conversion and Management, 51(12), 2846–2856.

[6] Wang, Z. Q., Zhou, N. J., Guo, J., & Wang, X. Y. (2012). Fluid selection and parametric optimization of organic Rankine cycle using low temperature waste heat. Energy, 40(1), 107–115.

[7] Tchanche, B. F., Papadakis, G., Lambrinos, G., & Frangoudakis, A. (2009). Fluid selection for a low-temperature solar organic Rankine cycle. Applied Thermal Engineering, 29(11-12), 2468–2476.

[8] Wang, R., Jiang, L., Ma, Z., Gonzalez-Diaz, A., Wang, Y., & Roskilly, A. (2019). Comparative Analysis of Small-Scale Organic Rankine Cycle Systems for Solar Energy Utilisation. Energies, 12(5), 829.

[9] Heng, S. Y., Asako, Y., Suwa, T., Tan, L. K., Sharifmuddin, N. B., & Kamadinata, J. O. (2019). Performance of a small-scale solar cogeneration system in the equatorial zone of Malaysia. Energy Conversion and Management, 184, 127–138.

[10] Saadon, S., & Mohd Nasir, N. A. (2020). Performance and Sustainability Analysis of an Organic Rankine Cycle System in Subcritical and Supercritical Conditions for Waste Heat Recovery. Energies, 13(12), 3035.

Journal of Physics: Conference Series **2053** (2021) 012020 doi:10.1088/1742-6596/2053/1/012020

Stochastic type-2 fuzzy modelling on GHG emission prediction for gas-fired power plant

Sazalina Zakaria[1,2*], Radin Diana R. Ahmad[1,2], Ahmad Rosly Abbas[1,2], Mohd Faizal Mohideen Batcha[1], Mohd Fauzi Zanil[3]

[1] Faculty of Mechanical and Manufacturing Engineering, University Tun Hussein Onn Malaysia (UTHM), Malaysia.
[2] Generation and Environment Department, TNB Research Sdn. Bhd., Malaysia.
[3] Department of Chemical and Petroleum Engineering, Faculty of Engineering & Built Environment, UCSI University, Malaysia.

Corresponding author's email :sazalina.zakaria@tnb.com.my

Abstract. This paper demonstrates the implementation of type-2 fuzzy logic model in prediction of greenhouse gas emission for gas fuelled power plant. The data collection and analysis on the GHG emission had been conducted with emphasis on the industrial scale with many uncertainty factors. The developed model will be used as a prediction tools to evaluate potential GHG emission for future study and mitigation planning. In this work, 100 sample sizes had been used as training dataset in the model framework which serve as the input information while 14 sample sizes are used for model validation to achieve model generalization. A novel Karnik-Mendel (KM) algorithm had been proposed with the Genetic Algorithm (GA) model optimization to achieve high performance and accuracy for the respective predictive model. In conclusion, the result showed that the model able to give an outstanding accuracy. On the validation analysis, the results showed developed model able to give satisfactory prediction with R^2 of 0.978 and 0.84968 for training and validation, respectively.

1. Introduction

Like in most other countries, greenhouse gas (GHG) emissions in Malaysia comes from the electricity generation, transportation, industrial processes, and waste sectors. The Malaysian energy sector is highly dependent on fossil fuels. This sector is a major emissions contributor in year 1994 to 2016. table 1 shows the emissions trends by the energy sector [1].

Table 1. Trends of Malaysia GHG emissions by energy sector, 1994-2016.

Year	GHG Emission [GgCO_2eq]	Share [%]
1994	92,049.66	68
2000	143,141.29	66
2005	198,514.01	73
2014	253,517.24	80
2016	251,695.02	80

Malaysia energy and emissions intensity have been steadily increasing. It is expected that the GHG emissions will continue to rise in tandem with the growing demand for fossil fuel. The increasing trend of fossil fuel consumption for electricity generation, particularly coal, contradict with the National goal to reduce the emissions intensity. Based on 2015 Paris Agreement Target, Malaysia intends to reduce GHG emissions intensity of GDP by 45% by 2030 relative to the emissions intensity of GDP in 2005.

The electricity generation GHG emissions is also projected to continue to grow in tandem with the increasing population and GDP. Analysis of the projected GHG emissions for the energy sector indicates that GHG emissions from this sector would continue to increase from 133,097 GgCO2e in 2014 to 191,990.661 Gg CO2e in 2030. Therefore, the data collection and analysis on the GHG emission had been conducted under this study with emphasis on the industrial scale with many uncertainty factors. The developed model will be used as a prediction tools to evaluate potential GHG emission for future study and mitigation planning.

The trend in process engineering is strive towards the system complexity which mainly due to the requirement of complex tasks and good accuracy [2]. Complex systems with certain variables varied by time are unpredictable and dynamic in the system. Normally, a conventional sensor will be installed in the production line which used as an indicator for the engineers to make proper judgement based on the real time value display by the indicator. On the other hand, a regressive model is used to calculate state variable that are normally is inaccessible in certain process. Determine parametric value based on model is attractive as in environmental process applications, it has many unavailability of states to measure.

Furthermore, stochastic behavior on the parametric state is commonly observed in prediction of GHG emission [3]. Stochasticity refers to the occurrence of random probability distribution or pattern that may be statistically analyzed yet may not be accurately predicted. Therefore, stochastic response is undesirable in terms of operability as it would be challenging to determine parametric value in environmental process application. Moreover, the randomness could be dreaded since it would be difficult to forecast an appropriate representation of parameter's value due to inconsistent dataset. Therefore, development of a stochastic based model to predict GHG emission is of great research interest in this paper. Thus, by providing the input and output dataset, these predictions are possible to be determined [4].

The selection of type-2 fuzzy logic as a proposed model's framework in the environmental process applications which potentially eliminated the unknown uncertainty factor and enhance the stability of the prediction. In general, the fuzzy logic, has two type of mechanism which named as type-1 fuzzy logic (T1FL) and type-2 fuzzy logic (T2FL). As compared to T1FL, the T2FL managed to handle higher degree of uncertainties with its three-dimensional nature of type-2 fuzzy sets (T2FS). In this paper, the uncertainties in the predictive model for GHG emission will be solved by using T2FL as this approach potential provide degree of freedom to model the process which has many uncertainties.

Type-1 fuzzy sets were introduced by Zadeh in 1965 to process and manipulate data and information affected by uncertainty or imprecision [5]. It had represented the vagueness and uncertainty of linguistic problem. Meanwhile, type-2 fuzzy set had been designed with the three-dimensional membership function of footprint of uncertainty. The additional block of type reducer had been introduced before the defuzzification process. This had indicated the difference between the T1FL and T2FL. The membership function of type-1 is crisp value while the membership function of type-2 has a region named as footprint of uncertainty which exist upper and lower limit of the primary membership. The uncertainty region exists between the upper and lower limit of primary membership value.

The internal architecture of type-2 fuzzy logic consists of four blocks which named as fuzzifier, rule base inference, type reduction and defuzzifier. Initially, the input variable had been fuzzified in the fuzzifier block into fuzzy input set. The fuzzifier block had map the crisp point into a type-2 fuzzy set. Next, the fuzzy input sets will activate the rule-base inference block in which the IF-THEN rule had been preformulate in the system. The inference engine combines rules and give a mapping function from input type-2 fuzzy sets to output type-2 fuzzy set. The operation of unions, intersections and composition will be used in the inference system to calculate the desired value for the output value [6]. The type-2 output from inference engine will proceed into next block which named as type reduction block. The type reduction block will reduce the type-2 output sets into type-1 output sets. Type reduction block plays a vital role in type-2 fuzzy inference system (T2FIS) for the computational burden to reduce T2FL into T1FL. The selection of type reduction can be considered as the bottleneck of T2FIS determine the overall T2FL model performance. Normal type reduction block is Karnik Mendel (KM) algorithm in which a point search process will occur in this block. The KM algorithm are well-known for converge monotonically and super-exponentially fast in two to six iterations. Researchers had been proposed to enhance KM type reduction with the objective of reduce the computational cost and computational time as well as increase the accuracy of predictive model [6].

The objective of this works is to develop an empirical model to predict GHG emission by using type-2 fuzzy logic system. The prediction focusses on the emission from gas-fired power plant. To carried out the model development, a type-2 fuzzy reduction algorithm with Extend Karnik Mendel (EKM) algorithm with new initialization (ANI) had been proposed. This modeling algorithm able to provide a less computational time of convergence with a few of iteration [7].

2. Methodology

2.1. Data Collection

In this paper, the gas density at standard temperature and pressure (STP) is defined in between 0.7 to 0.9 kg/m3. Figure 1 show the distribution of each site at density versus GHG emission factor. In this model, the range from 0.55 to 0.9 kg/m3 has been consider as a training and validation dataset.

This indicate the data in the inventory has ill-defined information which need to be corrected before proceeding to GHG emission factor model development. As tabulated in table 2, there were twelve inputs which were C1, C2, C3, iC4, nC4, iC5, nC5, C6, C7, C8, CO2 and N2 and one output layer corresponded to gas density.

2.2. Model development of Type-2 Fuzzy

The type-2 model with novel KM algorithm named as EKMANI had designed to estimate the output state variable with high accuracy and low computational time. In this paper, the T2FL model had been developed as in figure 2 to predict the GHG emission by using input information as in table 2. Finally, the performance evaluation on the proposed model had been done to evaluate its significant for accuracy and computational load performances.

The development starts with determine the fuzzy linguistic membership function with the respective input and output variables in the type-2 fuzzy toolbox in MATLAB [8]. The linguistic variable named as xi for i = 1 to 12 which were C1, C2, C3, iC4, nC4, iC5, nC5, C6, C7, C8, CO2 and N2 respectively. Then, it will be investigated with the respective membership functions from the previous computation value until satisfied best fit of R^2. The respective input variables will be labelled with the triangle interval in T2FIS. Meanwhile, the output variable will be developed by using the linear mathematical equation in TSK Fuzzy Inference System (FIS).

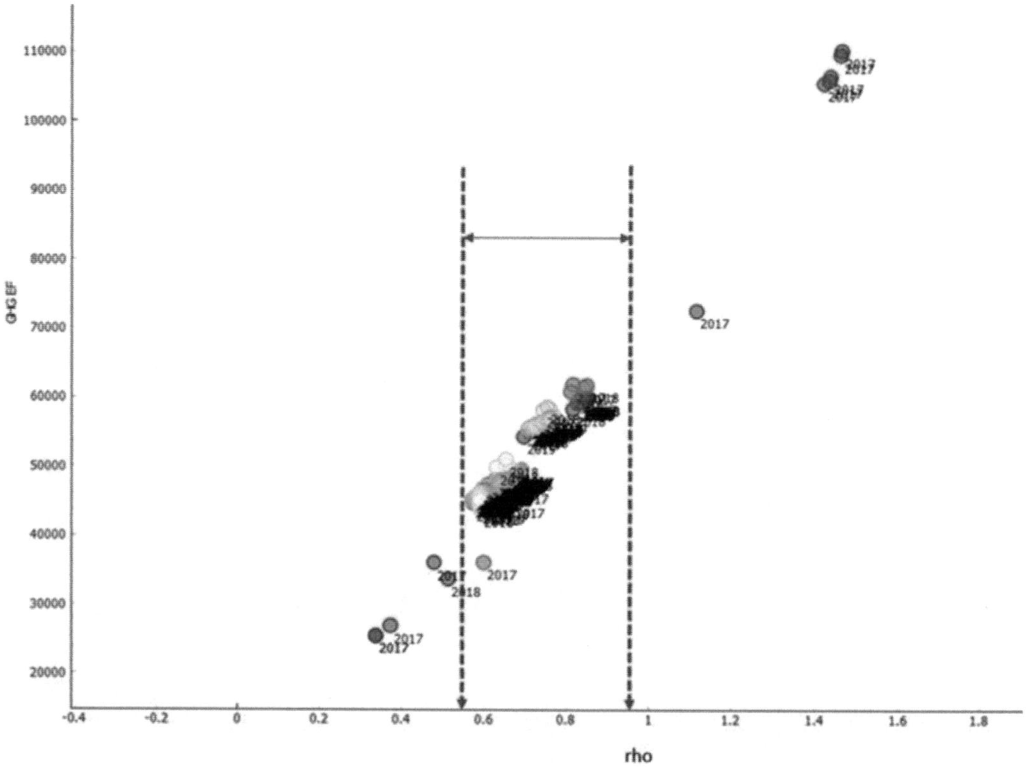

Figure 1. Distribution of density for gas fueled power plant.

Table 2. Greenhouse gas (GHG) emission inputs-output distribution of gas fuel composition.

	Unit	Centre	Dispersion	Min	Max
Inputs:					
C1 (CH4)	%	91.28	0.05	68.57	97.17
C2 (C2H6)	%	3.26	0.31	1.44	4.99
C3 (C3H8)	%	1.07	0.63	0.10	2.61
C6+ (C6H14)	%	0.05	1.82	0	0.39
nC4 (C4H10)	%	0.18	0.89	0.01	0.64
iC4 (C4H10)	%	0.18	0.83	0.02	0.60
nC5 (C5H12)	%	0.03	1.44	0	0.24
iC5 (C5H12)	%	0.06	1.07	0	0.24
N_2	%	1.05	1.36	0.06	19.02
CO_2	%	2.83	0.79	0.01	9.01
Density, rho	kg/Sm^3	0.66	0.12	0.57	0.85
Output:					
GHG EF	kgCO_2/TJ	51185.4	7539.8234	4636.9	56571.3

Next, the fuzzy inference rule will develop in T2FIS for the decision making for the fuzzy observer system with the proposed variable. Meanwhile, novel type reduction method of KM algorithm had been built in the MATLAB for the type-2 fuzzy toolbox [8]. After the development of FIS and linguistic variable with respective membership function, the fuzzy observer will be exported to Simulink for the parametric prediction process. Next, an optimization process on the process model will be carried out to optimize the output constants in the interval linear equation. Finally, the predicted data from the fuzzy observer will be discussed with result obtained from the simulation data and evaluate with the respective performance index.

The development starts with determine the fuzzy linguistic membership function with the respective input and output variables in the type-2 fuzzy toolbox in MATLAB [8]. The linguistic variable named as xi for i = 1 to 12 which were C1, C2, C3, iC4, nC4, iC5, nC5, C6, C7, C8, CO2 and N2 respectively. Then, it will be investigated with the respective membership functions from the previous computation value until satisfied best fit of R^2. The respective input variables will be labelled with the triangle interval in T2FIS. Meanwhile, the output variable will be developed by using the linear mathematical equation in TSK Fuzzy Inference System (FIS). Next, the fuzzy inference rule will develop in T2FIS for the decision making for the fuzzy observer system with the proposed variable. Meanwhile, novel type reduction method of KM algorithm had been built in the MATLAB for the type-2 fuzzy toolbox [8]. After the development of FIS and linguistic variable with respective membership function, the fuzzy observer will be exported to Simulink for the parametric prediction process. Next, an optimization process on the process model will be carried out to optimize the output constants in the interval linear equation. Finally, the predicted data from the fuzzy observer will be discussed with result obtained from the simulation data and evaluate with the respective performance index.

Fuzzy linguistic variable is the input and output variables for the GHG prediction with respective membership function with each linguistic label. It can be defined as the values that assigned to the respective words or labels for each variable. In type-2 fuzzy model, interval values will be assigned to the respective words or variable for the indication or degree of membership function for the label. The idea of fuzzy linguistic variable is essential for the development of fuzzy set theory as the fuzzy logic is primary associated with the reasoning for the vague term in the language. The input and the output set were labelled with respective terms and word with the upper interval and low interval [5].

The output variable of this paper named as GHG had been range from 4636.9 to 56571.3 kgCO2/TJ. The membership function for the output variable had labelled with low (min), normal (center), and high (max) with respective upper and lower interval of 5% (from 2006 IPCC Guidelines for National Greenhouse Gas Inventories) path of uncertainty. In the type-2 fuzzy model, the output variables will be represented in the interval linear form as in equation (1) and as in equation (2):

$$GHG_u = \sum_{i=1}^{12} \left(\overline{A}_i x_i\right) + \overline{B} \tag{1}$$

$$GHG_l = \sum_{i=1}^{12} \left(\underline{A}_i x_i\right) + \underline{B} \tag{2}$$

Where A and B is linear coefficient upper and lower (denoted from the upper and lower bar) which will be optimized by genetic algorithm. By referring to table 2, the input variables of TSL type-2 fuzzy model, are labelled as low (min), normal (center), and high (max). Then upper and lower membership had been graded as 5% upper and lower from membership range.

The IF-THEN rule is the rule of inference in which it is a knowledge-based system [9]. Total of 144 fuzzy inference rule (upper and lower). The IF-THEN rule in the type-2 fuzzy system is described as for instance GHG is low when the x1 is low, x2 is low … x12 is low. Meanwhile, the next scenario is GHG is low when the x1 is low, x2 is normal … x12 is low. The inference sequence is repeatedly until 144 fuzzy rules.

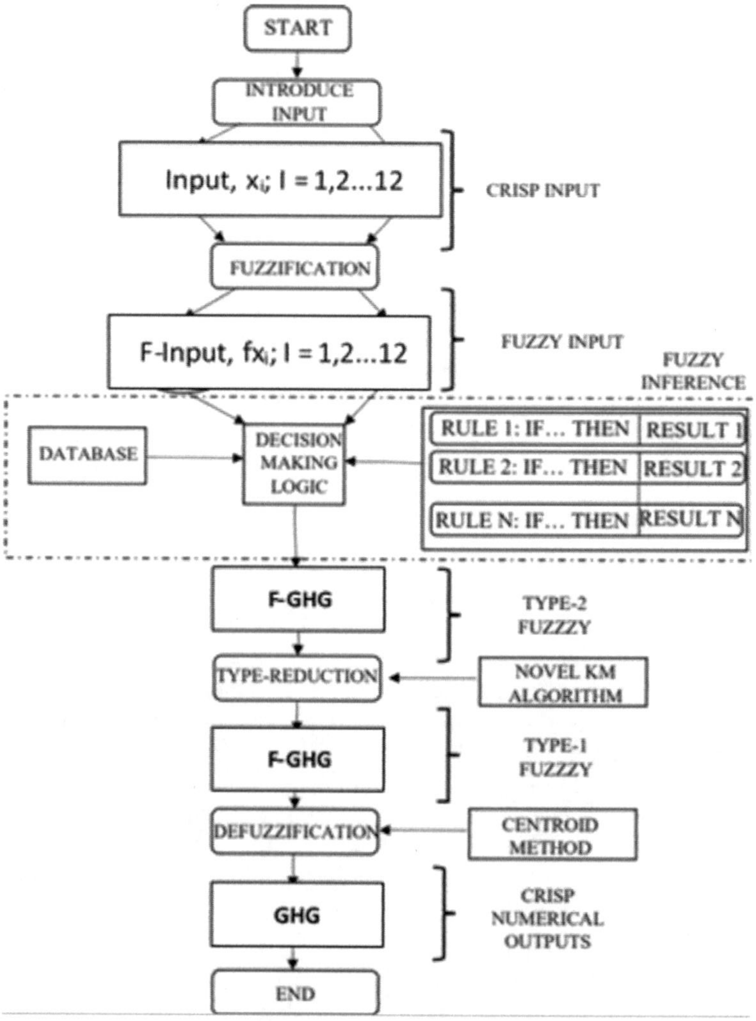

Figure 2. Model development flow chart.

The novel KM algorithm named as EKMANI had been developed in MATLAB as the new type reduction method as in table 3 to reduce the computational time and number of iterations required for the find the switch point. The EKMANI algorithm had improved for the initialization process from EKM algorithm. The summary of complete EKM algorithm with new initialization had been show as follow [8].

Table 3. EKMANI algorithm.

Step	Left Point	Right Point
1	x_i had been sorted by increasing order	x_i had been sorted by increasing order
2	Renumber the index of lower and firing intervals, $\underline{\omega}_i$ and ω^{-}_i respectively to maintain the original value to the corresponding rule consequent.	Renumber the index of lower and firing intervals, $\underline{\omega}_i$ and ω^{-}_i respectively to maintain the original value to the corresponding rule consequent
3	Compute: $k = 0.35\,N$	Compute: $k = 0.65\,N$
4	Set the k to the nearest integer. $a = \sum_{i=1}^{k} x_i\overline{\omega_i} + \sum_{i=k+1}^{N} x_i\underline{\omega_i}$ $b = \sum_{i=1}^{k} \overline{\omega_i} + \sum_{i=k+1}^{N} \underline{\omega_i}$	Find the switch point k where; $a = \sum_{i=1}^{k} x_i\underline{\omega_i} + \sum_{i=k+1}^{N} x_i\overline{\omega_i}$ $b = \sum_{i=1}^{k} \underline{\omega_i} + \sum_{i=k+1}^{N} \overline{\omega_i}$
5	Compute: $y = \frac{a}{b}$ In which satisfy the condition of $x_{k'} < y < x_{k'+1}$	Compute: $y = \frac{a}{b}$ In which satisfy the condition of $x_{k'} < y < x_{k'+1}$
6	The k' value should be identified in such that k' ε [1, N-1] and check whether k'= k.	The k' value should be identified in such that k' ε [1, N-1] and check whether k'= k.
7	Computed the value of s = sign (k'-k) and $a' = a + s \sum_{i=\min(k,k')}^{\max(k,k')} x_i(\overline{\omega}_i - \underline{\omega}_i)$ $b' = a + s \sum_{i=\min(k,k')}^{\max(k,k')} (\overline{\omega}_i - \underline{\omega}_i)$ $y'' = \frac{a'}{b'}$	Computed the value of s = sign (k'-k) and $a' = a - s \sum_{i=\min(k,k')}^{\max(k,k')} x_i(\overline{\omega}_i - \underline{\omega}_i)$ $b' = a - s \sum_{i=\min(k,k')}^{\max(k,k')} (\overline{\omega}_i - \underline{\omega}_i)$ $y'' = \frac{a'}{b'}$
8	Set y = y', a= a' and b = b' and k = k'. Proceed with step 4.	Set y = y', a = a' and b = b' and k = k'. Proceed with step 4.

3. Result and Discussion

The performance analysis had been done on the type-2 fuzzy logic with the objective to evaluate the efficiency of the implementation of observer in the environmental process application [10]. The different between predicted value with the actual (true) value for GHG emission had been done with the performance index. The predicted data will obtain from the result of the type-2 fuzzy logic model while the actual data will be obtained from the repository data fuel certificate. The predictive model will be evaluated for the predictive performance (table 4). The predicted data will evaluate with the term of coefficient of determination, R^2.

Table 4. Performance evaluator.

Performance evaluator	Calculation
Coefficient of determination	$R^2 = \frac{\sum_{i=1}^{N}(P_i - P_m)^2}{\sum_{i=1}^{N}(P_i - P_m)^2 + \sum_{i=1}^{N}(O_i - P_i)^2}$

The above performance analysis had indicated that O_i is true value, P_i is predicted value from type-2 fuzzy logic and P_m is predicted average value from type-2 fuzzy logic. R^2 coefficient better known as correlation coefficient measures the strength and direction of linear relationship between two variables on a scatterplot.

The 14 data points has been used in the model validation performance (table 5). The data points were chosen randomly from data inventory. Good model fit for both training and validation has been found as figure 3(a) and figure 3(b). The trade-off between training (0.978) and validation (0.84968) fitness is to guarantee the generalization of developed model. This will be benefited to the prediction of unseen input values for future work prediction. Similar performance has been reported on GHG prediction where for training R^2 above 0.8 is acceptable [11].

Table 5. Validation Dataset.

	Inputs												Output
#	C1	C2	C3	iC4	nC4	iC5	nC5	C6+	C7	C8	CO2	N2	GHG
1	92.61	3.24	1.50	0.30	0.35	0.14	0.10	0.17	0.00	0.00	1.43	0.16	56782
2	92.61	3.24	1.50	0.30	0.35	0.14	0.10	0.17	0.00	0.00	1.43	0.16	56763
3	92.73	3.32	1.47	0.29	0.33	0.14	0.10	0.07	0.00	0.00	1.40	0.15	56774
4	92.75	3.08	1.48	0.30	0.35	0.14	0.30	0.07	0.00	0.00	1.56	0.17	56773
5	92.73	3.43	1.44	0.29	0.33	0.14	0.10	0.07	0.00	0.00	1.34	0.14	56793
6	90.93	4.33	1.74	0.31	0.31	0.08	0.04	0.07	0.00	0.00	2.09	0.10	56761
7	90.54	4.67	1.88	0.36	0.33	0.07	0.04	0.05	0.00	0.00	2.01	0.06	56758
8	93.54	2.70	0.67	0.14	0.10	0.02	0.01	0.02	0.00	0.00	2.00	0.80	56781
9	92.73	3.38	1.45	0.32	0.39	0.03	0.01	0.02	0.00	0.00	1.07	0.59	56756
10	91.62	3.18	0.82	0.18	0.13	0.04	0.02	0.14	0.00	0.00	2.62	1.26	58355
11	93.89	2.92	0.13	0.03	0.02	0.01	0.01	0.15	0.00	0.00	1.41	1.44	56779
12	84.89	3.47	2.00	0.46	0.48	0.19	0.14	0.38	0.00	0.00	6.85	1.15	56746
13	84.71	3.69	2.27	0.52	0.55	0.21	0.15	0.36	0.00	0.00	6.39	1.14	56754
14	84.71	3.69	2.27	0.52	0.55	0.21	0.15	0.36	0.00	0.00	6.39	1.14	57081

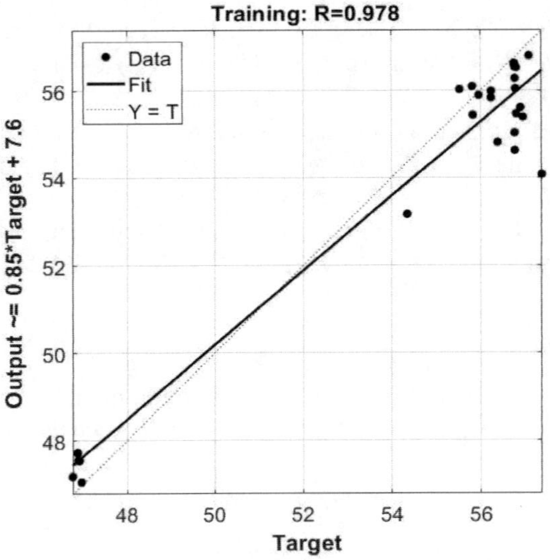

(a)

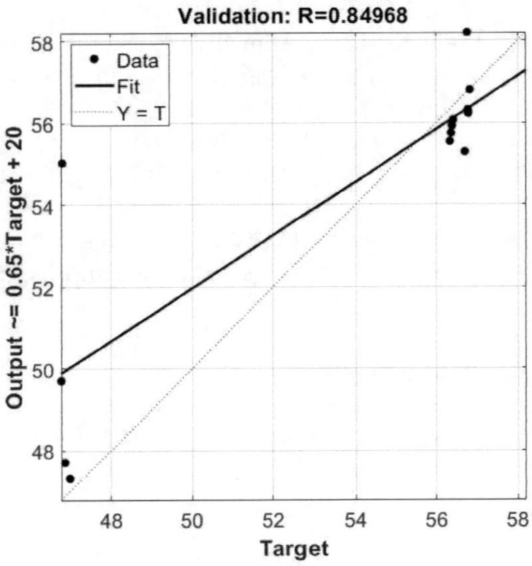

(b)

Figure 3. Model Performance evaluation: 3(a) Training, 3(b) Validation.

4. Conclusion

The model was developed empirically to predict the value of GHG emission. The scope of model covers the inputs-output range as in table 2 for specific 12 species in gas composition. In this paper, modeling was performed by using type-2 fuzzy logic with EKMANI algorithm. The model is a data driven correlation which developed based on regression technique and it requires many data points (dataset) which related to the input and output of process. In this paper, the dataset is taken only for those in density range of 0.55 to 0.9 kg/m3. The model development focus on the accuracy of the prediction. However, there is a trade-off between accuracy in the training error and validation error and it is a common problem in the empirical system modelling. The R-Square, R^2 value for training and validation is computed as 0.978 and 0.84968 to represent best fit of type-2 fuzzy logic output function coefficient, respectively. Thus, model can be assumed has small prediction error with respect to the validation data.

5. References

[1] Tudu S et. al 2021 Solution of generalised type – 2 Fuzzy boundary value problem *Alex. Eng. J.* **60** (2) 2725-39.

[2] To W M and Peter K C, 2017 GHG emissions from electricity consumption: A case study of Hong Kong from 2002 to 2015 and trends to 2030 *J. Clean. Prod.* **165** 589-98.

[3] Davor Z A et. al 2015 Modeling of energy consumption and related GHG (greenhouse gas) intensity and emissions in Europe using general regression neural networks. *Int J Greenh Gas Con.* **20** 244–53.

[4] Lama A et. al 2020 Greenhouse gas emission prediction on road network using deep sequence learning *Transp. Res. D.* **88** 102593.

[5] Tufan K, 2020 Interval Type-2 Fuzzy Logic System Toolbox *MATLAB File Exchanger.* [Online]. [Accessed 24 May 2020].

[6] Emanuel O R, Patricia, M. and Oscar, C. 2017 New methodology to approximate type reduction based on a continuous root-finding Karnik Mendel algorithm *Algorithms* **10** 77.

[7] Yeh CY, Jeng WH and Lee SJ 2011 An enhanced type-reduction algorithm for type2 fuzzy sets. *Ieee T Fuzzy Syst.* **19**(2) 227–40.

[8] Salaken SM 2016 Modification on enhanced Karnik-Mendel algorithm *Expert Syst Appl.* **65**

283-91.
[9] Erdal K M and Ahmadieh K 2016 Fundamentals of Type-1 Fuzzy Logic Theory *Fuzzy Neural Networks for Real Time Control Applications* (Oxford: Butterworth-Heinemann) Chapter 2 pp 13-24.
[10] Zanil M F et. al 2020 Forecasting greenhouse gas emissions from coal-based resource in power plant using a nonsupervisory artificial neural network *IOP Conf. Ser. Earth Environ. Sci.* 463.
[11] Davor Z A et. al 2014 Forecasting GHG emissions using an optimized artificial neural network model based on correlation and principal component analysis *Int J Greenh Gas Con.* **20** 244-53.

Acknowledgements
The authors would like to express their gratitude to the *Akaun Amanah Industri Bekalan Elektrik* (AAIBE) and TNB Research Sdn. Bhd. for their support in providing the research fund to carry out this project.

Generic review on the potential of nuclear-renewable hybrid system for sustainable power production in Malaysia

Afiq Ariffin[1], Nor Afifah Basri[1], Ahmad Termizi Ramli[1], Suhairul Hashim[1]

[1]Universiti Teknologi Malaysia, UTM, 81310 Johor Bahru, Malaysia

*Corresponding author's email : norafifah@utm.my

Abstract. Malaysia is diligently focusing its effort in developing renewable energy technology to diversify its energy sources and ensure security and sustainability in the energy sector. The development of renewable energy sector is expected to achieve 20 percent by the year 2025 to fulfil the aim of lowering greenhouse gases (GHG) produced by the energy industry. It is aligned with the commitment to the Paris Agreement signed in 2015 to reduce the emissions intensity to Gross Domestic Product (GDP) by 45% relative to 2005 levels by 2030. The rapid development of advance technology in the energy industry paved a new idea of integration of different energy resources for various industrial purposes, including power generation. The hybrid system between nuclear and renewable energy is an innovative application focusing to increase the efficiency and availability of electricity production, as well as providing additional services to industries from the energy generation process in the hybrid power plant. This paper presents the overview of the current scenario in the Malaysia energy industry, renewable energy status, the potential of nuclear-renewable hybrid system and the challenges of nuclear power development in Malaysia.

1. Introduction

Malaysia, energy sector is leading the other sectors (agriculture, waste, and industrial processes) for GHG emission record with 76% from total emission in 2011 [1]. Fossil fuel as the major contributor to GHG emission among energy sources are tremendously affecting global climate. Primary GHG consist of CO_2 and it has the most lethal effect on our environment [2]. According to Kiris and Kasap [3], CO_2 release for fossil fuel is 888 g/kW.h, 733 g/kW.h, and 499 g/kW.h for coal, oil, and natural gas, respectively.

Malaysia have signed and committed to Paris Agreement in response of mitigating the worldwide climate change threat. This Multilateral Environmental Agreement (MEA) was signed in 2015, and it requires Malaysia to reduce greenhouse gas emission by 45% by 2020 in relation to the 2015 GDP. Since the energy sector is identified as the main GHG emission contributors, the effort to reduce the national emission, along with decarbonisation strategy in the industry have been included in the National Development Plan such as Green Technology Master Plan (2017-2030) and Energy Efficiency Action Plan. Therefore, to be able to implement the SDGs and successfully comply with the Paris Agreement, Malaysia is expected to rely on renewable energy and other low carbon sources rather than fossil fuels.

Renewable energy (RE) development in Malaysia has been improving and contributing to a significant amount of energy in 2017 with 6947 ktoe which is triple the amount in year 2000 [4]. However, RE did not have the competency to replace the fossil fuel source as the main baseload in the energy mix. Hence, using RE alone to significantly reduce the GHG emission in the industry while providing energy is not realistic. Another form of low carbon source like nuclear is proven as an alternative to assist RE in producing energy through hybrid technology [2].This paper presents an overview of Malaysia energy scenario, the current development of RE, the potential of a hybrid system between nuclear and renewable energy and challenges for nuclear development in Malaysia. The aim is to stimulate further discussion and research on RE, nuclear energy and its hybrid technology.

2. Energy mix scenario in Malaysia

The primary energy production in Malaysia is contributed by 8 fuel types – gas, oil, hydropower, coal, biodiesel, biomass, solar and biogas. In 2018, the contribution of primary energy supply comes from natural gas (41%), crude oil and petroleum (30%), coal (22%) and hydropower and renewables (7%) with total amount of 99,873 ktoe [4]. From the primary energy supply, a total of 169,897 GWh electricity was generated from natural gas (39.1%), coal (43.4%) hydropower (15.6%), renewables (1.3%) and others (0.6%) as depicted in figure 1. The electricity generation capacity shows a steady increase from 156,000 GWh in 2016 and 162,184GWh in 2017. The amount of generated electricity is within the installed capacity with the total amount of 33,991 MW [5].

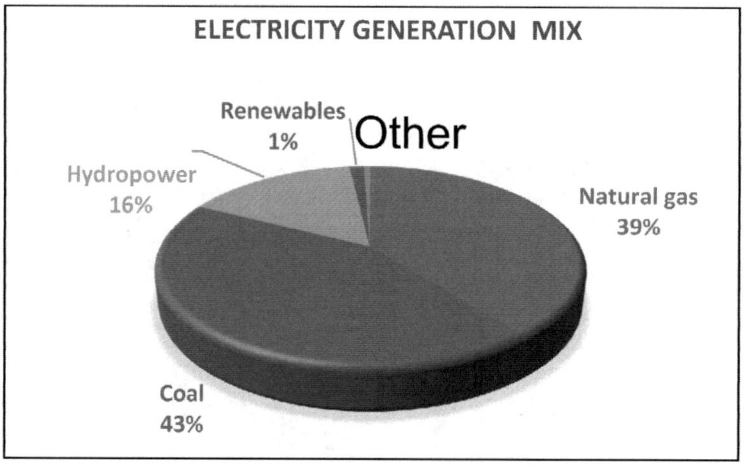

Figure 1. Electricity generation mix of 2018 [4].

There are 5 primary consumer sectors in the 2018 national energy consumption data – transportation, industry, non-energy use, residential and commercial, and agriculture. Transportation sector is leading the energy consumption with 38.5% followed by industry (28%), non-energy use (20%), residential and commercial (12.5%), and agriculture (1%). However, the electricity consumption by the same sector is monopolized by the industrial sector (49.8%), followed by commercial (29%), residential (20.5%), agriculture (0.4%) and transport (0.3%) sectors respectively with a total of 152,866 GWh. The electricity consumption is increasing from the 1990, in parallel with the development of economy and technology. The consumption data in 2018 has increased by 64.5% from 2008, which indicates large demand to the electricity sector.

Table 1 shows the available capacity, peak demand, and reserve margin for 3 regions in Malaysia in 2017 and 2018. Peak demand in 2018 stands at 18,338 MW in Peninsular Malaysia, 3,504 MW in Sarawak and 955 MW in Sabah [6]. This value increase by 3% from the previous year, 2017. The available capacity is higher than the peak demand with good reserve margin of more than 25% each year. However, the data shows major increase in demand but smaller increase in capacity, which caused the decreasing of reserve margin. Forecasted peak demand for Malaysia in 2030 will reach 22,815 MW, with projected growth of 1.8% per annum from 2020 [7]. On the other hand, researchers forecasted that reserve margin will decrease along with time because of retiring power plants and the increase in demand

[8]. Considering the existing and retiring power plants, the reserve margin will reach below 25% by 2030, which may lead to energy crisis if not planned properly.

Table 1. Malaysia's electricity generation capacity, demand and reserve margin in 2017 and 2018 [6].

Region	**Available capacity (MW)**		**Peak demand (MW)**		**Reserve margin (%)**	
	2017	**2018**	**2017**	**2018**	**2017**	**2018**
Peninsula Malaysia	23,711	24,227	17,790	18,338	35.5	32.1
Sabah	1,276	1,277	938	955	36.0	33.7
Sarawak	4,233	4,427	3,302	3,504	28.2	26.3

3. RE contribution in Malaysia energy mix

Renewable energy was first introduced in national energy policy on 1979. From then, several programs have been launched by the government in order to support the development of renewable energy in Malaysia. The initial target is to achieve 20% RE capacity mix by 2025, with a total of 3,758 MW new RE capacities should be installed starting 2020. In 2018, renewable energy (excluding hydropower) supplies only 1.3% from the total electricity generation. In 2019, there are 119 renewable energy power plant includes mini hydro, solar and biomass [9]. Upon completion of Malaysia's second Large Scale Solar (LSS) project in Bukit Selambau, Kedah, the RE capacity excluding large hydro plants will only stand at 403 MW, still miles away from the initial target. However, the national energy commission is confident to achieve the target with 11 new RE plants generation planned to operate from 2020 to 2030. If the new RE plants were operating according to the target, the projected capacity for RE will increase from 9% in 2020 to 23% in 2030, in tandem with reduction on fossil fuel capacity in the mix.

Among RE, hydropower contributed the highest percentage in RE mix with steady supply within 10% to 15% contribution in the energy mix. Contribution from solar power is increasing steadily with the completion of LSS project phase 1 and 2. However, research on biogas, geothermal and wind are continuously conducted to increase their contribution in the energy mix. From the current scenario, it can be concluded that only hydro and solar is able to be achieve commercialization in electricity industry in Malaysia and having the potential of hybrid with nuclear energy.

3.1. Solar

As an equatorial and tropical country, Malaysia is blessed with high solar radiation and have a promising future in solar energy development [10]. Several programs are introduced by government in encouraging solar PV (Photovoltaic) development. LSS is one of the recent programs that was introduced by Malaysia Energy Commission with cooperation of Tenaga Nasional Berhad (TNB). The purpose of LSS program is to develop large scale solar plant in Malaysia through a competitive bidding process. The first LSS project in Sepang, Selangor generates 50 MW with a system availability of over 99%. The second LSS project in Bukit Selambau, Kedah has started operating in September 2020, 114 days earlier than the initial target in December 2020 [11]. The third LSS project (LSS-3) was launched in June 2020 and scheduled to operate commercially by the end of 2021. LSS-3 with a capacity of 500 MW is expected to bring a significant effect in achieving the target of 20% contribution from RE by 2025 [12].

In the recent years, solar technology is rapidly developing, where many new technologies are introduced to increase solar competency in the energy industry. Chua et al. [13], in his research predicted that if solar energy remains competitive in Malaysia energy market, it will surpass other RE in electricity generation. However, solar energy has a very long way to achieve the same reputation as hydropower, especially in providing stable and sustainable amount of electricity.

3.2. Hydropower

Hydropower is the leading RE source in contributing electricity generation worldwide, including Malaysia. There are 10 large hydro power plant with combined capacity of 5534 MW in Malaysia [14]. Hydropower holds a share of 76% contribution among renewable energy in the electricity generation capacity and contributes 10% to 15% % of overall electricity generation [15]. The installed capacity of hydropower reached 2,413 MW in 2016 with a generation of 5,072 GWh in 2018 [15].

Hydropower is considered green energy due to low GHG production. However, building large dam for hydro power plant cause negative effect to the environment. Hence, smaller scale hydropower is introduced several years ago in

Malaysia decrease the environmental impact compared to large hydropower plant [15,16]. Small scale hydropower produces power from 2 MW to 24 MW with 149 potential sites in Malaysia [15]. Small hydropower plants have low operating cost and the technology is suitable as decentralized power supply to meet the local small-scale electricity demand of remote areas, rural towns, or localized industries [17]. In 2014, there are 58 mini hydropower station around Malaysia, an increment of 65% compared in 2010 [18]. As one of the low cost RE, mini hydro has the prospect to reduce the environmental loading of Malaysia's energy usage in coming years [8,18]. Due to its advantages, such as clean, low environmental impact, low cost, and indigenous energy source, hydropower could play a vital role in the sustainable and green energy mix in Malaysia [19].

4. Small Modular Nuclear Reactor (SMR)

SMR is a reactor that have an output power less or equal 300 MWe and possess modularity concept. Baldwin et al. [20] describes the term modularity as building a complex product or process from smaller subsystem that can be designed independently and function together as a whole system. A module is the installation unit that is manufactured, assembled and tested in factory or workshop and transported to the SMR Nuclear power plant (NPP) site in such a state of readiness that no further significant processing is required to be done before installation. The SMR NPP is made of multiple such units of installation.

Globally, there are about 50 SMR designs with various power output at different stages of development [21]. A typical SMR consists of two sections: a power module and a power generation assembly. The power module houses the reactor core with the fuel rod assembly where nuclear fission generates heat. Most SMR designs available today are generation IV reactors that will be commercially deployed in several years [24,25].

SMR has many advantages for newcomer country in nuclear power programme because it provides investment flexibility and scalability. SMR is highlighted with its small footprint and compact design where it used less area compared to large NPP. It enables the government to find more suitable sites in development-limited areas or to redevelop unused area such as brownfields from retired fossil fuel plants. Due to its smaller scale and underground reactor core design, SMR may have lower visual impacts than large reactors located above surface. Thus, with less visual impact than conventional NPP, SMR can possibly located closer to the public settlements, where it can supply the intended electricity to the area.

The cost of SMR is a lot cheaper than the conventional NPP, depending on the reactor type and its capacity [22,23]. In the years coming, when more SMR design is commercialized and the licensing process is verified and endorsed, the capital costs will be more affordable hence making it accessible for a country that lack in financial support. Construction time for SMR take approximately 3 years compared to conventional NPP that would take approximately 8 years to complete [22]. Short construction time and short period of installation hastens the integration into the existing power grids [23]. Consequently, the capital cost can be reduced because 80% of the NPP life-cycle costs are incurred during the building phase [24].

5. Integration of Nuclear and RE: Nuclear – Hybrid System

Nuclear-renewable hybrid technology is the combination of nuclear and renewable resources in producing electricity and heat energy. Integration and coordination of multiple clean energy generation sources offers us the opportunity to maximize utilization of all clean energy technologies ensuring that our systems are both efficient in their operation and cost effective. This hybrid system is highly suitable for country that produce its electricity using both nuclear and renewable energy. Table 2 shows the potential of nuclear-RE hybrid system and the possible outputs gained. Sabharwall et al. [25] compared three different type of cases which are nuclear only, nuclear-wind hybrid, nuclear-wind with hydrogen production hybrid and conclude that nuclear wind with hydrogen production is more attractive economically that the return on investment of the integration is faster than other model.

Table 2 Nuclear-renewable hybrid system [2].

Energy sources	Coupling method	Storage method	Possible outputs
Nuclear + Wind	Electrical	Hydrogen	Electricity, hydrogen
Nuclear + Solar	Thermal	Thermal	Electricity, heat
Nuclear + Biomass	Thermal	Chemical	Electricity, biofuels
Nuclear + Geothermal	Thermal	Thermal	Electricity, heat
Nuclear + Wind + Geothermal	Electrical and Thermal	Chemical	Electricity, chemical product, synfuel

According to Khalis and Bicer [26], SMR is the most suitable advanced reactor nowadays to be used for hybridization of nuclear and RE. Nuclear-renewable hybrid energy system (N-R HES) are synchronously controlled systems that dynamically apportion thermal and/or electrical energy to provide responsive generation to the power grid. N-R HES contains multiple subsystems [27]; nuclear heat generation source, a turbine that converts thermal energy to electricity, at least one renewable energy source, and industrial process that utilizes heat and/or power from the energy sources to produce a commodity-scale product.

There are three possible general categories introduced in 2016 for the integration system; tightly coupled HES, thermally coupled HES, and loosely coupled HES [27]. All thermal energies added together and all electric energies are combined together in tightly coupled system. Available thermal and electrical energy will be used in needed industry application suitable to this technology. For loosely coupled system, RE source and nuclear reactor generate electric separately and linked together at the electricity distribution level. The system may include electrical-to-thermal energy conversion equipment to provide thermal energy input to the industrial process. Lastly for the thermal coupled, nuclear and renewable subsystems are co-controlled to provide energy and ancillary services to the grid. These categories have more than one connection point to the grid and includes thermally integrated subsystems that are tightly coupled with the heat generation source [27].

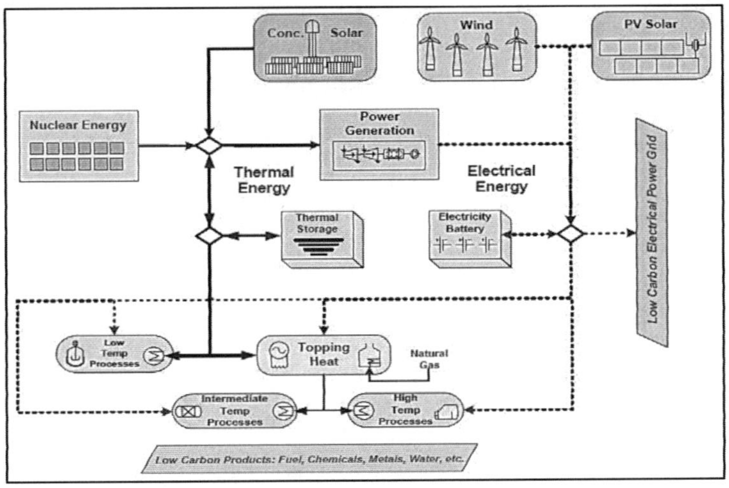

Figure 2. An example of hybrid energy system [2].

Figure 3 show recent patented hybrid nuclear-solar technology. It uses a nuclear reactor to heat a compressed working fluid that is expanded within turbines rotating compressors of both reactor plant and air pressurising plant. Heat exchangers removes the low-grade heat from the working fluid and pass it to moisture removal equipment on the air compressing plant's downstream side. Inter cooler heat exchangers are used to further cool the working fluid until it enters the compressors, lowering compressor power requirements [28].

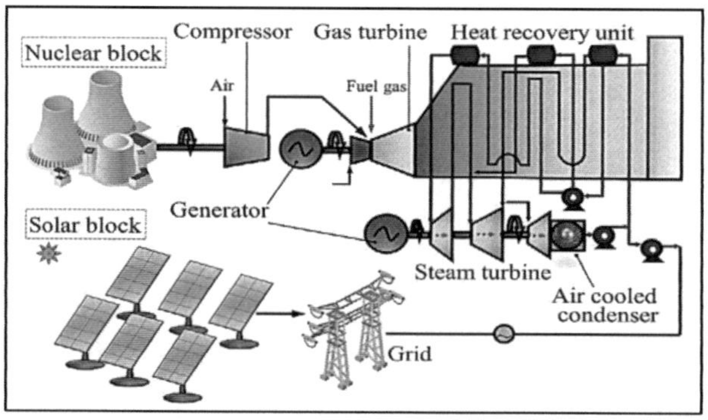

Figure 3. Hybrid nuclear solar energy system [28].

Integration of RE and nuclear energy comes with many advantages. Incorporation of the 2 technologies in 1 facility can improve their capability to become highly competitive in the energy sector, thus ensuring sustainable and green energy supply for future generation [2,23,29]. Coupling technology with nuclear allows the intermittent RE to have a higher grid penetration, thus increasing the overall efficiency to meet energy demand [2,29].On the other hand, the positive image of RE helps to improve the negative sentiments on nuclear operating side by side, which will eventually subjugate the challenges faced with their independent use can be incorporated with the sustainability of nuclear power generation in the hybrid or integration system to subjugate challenges faces with their independent use. The capability of producing a useful product such as hydrogen also makes this new technology indispensable in the future. Hybridization of RE and nuclear would also encourages industrial symbiosis [29].

6. Challenges towards nuclear power development in Malaysia

Malaysia has once considered nuclear as an alternative resource to replace fossil fuels for baseload capacity. However, social and financial restrictions have decelerated the effort towards realizing nuclear power development in Malaysia.

Public acceptance has been the critical factor in implementing nuclear energy for a newcomer country. Public acceptance for nuclear is particularly low because it is majorly affected by nuclear accident that happened throughout the history, and further exaggerated by the mass media and movies [30]. In Southeast Asia, Indonesia has the higher public acceptance in nuclear power while Malaysia appears to be struggling to win over public support post-Fukushima accident [31]. For example, Lynas corporation, a mining company that open its plant in Gebeng caused a heated political argument and protest among public regarding the radioactive materials in the residue of rare earth material extraction process. According to Atomic Energy Licensing Board (AELB) Malaysia, radioactivity of these residue is low and not compromising to the environment and the public nearby the plant. However, the public is still concerned about the management of the radioactive residue and consistently urged the government to send it elsewhere outside the country.

The waste management issue from Lynas have led to bigger discussion about the nation's ability to manage nuclear waste from a nuclear power plant. Although the amount of waste generated by nuclear power is very small relative to other thermal electricity generation technologies, nuclear waste disposal has been the greatest challenge in nuclear power generation [32]. For over 6 decades, there is no long-term repository licenced by any nation even it is widely agreed that High Level Waste (HLW) and long-lived Intermediate Level Waste (ILW) were to be stored in deep geological disposal [2].

A country could not elude from financial risk in developing any type of power generation. The construction of NPP would cost billions of dollars and is increasing by year. According to Suman et al [2]. the cost of constructing new NPP risen steeply compared to other low carbon alternatives. Adding to the problem, there is risk in the delay of construction and licensing that would escalate the capital cost of NPP [2]. Construction cost of nuclear power plant contribute majorly to the life cycle of the plant, while the fuel costs are a minor proportion of total generating costs. Hence, nuclear capital costs are greater than those for coal-fired plants and much greater than those for gas-fired plants [22].

Malaysia has taken its first step in producing green and sustainable energy by developing renewable energy such as solar, biomass, and hydropower. Due to many challenges hindering the development of nuclear in Malaysia, nuclear option is not prioritized for the time being. However, government has long accepted its capability to ensure national energy security and sustainability. With the increasing cost to import fossil fuel and limitation of power production from renewable sources, it is a matter of time when nuclear will be reconsidered as an energy source option in Malaysia.

7. Conclusion

Electricity consumption in Malaysia is projected to increase in parallel with population numbers that is expected to reach 41.5 million in 2040. There are only two options to decarbonise an ever-increasing demand in the electricity sector: nuclear power and renewable energy. Of these two options, only nuclear provides firmly dispatchable baseload electricity. On the other hand, decarbonising existing electricity system with 100% RE is unnecessarily expensive, and unsustainable for the long term. Even though Malaysia is expanding its RE production, more stable and larger amount of energy is needed to ensure the energy security for the future. With the emergence of nuclear renewable hybrid system, Malaysia should consider adopting this technology by combining small modular reactor with RE to generate electricity and other industrial purposes. There are certainly some technical and economic issues that needs deep consideration towards the implementation of the hybrid system. However, its potential to diversify the energy mix and to lower carbon emission in the future has been slowly proven, and certainly cannot be neglected.

8. References

[1] Tang KHD. Climate change in Malaysia: Trends, contributors, impacts, mitigation and adaptations. Sci Total Environ 2019;650:1858–71. https://doi.org/10.1016/j.scitotenv.2018.09.316.

[2] Suman S. Hybrid nuclear-renewable energy systems: A review. J Clean Prod 2018;181:166–77. https://doi.org/10.1016/j.jclepro.2018.01.262.

[3] Kiris S, Kasap Y. Environmental Effects An analysis of the energy production efficiency of countries. Energy Sources, Part A Recover Util Environ Eff 2019;00:1–13. https://doi.org/10.1080/15567036.2019.1696428.

[4] MEC. Peninsular Malaysia Electricity Industry Outlook 2019. 2019.

[5] Malaysia Energy Commission. Industri Pembekalan Elektrik Malaysia Performance & Statistical Information. 2018.

[6] Suruhanjaya Tenaga. Malaysia Energy Statistics Handbook 2018. Suruhanjaya Tenaga (Energy Comm 2019:1–86.

[7] Malaysia Energy Commission. Report on Peninsular Generation Development Plan 2019. vol. 2019. 2020.

[8] Basri NA, Ramli AT, Aliyu AS. Malaysia energy strategy towards sustainability: A panoramic overview of the benefits and challenges. Renew Sustain Energy Rev 2015;42:1094–105. https://doi.org/10.1016/j.rser.2014.10.056.

[9] TNB. Integrated Annual Report 2019. 2019.

[10] Khan R, Go YI. Assessment of Malaysia ' s Large-Scale Solar Projects : Power System Analysis for Solar PV Grid Integration 2019;1900060:1–17. https://doi.org/10.1002/gch2.201900060.

[11] Sataloff RT, Johns MM, Kost KM. Integrated Annual Report 2019. 2019.

[12] Tan CS, Maragatham K, Leong YP. Electricity energy outlook in Malaysia. IOP Conf Ser Earth Environ Sci 2013;16. https://doi.org/10.1088/1755-1315/16/1/012126.
[13] Chua SC, Oh TH, Goh WW. Feed-in tariff outlook in Malaysia. Renew Sustain Energy Rev 2011;15:705–12. https://doi.org/10.1016/j.rser.2010.09.009.
[14] Map Data in GEO - GEO Systems 2018. http://globalenergyobservatory.org/constructNetworkIndex.php (accessed April 19, 2021).
[15] Yah NF, Oumer AN, Idris MS. Small scale hydro-power as a source of renewable energy in Malaysia: A review. Renew Sustain Energy Rev 2017;72:228–39. https://doi.org/10.1016/j.rser.2017.01.068.
[16] Hossain M, Huda ASN, Mekhilef S, Seyedmahmoudian M, Horan B, Stojcevski A, et al. A state-of-the-art review of hydropower in Malaysia as renewable energy: Current status and future prospects. Energy Strateg Rev 2018;22:426–37. https://doi.org/10.1016/j.esr.2018.11.001.
[17] Ho LW. Wind energy in Malaysia: Past, present and future. Renew Sustain Energy Rev 2016;53:279–95. https://doi.org/10.1016/j.rser.2015.08.054.
[18] Abd Hamid MF, Ramli NA, Mat Napiah SNB. Factors Affecting Mini Hydro Power Production Efficiency : 2015:85–8.
[19] Yuksel I. Dams and hydropower for sustainable development. Energy Sources, Part B Econ Plan Policy 2009;4:100–10. https://doi.org/10.1080/15567240701425808.
[20] Upadhyay S, Sharma MP. A review on configurations, control and sizing methodologies of hybrid energy systems. Renew Sustain Energy Rev 2014;38:47–63. https://doi.org/10.1016/j.rser.2014.05.057.
[21] IAEA. Advances In Small Modular Reactor Technology Developments. 2018th ed. Viena: Nuclear Power Technology Development Section, Division of Nuclear Power of the IAEA Department of Nuclear Energy; 2018.
[22] Zohuri B. Small Modular Reactors as Renewable Energy Sources. 2018. https://doi.org/10.1007/978-3-319-92594-3.
[23] Devanand A, Kraft M, Karimi IA. Optimal site selection for modular nuclear power plants. Comput Chem Eng 2019;125:339–50. https://doi.org/10.1016/j.compchemeng.2019.03.024.
[24] Ho M, Obbard E, Burr PA, Yeoh G, Obbard E. A review on the development of nuclear power reactors. Energy Procedia 2019;160:459–66. https://doi.org/10.1016/j.egypro.2019.02.193.
[25] Sabharwall P, Bragg-Sitton S, Boldon L, Blumsack S. Nuclear Renewable Energy Integration: An Economic Case Study. Electr J 2015;28:85–95. https://doi.org/10.1016/j.tej.2015.09.003.
[26] Khalid F, Bicer Y. Energy and exergy analyses of a hybrid small modular reactor and wind turbine system for trigeneration. Energy Sci Eng 2019;7:2336–50. https://doi.org/10.1002/ese3.327.
[27] Bragg-Sitton SM, Boardman R, Rabiti C, Kim JS, McKellar M, Sabharwall P, et al. Nuclear-Renewable Hybrid Energy Systems: 2016 Technology Development Program Plan 2016.
[28] Keller MF. Hybrid integrated energy production process. 2005.
[29] Sabharwall P, Bragg-sitton S, Boldon L, Blumsack S. Integration : An Economic Case Study. Electr J 2015;28:85–95. https://doi.org/10.1016/j.tej.2015.09.003.
[30] Dahlan NY, Ibrahim A, Rajemi MF, Baharum F. Analysis of the Impact of Nuclear Power Plant on Malaysia ' s Power Systems : Costs , CO 2 Emission and System Reliability 2020:206–11.
[31] Putra NA. The dynamics of nuclear energy among ASEAN member states. Energy Procedia 2017;143:585–90. https://doi.org/10.1016/j.egypro.2017.12.731.
[32] IAEA. Basic infrastructure for a nuclear power project. IAEA Tecdoc 1513 2006.

Acknowledgements
This research is funded by Universiti Teknologi Malaysia under Research University Grant (GUP: Q.J130000.2651.16J77).

Numerical studies of fluid flow and heat transfer in microchannel heat sink with trapezoidal cavities, ribs and secondary channel

S A Razali[1], N A C Sidik [2]* and S N A Yusof [2]

[1] Faculty of Mechanical Engineering, Universiti Teknologi Malaysia, 81310 Skudai, Johor, Malaysia
[2] Malaysia – Japan International Institute of Technology (MJIIT), Universiti Teknologi Malaysia, Jalan Sultan Yahya Petra, 54100 Kuala Lumpur, Malaysia

*Corresponding author's email : Azwadi@utm.my

Abstract. In order to improve reliability and prevent premature failure, heat produced by electronic devices must be transfer efficiently. The microchannel heat sink appears to be the most reliable cooling technology to improve heat transfer performance. A new Microchannel Heat Sink (MCHS) design has been proposed with trapezoidal cavities, ribs and secondary channels. Fluid flow and heat transfer characteristics for Reynold numbers from 100 to 500 are numerically studied and analysed. Four microchannel heat sinks with various related geometry have been considered in this study, for instance, microchannel with rectangular ribs (TRAC-RR) and microchannel with trapezoidal cavities (TRAC). The finite volume method (FVM) is used to solve the governing equations, and the computations are performed using the SIMPLE algorithm. The present design's overall performance is evaluated using the friction factor, the Nusselt number, and performance evaluation criteria (PEC). The results show that the TRAC-RR-SC MCHS is the best design for the proposed design compared to the other four geometries, with a maximum PEC of 1.78. Additionally, the secondary flow field analyses visually show the hydraulic and thermal performance enhancements due to interruption, flow mixing and redevelopment of the thermal boundary layer.

1. Introduction

With the emergence of technology in recent years, microelectronic devices and chips are shrinking in size to maximise performance, which results in extremely high heat generation. Microdevices thermal management has become a hot subject that has intrigued the interest of a rising number of academics. MCHS single-phase liquid cooling has been known as one of the most efficient methods for microdevices cooling, with advantages such as compactness and ease of integration with microchips or high-density data storage systems [1]. The simple plain channel is incapable of meeting these need as the heat load rises and the temperature of microdevices is very limited.

Bargles [2] has classified a large number of augmentation techniques. There are three types of augmentation techniques: active methods, passive methods, and a combination of the two. On the other hand, passive methods have a broader application than active methods because they do not require external power. These methods include the use of MCHSs with grooves/cavities to improve heat transfer. Chai et al. [3] experimentally and numerically compared the convective heat transfer of MCHSs with various cavities. Thermal resistance, heat transfer, and pressure drop were all measured in the new microchannels and compared with the rectangular microchannel. The thermal boundary layer and secondary flow improved the heat transfer. The combined cavity and ribs structure in MCHSs demonstrated its advantages for further heat transfer enhancement. Numerous researchers discovered that MCHSs with ribs and cavities on the sidewalls performed better due to jetting effects and vortexes [4,5]. Li et al. [6] Compared the microchannels without ribs or cavities, and results showed that the suggested design enhanced flow mixing by encouraging chaotic advection and jetting. Moreover, the microchannel presented shows a more uniform velocity and temperature at the heat sink substrate reduced which give better heat transfer performance. Japar et al. [7] explored the flow and heat transfer structure in MCHS with fins, cavities, and secondary channels, providing a low pumping power consumption with a high heat transfer rate. Numerous studies have been conducted to optimise the geometrical structure of microchannel heatsinks [8-10]. However, only few investigations existed on the effect of a combination of multiple passive technique, specifically combination of trapezoid cavities, rectangular ribs, and secondary channels have been reported.

Therefore, this paper aims to emphasise studying the effectiveness of microchannel heat sink hybrid designs with a secondary channel for high heat transfer rates and utilises the secondary channel's characteristics, allowing for a greater flow area and reducing the pressure drop. The performance evaluation criteria in this investigation are established on a comparison of the suggested design's performance with straight rectangular microchannel performance.

2. Physical model

The geometric of the TRAC-RR-SC microchannel heat sink's computational domain has been displayed in figure 1 (a), which is a symmetrical part of the whole MCHS. Figure 1 (b) and table 1 show the detailed structures of TRAC-RR-SC MCH. Four different configurations of microchannels (RC-RR, TRAC, TRAC-RR, TRAC-RR-SC) are developed to study the effect of cavities, rib and secondary channels on the fluid flow and heat transfer performance.

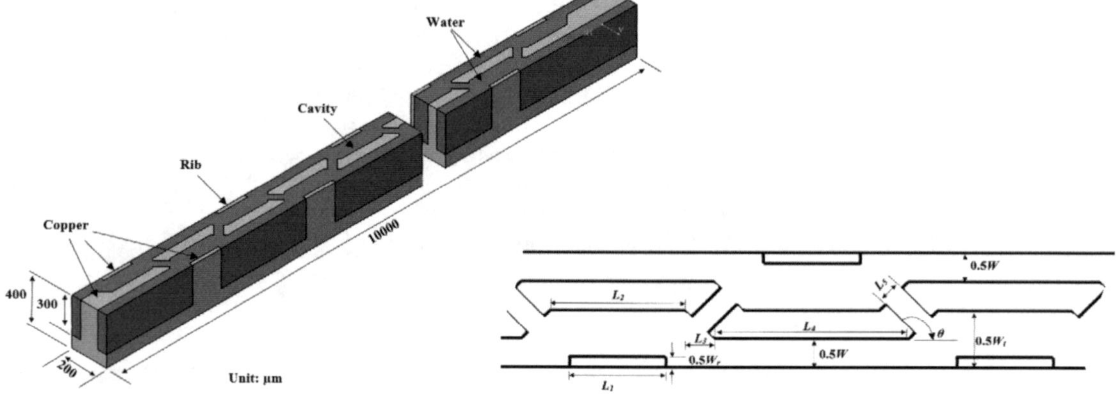

(a) Isometric View (b) Top View

Figure 1. TRAC-RR-SC (a) Geometric of the computational domain (b) Structures of MCHS.

Table 1. List of a geometric dimensions of TRAC-RR-SC microchannels.

Parameters	Wt	W	Wr	L1	L2	L3	L4	L5	θ
Values, μm	300	150	56	250	350	75	500	60	45

3. Numerical Methods

3.1. Governing Equation

Continuity, momentum, and energy-governing equations in Cartesian coordinates in the nondimensional form are written as follows:

$$\frac{\partial u}{\partial x}+\frac{\partial v}{\partial y}+\frac{\partial w}{\partial z}=0 \tag{1}$$

$$u\frac{\partial u}{\partial x}+v\frac{\partial u}{\partial y}+w\frac{\partial u}{\partial z}=-\frac{1}{\rho_f}\frac{\partial \rho}{\partial x}+\frac{\mu_f}{\rho_f}\left(\frac{\partial^2 u}{\partial x^2}+\frac{\partial^2 u}{\partial y^2}+\frac{\partial^2 u}{\partial z^2}\right) \tag{2}$$

$$u\frac{\partial v}{\partial x}+v\frac{\partial v}{\partial y}+w\frac{\partial v}{\partial z}=-\frac{1}{\rho_f}\frac{\partial \rho}{\partial y}+\frac{\mu_f}{\rho_f}\left(\frac{\partial^2 v}{\partial x^2}+\frac{\partial^2 v}{\partial y^2}+\frac{\partial^2 v}{\partial z^2}\right) \tag{3}$$

$$u\frac{\partial w}{\partial x}+v\frac{\partial w}{\partial y}+w\frac{\partial w}{\partial z}=-\frac{1}{\rho_f}\frac{\partial \rho}{\partial z}+\frac{\mu_f}{\rho_f}\left(\frac{\partial^2 w}{\partial x^2}+\frac{\partial^2 w}{\partial y^2}+\frac{\partial^2 w}{\partial z^2}\right) \tag{4}$$

$$u\frac{\partial T_f}{\partial x}+v\frac{\partial T_f}{\partial y}+w\frac{\partial T_f}{\partial z}=-\frac{k_f}{\rho_f c_{pf}}\left(\frac{\partial^2 T_f}{\partial x^2}+\frac{\partial^2 T_f}{\partial y^2}+\frac{\partial^2 T_f}{\partial z^2}\right) \tag{5}$$

$$0=k_s\left(\frac{\partial^2 T_s}{\partial x^2}+\frac{\partial^2 T_s}{\partial y^2}+\frac{\partial^2 T_s}{\partial z^2}\right) \tag{6}$$

These assumptions are made to analyse the fluid flow and heat transfer in TRAC-RR-SC microchannel; 3D fluid flow, laminar, incompressible flow and steady-state, Newtonian, constant thermo-physical properties, velocity zero across solid boundaries, and negligible gravity and radiation heat transfer.

3.2 Boundary Condition and Grid Independence

A three-dimensional solid-fluid heat transfer numerical model has been created using Catia V5, and the numerical simulation has been performed using Fluent 17.0. In this investigation, copper MCHS is used with a constant wall heat flux of 1000 kW/m2 supplied to the substrate's bottom, and with an inlet temperature of 300 K, water is used as the working fluid. The average velocity at the entrance of the microchannel is fixed as 0.5–2.5 m/s for analysis. Figure 2 shows that Fluent 17 has generated a uniform grid throughout all zones.

The finite-volume method is used to discretized governing equations for all domains. The SIMPLEC algorithm is applied to resolve the coupling of velocity and pressure. Meanwhile, the second-order upwind scheme is set for the convective terms. Also, the second-order central difference scheme is selected for the diffusion term. The residual convergence criteria are set to be 10^{-6} for continuity and 10^{-9} for energy. To confirm mesh independence for the microchannel, the grid number of microchannel TRAC-RR-SC was varied between 0.6 million and 1.4 million when the Reynolds number was 500. The difference in relative error between the finest grids (J1) and the remainder of the grids (J2) is represented as (7)

$$e\%=\left|\frac{J_2-J_1}{J_1}\right|\times 100 \tag{7}$$

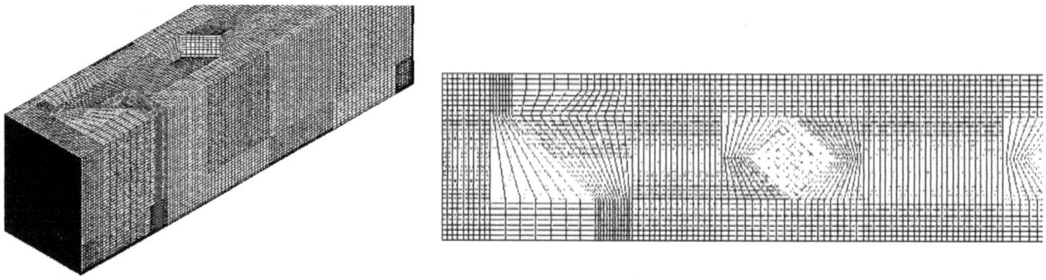

Figure 2. Three-dimensional grid diagram model.

According to table 2, the variations in Nusselt number and pressure drop are less than 0.91 percent. As a result, a grid number of 1.2 million is chosen to save computation time.

Table 2. Grid independence test.

Mesh	Nusselt.no	e%	Pressure drop	e%
1.4M	19.55	baseline	40267.46	baseline
1.2M	19.37	0.91	40216.34	0.12
1.0M	19.30	1.27	40152.01	0.28
0.8M	18.63	4.70	40115.25	0.37
0.6M	18.40	5.85	40095.73	0.42

3.3. Data Acquisition

The following equation calculates the Reynolds number (Re) and hydraulic diameter (Dh);

$$\mathrm{Re} = \frac{\rho u_m D_h}{\mu} \tag{8}$$

$$D_h = \frac{2H_c W_c}{H_c + W_c} \tag{9}$$

Where *Dh* is the hydraulic diameter in the microchannel. The friction factor is calculated from the pressure drop values using the equation below to assess hydraulic performance;

$$f_{app} = \frac{2D_h \Delta P}{L_t \rho u_m^2} \tag{10}$$

Where *Lt* and *ΔP* are the total lengths of microchannel and pressure drop across the microchannel, the average heat transfer coefficient is calculated using the following formula:

$$h_{ave} = \frac{q_w A_{film}}{A_{con.}\left(T_{w,ave} - T_{f,ave}\right)} \tag{11}$$

$$Nu_{ave} = \frac{h_{ave} D_h}{k_f} \tag{12}$$

Where D_h, k_f, q_w, A_{film}, $A_{con.}$, $T_{w,ave}$ and $T_{f,ave}$ are the hydraulic diameter, the fluid's thermal conductivity, heat flux, the heated region, the convection heat transfer region, channel walls' average temperature and average fluid temperature.

A thermal-hydraulic performance evaluation criteria (PEC) index is performed to measure overall heat transfer performance while taking the pressure penalty into considerations which is defined as:

$$PEC = \frac{Nu / Nu_o}{(f / f_o)^{1/3}} \tag{13}$$

4. Results and Discussion

4.1 Validation

In order to analyse the accuracy of the numerical solution procedure, the present plain channel is firstly simulated and compared with the correlation given by Steinke and Kandlikar [8] and Darcy friction factor for the apparent friction factor. Figure 3(a) shows the plot of friction factor for Reynolds range of 100–500. The correlations are given by Steinke and Kandlikar [11] is expressed as (14-16), respectively.

$$f_{app} = f + \frac{\alpha(x) D_h}{L_t} \tag{14}$$

$$P_o = f\,\mathrm{Re} = 96[1 - 1.3553(AR) + 1.946(AR)^2 - 1.7012(AR)^3 + 0.9564(AR)^4 - 0.2537(AR)^5] \tag{15}$$

$$a(x) = 0.6796 + 1.2197(AR) + 3.3089(AR)^2 - 9.5921(AR)^3 + 8.9089(AR)^4 - 2.9959(AR)^5 \tag{16}$$

The Darcy friction factor is written as (17):

$$f_D = \frac{64}{\mathrm{Re}} \tag{17}$$

Meanwhile, the pressure drop is compared with correlations published by Steinke and Kandlikar [11] and displayed in figure 3(b). The pressure drop of the microchannel is defined as;

$$\Delta p = \frac{2(f\,\mathrm{Re})\mu u_m L_t}{D_h} + \frac{a(x)\rho u_m^2}{2} \tag{18}$$

A good agreement is observed between the present model and theoretical data, as displayed in figure 3. It is clearly shown that the model data has very good matching with theoretical data. Thus, this numerical method can assess the performance of all designs in this current work.

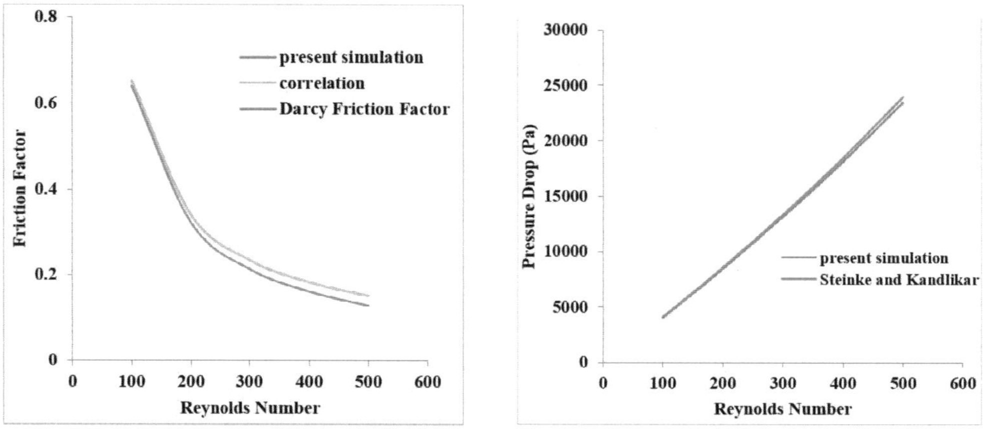

Figure 3. Comparison between the present model and theoretical data of (a) friction factor with correlation and (b) pressure drop with Steinke and Kandlikar [11].

4.2 Effects of trapezoidal cavities, ribs and secondary channel on heat transfer and fluid flow

Microchannel geometric configuration has a significant effect on the efficiency of fluid flow and heat transfer. The velocity and streamline are specified in the x–z plane (y = 0.25 mm) at Re = 500, respectively. As shown in figure 4, it can be obvious that the formation of cavities, ribs and secondary channel distinguishes the flow area from the rectangular channel (RC) where the streamlines are parallel. Fluid velocity constantly fluctuates and streams along the microchannel length because of the sudden expansion and blocking in the rib, cavities, and secondary channels.

The maximum velocity inside RC and trapezoidal cavities (TRAC) appear at the channel region's center. In contrast, the lowest flow velocity is discovered in the trapezoid cavities of TRAC channel, where the stagnation zone is located. The uniform velocity inside microchannels occurs at trapezoidal cavities-rectangular rib-secondary channel (TRAC-RR-SC). The velocity gradient is also elevated near the cavity, reflecting a stronger mix in the channel of near-wall hot fluid and center cold fluid. In the meantime, flow velocity has decreased in the downstream regions of the ribs.

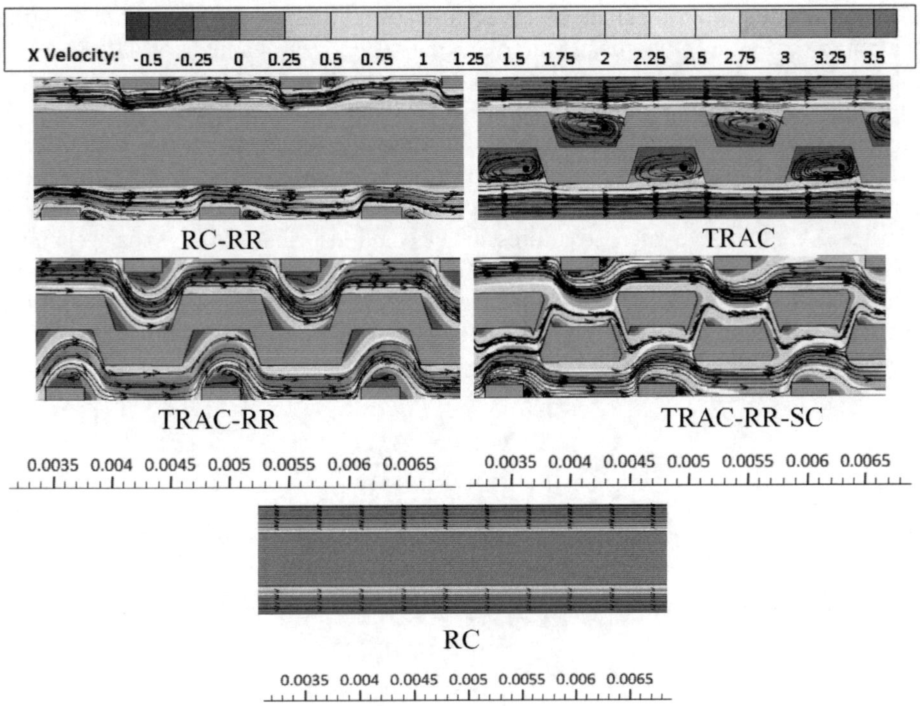

Figure 4. The velocity and streamline distributions for different structures on the x–z plane (y = 0.25 mm) at Re = 500.

4.3 Effect of trapezoidal cavities, ribs and secondary channel on microchannel performance

In the present investigation, the performance analysis parameters are focused on the differentiation of the suggested design performance with straight rectangular microchannel performance. The deviation of Nu/Nu_o and f/f_o with Reynolds number is shown in figures 5(a) and (b). The subscript (o) denotes rectangular microchannel RC. Basically, a higher Nusselt ratio will increase the performance, while a higher friction factor ratio will decrease the performance and vice versa.

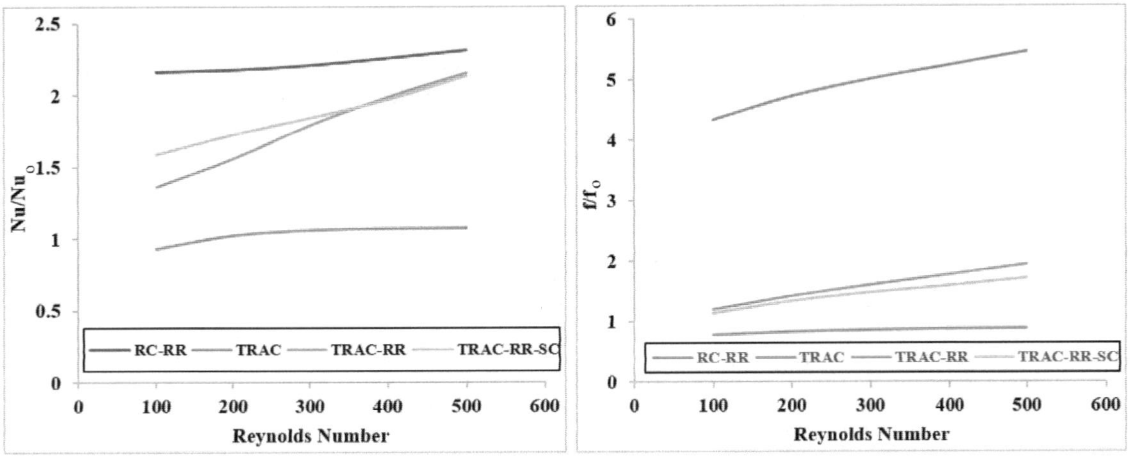

Figure 5. Variation of (a) Nu and (b) *f* with Reynolds Number for various geometries.

The best heat performance with a Nusselt ratio of 2.3 can be ascertained in rectangular-rectangular rib (RC-RR). Still, it also has the highest friction factor ratio due to the blocking rib configuration effect.

Meanwhile, TRAC shows the lowest thermal performance and the lowest friction factor ratio as a result of the stagnation zone found in cavities. Moreover, the secondary channels of TRAC-RR-SC greatly minimize the frictional loss incurred by ribs. It should be acknowledged that the friction loss average is 70% lower than RC-RR.

The overall performance of microchannels of different geometric configurations is now being compared using performance evaluation criteria (PEC). As mentioned in equation (13), this quantity indicates a thermal increase in hydrodynamic performance. The result shows PEC value is higher than one, which indicates that passive microstructures such as cavities, ribs, and secondary channels improve the heat sink's overall efficiency compared to simple microchannels. In figure 6, PEC variance is given within the Reynolds numbers 100 < Re < 500 range for all configurations. Moreover, for trapezoidal channel-rectangular rib (TRAC-RR) and TRAC-RR-SC configurations, this performance criterion gradually increases as the Reynolds number increases.

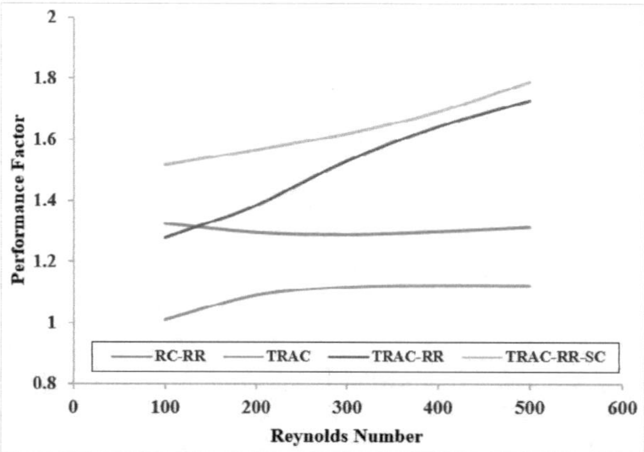

Figure 6. Variation of PEC with Reynolds Number for various geometries.

5. Conclusion

The effects of combining passive heat transfer methods, particularly cavities, ribs, and secondary channels on microchannel overall performance, were studied in this paper. Additionally, the fluid flow and heat transfer characteristics of the microchannel heat sink were numerically investigated. The main findings can be summarized that by combining passive heat transfer methods such as cavities, ribs, and secondary channels, hybrid techniques can significantly improve inclusive performance in microchannel heat sinks compared using individual techniques. The results show that the TRAC-RR-SC MCHS is the best design for the proposed design compared to the other four geometries, with a maximum PEC of 1.78.

6. References

[1] Li S, Zhang H, Cheng J, Li X, Cai W, Li Z, and Li F 2019 A state-of-the-art overview on the developing trend of heat transfer enhancement by single-phase flow at micro scale *Int J Heat Mass Transf* **143** 118476

[2] Bargles AE 1985 *Techniques to augment heat transfer. Handbook of Heat Transfer Applications* Ed W.M. Rohsenow, J.P. Hartnett, E.N. Ganic (New York: McGraw-Hill) Chapter 3 p 3-80

[3] Chai L, Xia G, Wang L, Zhou M, and Cui Z 2013 Heat transfer enhancement in microchannel heat sinks with periodic expansion-constriction cross-sections *Int J Heat Mass Transf* **62(1)** 741–51

[4] Bayrak E, Olcay AB, and Serincan MF 2019 Numerical investigation of the effects of geometric structure of microchannel heat sink on flow characteristics and heat transfer performance. Int J Therm Sci **135** 589–600

[5] Zhai YL, Xia GD, Liu XF, and Li YF 2014 Heat transfer in the microchannels with fan-shaped reentrant cavities and different ribs based on field synergy principle and entropy generation analysis *Int J Heat Mass Transf* **68** 224–33
[6] Li YF, Xia GD, Ma DD, Jia YT, and Wang J 2016 Characteristics of laminar flow and heat transfer in microchannel heat sink with triangular cavities and rectangular ribs *Int J Heat Mass Transf* **98** 17–28.
[7] Japar WMAA, Sidik NAC and Mat S 2018 A comprehensive study on heat transfer enhancement in microchannel heat sink with secondary channel *Int Commun Heat Mass Transf* **99** 62–81.
[8] A. Sahak AS, Che Sidik NA, and Akmal Yusof SN 2020 A Brief Review of Particle Dispersion of Cavity Flow *J. Adv. Res. App. Sc. Eng. Tech* **20(1)** 27-41
[9] Aziz Japar WMA, Che Sidik NA, Aid SR, Asako Y, and Lit Ken T 2018 A Comprehensive Review on Numerical and Experimental Study of Nanofluid Performance in Microchannel Heatsink (MCHS) *J. Adv. Res. Fluid Mech. Therm. Sc.* **45(1)** 165-76.
[10] Aziz Japar WMA, Che Sidik NA, Saidur R, Asako Y, and Mohd Azmi MI 2019 Temperature Minimization on the Substrate of a Heat Sink By Rib-Groove Microchannel Heat Sink with Effective Energy Consumption: Groove Geometry Parameter Effects. *CFD Lett.* **11(9)** 1-13.
[11] Steinke ME, and Kandlikar SG 2006 Single-phase liquid friction factors in microchannels *Int J Therm Sci* **45(11)** 1073–83.

Acknowledgements
The authors thankfully acknowledge the supports funder by the Ministry of Higher Education under FRGS, Registration Proposal No :FRGS/1/2019/STG07/UTM/02/18 & UTM R.K130000.7843.5F273.

Nitrogen Gas Adsorption Filter (NgAF) to enhance producer gas quality

Mohd Mahadzir Mohammud [1], Nor Azirah Mohd Fohimi [1*], Muhammad Arif Ab Hamid Pahmi[1,2], and Ariffatul Amirah Hairun Anuar[3]

[1]Faculty of Mechanical Engineering, Universiti Teknologi MARA, Cawangan Pulau Pinang, 13500 Permatang Pauh, Pulau Pinang, Malaysia.
[2]Creative and Innovation Research Group in Automative and Aviation (CIRAA), Universiti Teknologi MARA, Cawangan Pulau Pinang, 13500 Permatang Pauh, Pulau Pinang, Malaysia.
[3]Medivest Sdn. Bhd. Service Centre Hospital Mersing, Jalan Ismail, 86800 Mersing, Johor, Malaysia.

*Corresponding author email: noraz330@uitm.edu.my

Abstract. Biomass gasification is a thermochemical conversion process of solid biomass into a gaseous fuel called producer gas that can be used to generate power and electricity. The producer gas consists of around 47% of Nitrogen (N_2), 24% Carbon Monoxide (CO), 16% Carbon Dioxide (CO_2), 12% Hydrogen (H_2), and 1% Methane (CH_4). However, Nitrogen (N_2) content in the producer gas reduces its heating values as N_2 acts as a diluent because of the low calorific value (LCV) of gas. This study aims to design a Nitrogen gas filter for capturing nitrogen gas from producer gas to increase the heating value of producer gas as fuel in combustion. The method to increase the heating value of producer gas will increase the number of combustible gases or reduce the composition of non-combustible gases in producer gas. The use of material name zeolite with its microporous structures able to adsorb nitrogen molecules and act as catalysts to chemical reactions. Zeolites 5A have a small pore highly efficient to adsorb nitrogen gas because pore diameter is relatively similar to the size of nitrogen molecules. The quality of the producer gas depends on the design and operating parameters of the zeolite catalyst. Nitrogen Gas Adsorption Filter is a new method that has to be designed to improve the previous producer gas quality. Nitrogen Gas Adsorption Filter consists of a cylindrical shape body packed with crushed zeolites 5A. When this method of adsorption process is applied, the heating value of the producer gas is increased by observing the quantity of blue flame colour produced by NgAF.

1. Introduction

In 2017, the world primary energy consumption increased by 2.2%. This growth was driven by natural gas consumption which has increased by 3.39 trillion cubic feet since 2010 [1]. Biomass gasification is a thermochemical conversion process of solid biomass into a gaseous fuel called producer gas that can be used to generate power and electricity. Producer gas is a mixture of gases typically composed of 24% of Carbon Monoxide (CO), 12% Hydrogen (H_2), 1% Methane (CH_4), 16% Carbon Dioxide (CO_2), and 47% Nitrogen (N_2) [2]. CO, H_2, and CH_4 are combustible gases [3], whereas N_2 and CO_2 are non-combustible gases [4]. These two non-combustible gases act as a diluent. High concentrations of CO_2

and N_2 reduce the heating value of the gas and may cause other technical problems and are thus removed to meet pipeline and transport specifications [5]. In this paper, the study is only focused on N_2 gas. Improving the quality of the producer gas with the absorption process of CO_2 has been studied and done by Mahadzir et al. [6.7].

Nitrogen gas has the highest composition in producer gas compared to carbon dioxide. The known method to increase the heating value of producer gas was by manipulating the gasification conditions, which will increase the number of combustible gases or reduce the composition of non-combustible gases in producer gas [8,9]. N_2 gas can be reduced by the process of adsorption. Adsorption is a major process for the separation and purification of gases in different industries. Moreover, the selection of appropriate adsorbents in this procedure is an important factor, where the ideal adsorbent should have some specifications including high adsorption capacity, selectivity, hydrothermal stability, and facile regeneration [10].

Zeolites are the most famous adsorbents that are generally used for N_2 separation. The N_2 adsorption properties of different zeolites have been investigated in [11,12,13] including "Zeolite X", "Zeolite A", "Clinoptilolite", "Chabazite", "mordenite", etc. Zeolites have been discovered to have the ability in adsorbing gases into their microporous structure. It can adsorb organic and mineral molecules in the gas phase without any modification of their structure. This adsorption is due to their high specific surface (40 to 800 m^2/g) to some hydrophobic-hydrophilic surface effects and their structure. In industry, zeolites have been used widely for medical applications and the separation of gases. There are commercial units of zeolites that can provide oxygen of 95% purity to be used in hospitals for medical purposes. The gas separation process using zeolites enables oxygen to be separated from the air almost in a state of purity because zeolites preferentially adsorb N_2 which is 80% of the air due to it has a much larger quadrupole moment [14].

In the industry sector, zeolites have been used widely for medical applications and the separation of gases. There are commercial units of zeolites that can provide oxygen of 95% purity to be used in hospitals for medical purposes. The gas separation process using zeolites enables oxygen to be separated from the air almost in a state of purity because zeolites preferentially adsorb N_2 which is 80% of air. After all, it has a much larger quadrupole moment [8].

This paper presents a new approach is to design a Nitrogen gas filter that can reduce the composition of N_2 gas from producer gas in order to increase the heating value of producer gas as fuel in combustion. The producer gas has been produced by the gasification process using a downdraft gasifier [15]. According to the best knowledge of authors, Zeolites 5A as adsorbents in N_2 adsorption. Zeolites 5A as adsorbents are designed to be used in NgAF, in which nitrogen adsorption with high capacity is the main challenge. The conclusion has been determined where the NgAF has been presented. Other researchers so far have not used zeolites as bed material for the adsorption process in the gasification process. Therefore, there is still room for the author to study in this area.

2. Methodology

2.1. Data Collections: Nitrogen Gas Adsorption by Zeolites

Zeolites have small pores that can adsorb small polar molecules. Its internal microporosity gives rise to the high surface area and provides sites for the adsorption of molecules [8,16]. Zeolites have a higher tendency to adsorb N_2 over O_2 because of the permanent quadrupole moment which interacts with cations in zeolite structure [10]. Generally, the common adsorption processes of air separation are divided into two categories [17]. The first category is consisting of processes that make use of zeolites as N_2 adsorbents under the equilibrium conditions and O_2 is a processed product. The second one contains processes that utilize Carbon Molecular Sieves (CMSs) as O_2 adsorbents. Based on kinetic separation in this kind of category, O_2 is adsorbed owing to its faster permeation and higher selectivity. Moreover, N_2 is produced as a product in such processes. In the past, numerous studies of gas adsorption by zeolites have been done using various types of zeolites. Some that are closely related to nitrogen adsorption is simplified in table 1:

Table 1. Previous studies of gas adsorption selectivity.

Researchers	Gas pair	Zeolites Adsorbent	Adsorption Selectivity
[18]Ocean C et. al	CO_2/N_2	NaK-ZK-4 Zeolites	CO_2 over N_2
[19]Yaoguo F et. al	N_2/O_2	Faujasite (FAU) zeolites	N_2 over O_2
[20]Hamida P et. al	N_2/O_2	M_xNa_{96-x} LSX zeolites	N_2 over O_2
[21]Samira S et. al	CO_2/N_2	FAU- and LTA-type zeolite nanocomposites	CO_2 over N_2
[22]Jiangfeng Y et. al	CH_4 /N_2	ZK-5 zeolite	CH_4 over N_2
[23]Patrícia A P M et. al	CO_2/N_2	5A zeolite	CO_2 over N_2

In a study of separation of O_2 using zeolite 5A, the air is allowed to pass at 10 to 30 cm^3(STP)/min through a sample of about 18g of zeolite 5A. The equilibration was reached in the 10 minutes allowed for each adsorption point. For air passing at 20 cm^3/min into the bed of 5A zeolite beads, from a time of 2 min to about 12 min, pure O_2 was recovered at the outflow, showing the essential of N_2 adsorption, equivalent to 2.4 cm^3/g of zeolites. According to this study, the amount of O_2 recovered from the air is independent of the flow rate range 15 to 30 cm^3/min, indicating that the gas transport to the surface of zeolite particles was not the limiting step in adsorbent utilization [24]. Furthermore, in this study, following the adsorption run, desorption takes place at 25°C results in outflow gas had a composition close to that of air. This means that zeolites are reusable for another N_2 adsorption process once the gas has been desorbed from its microspores.

The adsorption capacity in the adsorption bed depends on the factors such as pressure, temperature, and flow rate. The adsorption and desorption cycle of a Pressure Swing Adsorption (PSA) system operates by pressure increasing and decreasing. As pressure increases, the adsorption rate of more strongly adsorbed components increases [11]. On the other hand, at near room temperature and atmospheric pressure, N_2 adsorption from the air is 2.4 cm^3/g in zeolites 5A bed in 12 minutes duration. The adsorbing capacity of zeolites is expected to be the same in the N_2 adsorption process from producer gas. If so, 1kg of zeolites 5A will be able to absorb up to 2400 cm^3 of N_2 gas. On average, the gasification of 1 kg of biomass produces about 2.5 m^3 of producer gas of which 47% of the volume is the composition of nitrogen gas [25]. This is equivalent to 1.18 m^3 of nitrogen produced from 1 kg of gasified biomass. Taking this data, about 50kg of zeolites is needed to entirely adsorb nitrogen in producer gas from 1 kg of biomass.

2.2. *Design of Nitrogen Gas Adsorption Device*

The dimension of NgAF is restricted in a lab scale to compromise and to ensure it is compatible with the designs of existing downdraft gasifier, heat exchanger and cooling cum cleaning systems in this project [15].

The flow of producer gas in this study is the high-temperature producer gas from gasifier will flow through a heat exchanger to bring down its temperature to ambient temperature (60 – 30°C). The producer gas is then will pass through a cleaning cum cooling system where tar and moisture content will be removed from the gas. After the cleaning and cooling process takes place, this is the section where producer gas will be going through NgAF to capture N_2 gas content and increase the heating value of producer gas.

In industry, there are commercial units of N_2 absorber using zeolites that carry out air separation process mostly to produce oxygen in an almost pure state. This commercial unit is known as an oxygen concentrator. The concentrator consists of two adsorption zeolite columns, a dual-head air compressor, five three-way solenoid valves, a back-pressure regulator, an oxygen sensor a surge tank, a pressure sensor, and a cooling fan to avoid the temperature rise of the air compressor [26]. Figure 1 shows the schematic of the oxygen concentrator using a two-column PSA cycle. The working principle of the oxygen generator is that it adopts air as raw materials with pressure swing adsorption (PSA) technology and the N_2 from the air can be selectively adsorbed by the zeolite molecular sieves.

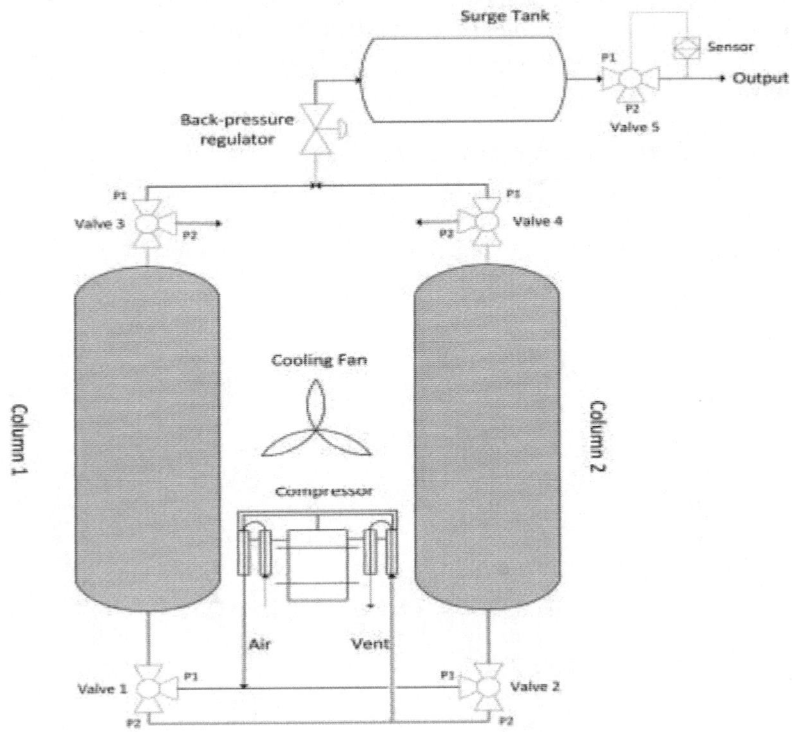

Figure 1. Schematic of the Oxygen Concentrator [20].

2.3. *Experimental Set-Up*

The NgAF was installed as in figure 2 and ready to be operated to test its efficiency in improving the heating value of the producer gas.

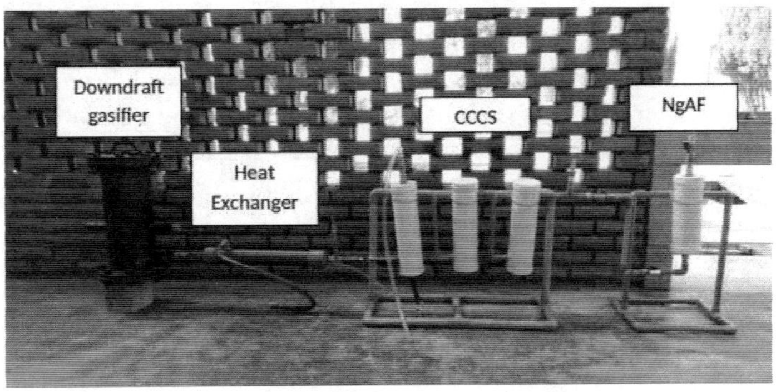

Figure 2. Devices Installed According to Their Stages

The NgAF was properly connected to the Cleaning Cum Cooling System (CCCS) to ensure gas from CCCS's outlet flow into NgAF inlet. 1 kg of Zeolite sample was placed into the NgAF filter body. The top end cap of the filter body was closed tightly to ensure no leakage. The valve at the inlet was opened to allow producer gas to enter the filter body while the valve at the outlet was fully closed to allow adsorption to take place. The outlet valve was opened every 30 minutes. The outlet producer gas was ignited to observe the combustion efficiency. The zeolite sample was weighted after the gas filtration was finished. The results mainly focused on the combustion efficiency in terms of the colour of flame produced during ignition. The flame produced before and after the gas pass through NgAF were compared to analyze the intensity of blue flame produced.

3. Result and Discussion

3.1. Nitrogen gas adsorption filter (NgAF)

The designation of NgAF in this paper is partially inspired by the design of an oxygen concentrator in Figure 1. The NgAF as illustrated in figure 3 used different working principles and approaches to capture nitrogen gas from producer gas. Just like the oxygen concentrator, NgAF is designed with two cylindrical filter columns containing zeolites.

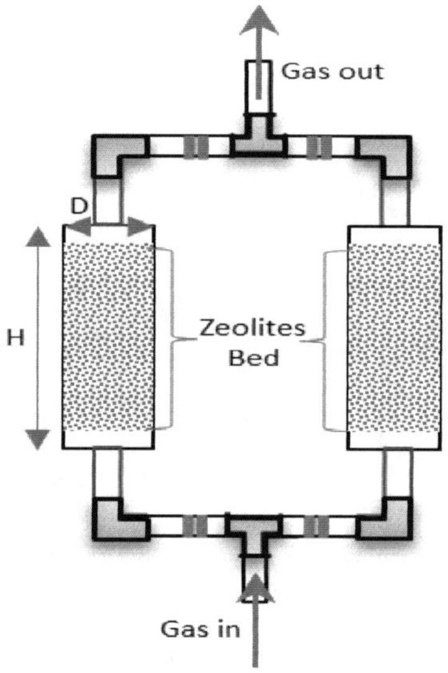

Figure 3. Schematic Diagram of NgAF.

However, the NgAF does not use the PSA technology but the adsorption process will take place at 25-35°C temperature and 100kPa pressure by allowing producer gas to pass through the zeolites bed for a certain period. The NgAF is in a lab-scale dimension to make it compatible with the existing downdraft gasifier [15]. Polyvinyl chloride, PVC pipe is selected to be used as the body column because it has a smooth surface which causes frictional resistance. Other than that, it starts to decompose when the temperature reaches 140 °C, with melting temperature starting around 160 °C. figure 4a and 4b show the detailed views of NgAF.

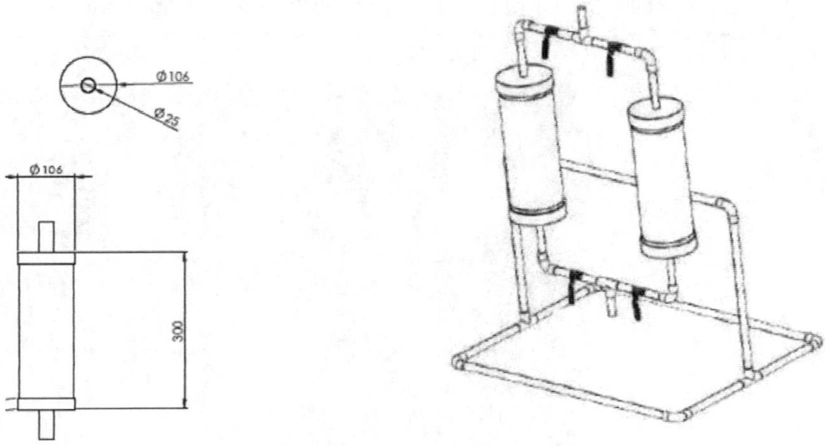

(a) (b)

All dimensions are in meter (mm)

Figure 4: Detail design of NgAF (a) The details of dimension (b) Isometric view.

The NgAF consists of two cylindrical columns and valves at each column inlet and outlet. Compared to the industrial oxygen concentrator, the NgAF takes producer gas as raw material to pass through the zeolite bed. Zeolite 5A is used in this study. It has a higher tendency to adsorb nitrogen gas molecules into its microspore is due to the relative similarity between its pore size (0.5 nm) and the size of nitrogen molecules (0.364 nm) that allows nitrogen molecules to be captured. Zeolite 5A molecular sieve has a surface area of 571 m^2/g and a density of 1.16 g/cm^3. To adsorb 0.59 m^3 of nitrogen gas from producer gas produced by gasification of 0.5 kg biomass residue, 25 kg of Zeolites 5A is required with 12.5 kg separated in each NgAF column. 12.5 kg of zeolites is equivalent to 10775cm^3. Thus, a cylindrical body with a volume of 11000 cm^3 is to be produced for the NgAF. The difference in zeolite volume and column volume is considering the air space in the zeolite bed. For the column body, PVC pipe with a large diameter is to be used. Thus, a PVP pipe of diameter D = 18 cm is selected.

3.2. Quality of producer gas, before and after NgAF

The producer gas was ignited with fire to test the presence of combustible gases that are methane (CH_4), carbon monoxide (CO) and hydrogen (H_2). The flames produced by the producer gas before and after passing through NgAF are compared in Table 2. The flames produced after passing through zeolite bed in NgAF contain blue flame even though it is hard to be seen. Referring to table 2, the observation of blue flame can be seen most clearly in the result of Experiment 3.

Table 2. Flames of producer gas before and after NgAF.

Experiment	Before NgAF	After NgAF
1		
2		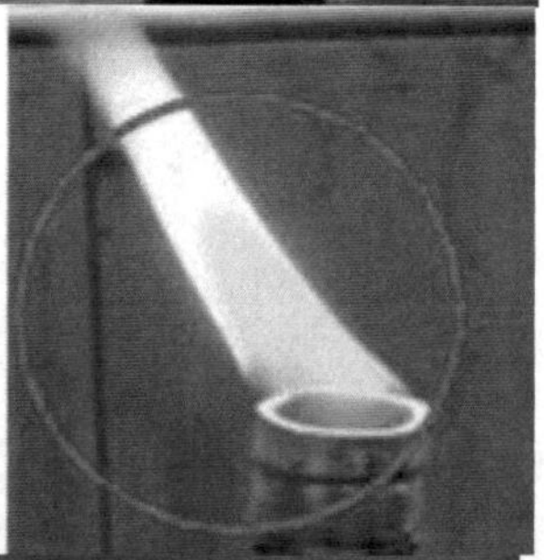
3	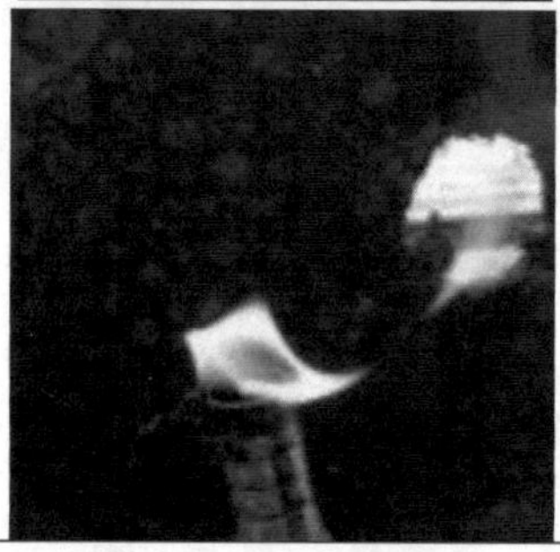	

The production of blue flame means that the combustion of producer gas is more complete compared to the combustion before the gass passing through NgAF. More complete combustion was produced because the producer gas out from the NgAF has high density and concentration compared to the gas before pass through the system. This is because before the gas enters NgAF, it contains a high weight percentage of non-combustible nitrogen gas that acts as a diluent to the producer gas. Zeolite bed in NgAF captured the nitrogen molecules into its micropores help in increasing the heating value of producer gas thus lead to the appearance of blue flame at the outlet of NgAF.

4. Conclusion

The design of Nitrogen gas Adsorption Filter (NgAF) has been designed. This is an alternative method to increase the heating value of the producer gas where the N_2 gas can be reduced using zeolites. The NgAF is designed based on the analysis and results obtained by other researchers. When this method of adsorption process is applied, the heating value of the producer gas is expected to increase the heating value by observing the colour of blue flame produced. The proposed design is the outcome of this writing, and it is referring to the paper reviewing and studies.

References

[1] BP statistical review of world energy 2018

[2] Hossaina M S, Islama M R, Rahmana M S, Kadera M A and Haniub H 2017 Biofuel from co pyrolysis of solid tire waste and rice husk *Energy Procedia* **110** 453 – 458

[3] Wei L, Jun W, Zhongyi F, Qing H, Nan Z, Mao X, Min S, Zhiyong Q and Zhi Z 2021 In-situ catalytic pyrolysis of waste tires over clays for high quality pyrolysis products *International Journal of Hydrogen Energy* **46** 6937 – 6944

[4] Sammy S 2012 Gasification, producer gas and syngas *Agriculture and Natural Resources* 8

[5] Crawley G M 2013 The world scientific Handbook of Energy, *World Scientific Publishing Co. Pte. Ltd*

[6] Mahadzir M M and Zainal Z A 2011 The ability of CaO-sand mixture to absorb CO_2 composition of the simulated gas and producer gas using bubbling fluidized bed reactor *International Journal of Arts and Sciences* **4** 37 – 41

[7] Mahadzir M M, Abdul R H and Zainal Z A 2014 Simulation and experiment of 50% calcium oxide-sand mixture in a bubbling fluidized bed CO_2 absorption *reactor Journal of Software Engineering* **8** 345-360

[8] Chakrabarti H C 2012 Biomass Gasification-Basics and Considerations *Minetech* 1–39.

[9] Bhavanam A and Sastry R C 2011 Biomass Gasification Processes in Downdraft Fixed Bed Reactors : A Review *International Journal Chemical Engineering Applies* **2** 425–433

[10] Kennedy D A and Tezel F H 2018 Cation exchange modification of clinoptilolite – screening analysis for potential equilibrium and kinetic adsorption separations involving methane, nitrogen, and carbon dioxide *Microporous and Mesoporous Materials* **262** 232-250

[11] Minghui F, Jihong S, Shiyang B and Hamida P 2015 Size effects of extraframework monovalent cations on the thermal stability and nitrogen adsorption of LSX zeolite *Microporous and Mesoporous Materials* **202** 44-49

[12] Rasoul N S, Shahram S, Farshad R 2018 Natural gas upgrading by selective separation on zeotype adsorbents *Journal of Natural Gas Science And Engineering* **54** 37-46

[13] Ehsan J S, Danial J F, Hadiseh K M and Nadia 2016 Comparative study of zeolite 5A and zeolite 13X in air separation by swing pressure adsorption *Korean Journal of Chemical Engineering* **33** 1391 -1401

[14] Christopher J R 2010 Properties and applications of zeolites *Science Progress* **93** 223–284

[15] Mahadzir M M, Aida H F and Ismail N I 2017 Studies and design downdraft gasification for teaching and learning in UiTM Penang branch *Proceedings of Mechanical Engineering Research Day* 211-212.

[16] Zheng Y, Li X and Dutta P K 2012 Exploitation of unique properties of zeolites in the development of gas sensors *Sensors* **12** 5170–5194

[17] Iman A K, Ehsan J S, Mehdi B and Mehdi F (2017) Dynamic Modeling of Nitrogen Adsorption on Zeolite 13X Bed *Fluid Mechanics Research International Journal* **1** 1–6

[18] Ocean C, Zoltan B, Nicolas F, Panagiotis K, Dariusz W and Niklas H 2020 Selective Adsorption of CO_2 on Zeolites NaK-ZK-4 with Si/Al of 1.8 – 2.8 *ACS Omega* **5** 25371 - 25380

[19] Yaoguo F, Yingshu L, Ziyi L, Quanli Z, Xiong Y, Chunyu Z, Chuanzhao Z, Haoyu W, Ralph T Y (2020) Insights into adsorption separation of N2/O2 mixture on FAU zeolites under plateau special conditions: A molecular simulation study *Separation and Purification Technology* **251** 117405

[20] Hamida P, Jihong S, Raza U 2020 Kinetic evaluation of dehydration in M_xNa_{96-x} LSX zeolites and resulting effects on selective adsorption of N_2 and O_2 *Microporous and Mesoporous Materials* **301** 110223

[21] Samira S and Mansoor A 2017 Characterization of CPs/Ca-exchanged FAU- and LTA-type zeolite nanocomposites and their selectivity for CO_2 and N_2 *adsorption Journal of Physics and Chemistry of Solids* **110** 116 -128

[22] Jiangfeng Y, Xuan T, Jiaqi L, Jun W, Hua S, Luogang W, Jinping L and Shuguang D 2021 Down-sizing the crystal size of ZK-5 zeolite for its enhanced CH_4 adsorption and CH_4/N_2 separation performances *Chemical Engineering Journal* **406** 126599
[23] Patrícia A P M, Ana M R, Kristin G, Alexandre F P F, Alírio E R 2017 Separation of CO_2/N_2 on binderless 5A zeolite *Journal of CO_2 Utilization* **20** 224 - 233
[24] Ertan A (2004) CO_2, N_2 and Ar Adsorption on Zeolites *MSc Thesis, İzmir Institute of Technology*
[25] Rajvanshi A K 2014 Biomass gasification *Alternative Energy Agriculture* **4** 1–21.
[26] Pan M, Omar H, and Rohani S 2017 Application of Nanosize Zeolite Molecular Sieves for Medical Oxygen Concentration *Nanomaterials* **7** 195

Acknowledgements
The authors would like to express their appreciation to Universiti Teknologi MARA (UiTM), Pulau Pinang, Malaysia, for giving the opportunity and financial supporting to do this biomass research study.

AUTHOR INDEX